Value Creation

Strategies for the Chemical Industry

Edited by
Florian Budde, Utz-Hellmuth Felcht,
and Heiner Frankemölle

2nd completely revised and enlarged edition

WILEY-VCH

WILEY-VCH Verlag GmbH & Co. KGaA

Editors

Dr. Florian Budde
Director
McKinsey & Company, Inc.
Taunustor 2
60311 Frankfurt
Germany

Prof. Dr. Utz-Hellmuth Felcht
Chairman of the Board of Management (CEO)
Degussa AG
Bennigsenplatz 1
40474 Dusseldorf
Germany

Dr. Heiner Frankemölle
Director
McKinsey & Company, Inc.
Magnusstrasse 11
50672 Cologne
Germany

Library of Congress Card No.: applied for

British Library Cataloguing-in-Publication Data
A catalogue record for this book is available from the British Library.

Bibliographic information published by Die Deutsche Bibliothek
Die Deutsche Bibliothek lists this publication in the Deutsche Nationalbibliografie; detailed bibliographic data is available in the Internet at <http://dnb.ddb.de>.

Printed in the Federal Republic of Germany.
Printed on acid-free paper.

Typesetting Kühn & Weyh, Satz und Medien, Freiburg
Printing Strauss GmbH, Mörlenbach
Bookbinding Litges & Dopf Buchbinderei GmbH, Heppenheim

ISBN-13: 978-3-527-31266-5
ISBN-10: 3-527-31266-8

Contents

Value Creation: Strategies for the Chemical Industry. 2nd Edition. F. Budde, U.-H. Felcht, H. Frankemölle (Eds.)

ISBN: 3-527-31266-8

Preface

The first edition of "Value Creation – Strategies for the Chemical Industry" was very well received. The effort required to compile a book on modern management approaches especially for chemical companies had been quite daunting; however, we were extremely gratified to receive positive feedback from many in the industry. After much deliberation, this response also decided us to publish a second and completely revised edition.

The second edition has seen some major changes. While the first edition was very much a "McKinsey book", this one incorporates a large number of contributions from authors in the industry, many of them board members of chemical or related companies, others leading representatives of institutions with strong ties with the chemical industry. The most prominently visible sign of the new direction is the change in the composition of the editors. It is the hope of the three of us that this mix of management experience combined with consulting insights will make for even more interesting reading.

Important changes in the book's content have also been driven by developments in the chemical industry's operating environment. The first edition went to press in early 2001, just as the Internet bubble was about to burst. The chemical industry had just seen itself dismissed as one of those "bricks and mortar" dinosaurs that would have no place in the new era. This perspective has now changed considerably. The industry's image – in the stock markets, but also in general – has improved significantly in the last five years.

That does not mean, however, that the external environment has become any easier. In particular, the growing economic importance of Asia (and especially China) both as a location for production and as a market will have a lasting impact on the chemical industry landscape, and many chemical companies are under enormous pressure to adapt. We are currently also seeing major shifts in feedstock prices, again resulting in a movement of some chemical production, here mostly towards the Middle East.

Other trends have remained unchanged, such as the eternal pressure to increase productivity to compensate for the price-cost squeeze, or to compensate for the ongoing commoditization of many of the industry's products and services by introducing innovations. In addition, we can see a continuing high level of

Value Creation: Strategies for the Chemical Industry. 2nd Edition. F. Budde, U.-H. Felcht, H. Frankemölle (Eds.)

ISBN: 3-527-31266-8

M&A activity, with private equity companies playing an increasingly important role in acquisitions in recent years.

Amid all these developments, value creation has remained a key objective, although the paths to it have, of course, been altered and augmented by the shifts in the business environment.

The editors have attempted to incorporate the changes into the agenda of this second edition, which falls into three main parts. Starting with an overview of the chemical industry, the book discusses the elements that are relevant to strategic decision-making in the chemical industry in general and in a number of key industry segments. In the next section, all the key elements of the chemical business system are addressed, reflecting the importance of functional excellence as a major driver of the industry's performance. The final set of chapters deals with special topics of interest to chemical companies, including current hot topics such as China and biotechnology.

We make no claim that our list of topics is comprehensive from the point of view of the chemical manager, who has to deal with many more issues than we could address here (e.g., environmental issues), but most of those that we have not included are well covered elsewhere.

Like the last edition, this book assumes some knowledge of business administration, without which it does not always make easy reading. Nevertheless, we hope it may be of interest not only to managers of chemical companies, but also to many others who have dealings with the chemical industry, from students considering it as a career option to investors wishing to assess the value creation potential of a particular investment. It might also have a relevance for managers in some other industries that share common traits with the chemical industry, such as pulp and paper, metals, or energy.

We hope that readers find this book both enjoyable and useful, and that they will come to share some of the enthusiasm we feel for an industry that contributes so much more than any other to the basis of modern life.

The Editors *Florian Budde, Utz-Hellmuth Felcht, Heiner Frankemölle*
July 2005

Acknowledgements

First and foremost, our warmest gratitude goes to all authors for the creativity, commitment, and enthusiasm that went into the writing of this book.

Our deepest thanks go to the top managers and senior executives from the chemical industry and chemicals-related institutions who were kind enough to contribute chapters. Their readiness to take part has enabled us to present a much richer, deeper, and broader picture of management concerns in the chemical industry than in the previous edition.

Closely behind this group, we also owe heartfelt thanks to the authors from McKinsey, and to all others who contributed to the material but were not directly involved in producing the written version.

We would also like to express a debt of gratitude to a number of people who provided enormous support behind the scenes on individual chapters: Eric Bartels, Ralf Dingeldein, Tobias Engel (Merck KGaA), Philip Eykerman, Khosro Ezaz-Nikpay, Helena Fransen, Clemens Grambow (Degussa), Elizabeth Gryger, Stefan Hilger (VCI), Parry Norling, Jens Riese, Ulf Schrader, Lionel Stanbrook (Syngenta), and Ulrich von Deessen (BASF).

Next, a big vote of thanks goes to the members of our extended team:

- Project leader Karsten Hofmann and project managers Sabine Deppe, Neville Gorman, Maren Hansen, and Marco Neuthard for their guidance along the project path and their expert and efficient liaison with all parties in the chain, from authors to publishers.
- Manuscript editors David Hunter, Annette Lehnigk, Jürgen Raspel, and Jonathan Turton for their creative and critical input and their contributions to the clarity and crispness of many chapters.
- Andreas Appelhagen, Birgit Fehmel, Annett Jensen, Shirley Li, Loula Merkel, Sam Samdani, Torsten Teichmann, and Sari Varpa of McKinsey's Chemicals Research and Information team for their professional and accurate provision of facts, figures, and background information and their proactive advice in finding the appropriate supporting evidence for many of the ideas the authors put forward.
- The graphic artists of McKinsey's Visual Graphics India team for helping to ensure that the illustrations delivered their messages clearly and accurately.
- Skilled translators Matthew Hitchens and Francine Barley.

Value Creation: Strategies for the Chemical Industry. 2nd Edition. F. Budde, U.-H. Felcht, H. Frankemölle (Eds.)

ISBN: 3-527-31266-8

Of all the great contributors behind the scenes, one individual stands out: Helen Robertson, our lead manuscript editor. Without her great commitment and relentless drive for excellence, this book would not have been possible.

Finally, we would like to say how greatly we appreciate the support and assistance of Karin Sora and Romy Kirsten of Wiley-VCH. Thank you, above all, for your patience and understanding during the manuscript production process.

We are confident that the end product will make all of you feel that your efforts have been worthwhile.

The Editors *Florian Budde, Utz-Hellmuth Felcht, Heiner Frankemölle*

List of Contributors

Alejandro Alcalde Rasch is a principal at McKinsey from the Dusseldorf office. He is a member of the global energy and materials sector and has worked in several different process industries, with an emphasis on chemicals and steel. His areas of expertise are in operations, where he has led several site services, maintenance, and manufacturing projects, and in marketing and sales. He holds a Dipl.-Kfm. and a PhD in business administration and management from the University of Duisburg.

Dr. Alejandro Alcalde Rasch, McKinsey & Company, Inc., Königsallee 60C, 40027 Dusseldorf, Germany

Scott Andre is a senior expert in McKinsey's Houston office. He provides expert support to chemical and petroleum clients, where he focuses on strategy, organization, and marketing in petrochemicals and petroleum refining. Before joining McKinsey, Scott Andre worked for ten years with Dow Corning in product and process development, economic evaluation, and new product commercialization. Scott has an MBA from Harvard Business School and a BS in chemical engineering from the University of Michigan.

Scott Andre, McKinsey & Company, Inc., Two Houston Center, 909 Fannin, Suite 3500, Houston TX 77010, United States

Thomas Augat is an expert in McKinsey's corporate finance practice based in Munich. He leads the German corporate finance research team and has worked mainly in the chemical and utilities industries with a particular focus on capital market diagnostics, portfolio management, and the redesign of planning and reporting processes. He holds a Dipl.-Kfm. from the University of Bayreuth, is a CFA charterholder, and worked for two years as research fellow at the chair of Finance and Banking of the Fernuniversität Hagen.

Thomas Augat, McKinsey & Company, Inc., Prinzregentenstrasse 22, 80538 Munich, Germany

Value Creation: Strategies for the Chemical Industry. 2nd Edition. F. Budde, U.-H. Felcht, H. Frankemölle (Eds.)

ISBN: 3-527-31266-8

Rolf Bachmann is a principal in McKinsey's Zurich office, having joined the firm in 1992. He has served various companies in specialty chemicals, pharma and life sciences, and the basic and advanced materials industries, and has been involved in several industrial biotechnology projects. He is a leading member of McKinsey's chemicals practice and leads the practice's industrial biotech initiative. Rolf has an MBA and a PhD in economics from the University of Zurich.

Dr. Rolf Bachmann, McKinsey & Company, Inc., Alpenstrasse 3, 8065 Zurich, Switzerland

Eric Bartels is a principal at McKinsey. Since joining the Cologne office in 1998, he has worked primarily in process industries, focusing on chemicals and electric power. He is a member of the leadership team of McKinsey's global chemicals, organization, and post-merger management practices. He holds a master's degree in computer science and mathematics from the University of Bonn and a PhD in mathematics from the University of Oxford.

Dr. Eric Bartels, McKinsey & Company, Inc., Magnusstrasse 11, 50672 Cologne, Germany

Sönke Bästlein is a director at McKinsey. He initiated and heads McKinsey's European Asia House in Frankfurt. Sönke Bästlein serves clients in the chemical industry globally, in engineering and construction (large-scale projects) and in private equity (commercial due diligences). He has led knowledge initiatives on customer portfolio management and value contracting (capital productivity). Before McKinsey, Sönke Bästlein worked for five years with Uhde in chemical plant engineering. He studied process engineering at RWTH Aachen and earned a PhD in business administration from the Johann Wolfgang Goethe-University in Frankfurt.

Dr. Sönke Bästlein, McKinsey & Company, Inc., Asia House, Schumannstr. 59, 60325 Frankfurt/M., Germany

Werner Becker is senior advisor for technology and regulatory affairs in the Liquid Crystals division of Merck KGaA in Darmstadt, Germany. He is responsible for technical customer services and product management regarding liquid crystals. His remit also includes regulatory issues, registration procedures, and observing national and international legislation. He has published several booklets on liquid crystals. He graduated from Saarland University with a diploma in chemistry and a doctoral thesis in electrochemistry.

Dr. Werner Becker, Merck KGaA, Liquid Crystals Division, 64271 Darmstadt, Germany

Aldo Belloni is a member of the Executive Board of Linde AG. He is in charge of all the group's gas and engineering activities. Since joining Linde in 1980, he has worked on several assignments in petrochemicals, natural gas, and synthesis gas. In the early 1990s he was in charge of Linde's US engineering operation. He has sat on the Executive Board since 2000, and assumed full responsibility for the Gas business in 2003. He holds a degree in chemical engineering from Politecnico di Milano.

Dr. Aldo Belloni, Linde AG, Abraham-Lincoln-Str. 21, 65189 Wiesbaden, Germany

Robert Berendes is head of Business Strategy and Planning at Syngenta. Previously, he was a principal with McKinsey & Company. Prior to that, he worked at Procter & Gamble as a process engineer. Robert Berendes holds a master's degree in chemistry from the University of Cologne and a PhD in biophysics from the Max Planck Institute for Biochemistry/Technical University in Munich.

Dr. Robert Berendes, Syngenta International AG, Schwarzwaldallee 215, 4002 Basel, Switzerland

Achim Berg is a consultant with McKinsey based in Frankfurt. He has worked for various clients in the chemical industry in Europe, the United States, and Asia. His main areas of emphasis within McKinsey are buyout transactions, strategy, corporate reorganization, and operational improvement programs. He holds a master's degree and a PhD in business administration and economics from Witten/Herdecke University.

Dr. Achim Berg, McKinsey & Company, Inc., Taunustor 2, 60311 Frankfurt, Germany

Leonhard Birnbaum is a principal in McKinsey's Dusseldorf office. He is a member of the global energy and materials practice, serving clients across process industries. His main focus within McKinsey is on fundamental strategic and operational transformation work. He holds a master's degree and a PhD in chemical engineering from the University of Karlsruhe.

Dr. Leonhard Birnbaum, McKinsey & Company, Inc., Königsallee 60C, 40027 Dusseldorf, Germany

Florian Budde is a director of McKinsey, based in the Frankfurt office and with two years' experience in the Seoul office. He is co-leader of the European chemicals practice and has served a large number of chemical clients on strategic, organizational, and operational issues, with an emphasis on capital market-oriented corporate strategy. He holds a PhD in physical chemistry from the Fritz-Haber-Institut der Max-Planck-Gesellschaft/Freie Universität Berlin, and has worked as a researcher in the Physical Science department of IBM's T. J. Watson Research Center in New York.

Dr. Florian Budde, McKinsey & Company, Inc., Taunustor 2, 60311 Frankfurt/M., Germany

Andrea Cappello is a consultant in McKinsey's Milan office. He is a member of the European supply chain management practice and has worked in various fields with a particular focus on chemicals, consumer goods, and retail. Before joining McKinsey, he worked at SAP on supply chain management, manufacturing, and product development. He holds a degree in electronic engineering, with a specialization in automatic control, from the Politecnico di Milano.

Andrea Cappello, McKinsey & Company, Inc., Piazza del Duomo 31, 20122 Milan, Italy

Joël Claret is a director of McKinsey and works in Geneva. He is co-leader of the firm's global chemicals practice, working primarily in the specialty chemicals, energy, and basic materials industries. He has served clients in particular in the areas of strategy and organizational development, innovation, and marketing and sales improvement, as well as broad-scale change management. He holds a degree in electrical engineering from the Swiss Federal Institute of Technology, Lausanne, and Carnegie Mellon University, as well as an MBA from INSEAD.

Joël Claret, McKinsey & Company, Inc., Avenue Louis Casai 18, 1209 Geneva, Switzerland

Thomas M. Connelly, Jr. is senior vice president and chief science and technology officer at DuPont. He joined DuPont in 1977 as a research engineer in the elastomer chemicals department at the DuPont Experimental Station in Delaware. He has served in a variety of research and technical leadership roles in the USA, Asia, and Europe. He was named to his current position in 2000.

Thomas Connelly graduated with highest honors from Princeton University with degrees in chemical engineering and economics. He received his doctorate in chemical engineering from the University of Cambridge.

Dr. Thomas Connelly, DuPont Company, P.O. Box 80328, Wilmington, Delaware, 19880-0328, USA

Sabine Deppe is a consultant in McKinsey's Cologne office. She is a member of the chemicals and purchasing and supply management practices. Her work focuses on operational and organizational improvements in chemicals, especially in the context of broader performance transformation programs. She holds a PhD in chemistry from the University of Göttingen.

Dr. Sabine Deppe, McKinsey & Company, Inc., Magnusstr. 11, 50672 Cologne, Germany

Ralf Dingeldein is a consultant based in McKinsey's Asia House office in Frankfurt. He is a core group member of the firm's global chemicals and strategy practices, and has worked in various fields of the chemical industry. His main areas of emphasis within McKinsey are the Chinese market, business strategy, sales and distribution, as well as private equity transactions. Ralf holds a Dipl.-Wirtsch.-Ing. from the TH Karlsruhe and a PhD in business administration from the University of Essen.

Dr. Ralf Dingeldein, McKinsey & Company, Inc., Asia House, Schumannstr. 59, 60325 Frankfurt/M., Germany

Klaus Engel is the managing director of Brenntag Management GmbH and chairman and chief executive officer of the Brenntag Executive Board. He has over 20 years' experience in the chemicals and chemical distribution businesses and has held top management functions within the industry for years. He holds a PhD in natural sciences/chemistry from the Ruhr-University Bochum.

Dr. Klaus Engel, Brenntag Management GmbH, Stinnes-Platz 1, 45472 Mülheim a.d. Ruhr, Germany

Philip Eykerman is a principal in McKinsey's Antwerp office whose work focuses on the chemicals and utility sectors. Before joining McKinsey in 1997, Philip was a process/project engineer at Fluor Daniel with responsibility for proposals, conceptual studies, and basic engineering projects in the oil-refining sector. He holds a degree in chemical engineering from the Catholic University in Leuven and a master's degree in oil-refining technology from the Institut Français du Pétrole, and has had internships with Elf and Neste.

Philip Eykerman, McKinsey & Company, Inc., Cockerillkaai 26, 2000 Antwerp, Belgium

Khosro Ezaz-Nikpay is a principal at McKinsey. He is based in London from where he leads the European purchasing and supply management practice. He has carried out extensive work in this field, primarily with consumer goods, automotive, and process industry clients, and has also worked on strategy, post-merger management, and operational improvement. He holds a BS from UC Berkeley and a PhD in chemistry from Harvard University.

Dr. Khosro Ezaz-Nikpay, McKinsey & Company, Inc., No. 1 Jermyn Street, London SW1Y 4UH, UK

Gary Farha is a director in McKinsey's Washington DC office. He is co-leader of the North American chemicals practice and the global innovation practice. He has worked on a variety of strategic, organizational, and operational efforts throughout the world and has led a number of practice-building initiatives within the energy and chemicals sectors. Before joining McKinsey, he held several financial positions with Kansas Gas and Electric Company and worked for Boeing in supply management. He holds an MBA from Michigan State University and is a chartered financial analyst.

Gary Farha, McKinsey & Company, Inc., 600 14th Street N.W., Suite 300, Washington, DC 20005, USA

Utz-Hellmuth Felcht is chairman of the board of management of Degussa. He is also a member of the supervisory boards of RAG Coal International and SGL Carbon, and of the board of management of RAG. Utz-Hellmuth Felcht also serves as vice-president of the German Chemical Industry Association, and is a member of the board of the German Association for Chemical Technology and Biotechnology. In 2003, he was awarded the Order of Merit of the Federal Republic of Germany. He obtained his PhD from the University of Kaiserslautern and is an honorary Professor at the Technical University of Munich.

Prof. Dr. Utz-Hellmuth Felcht, Degussa AG, Bennigsenplatz 1, 40474 Dusseldorf, Germany

Heiner Frankemölle is a director of McKinsey based in Cologne. He leads the firm's global chemicals practice, and has worked in various fields of the process sector with a particular focus on specialty chemicals, petrochemicals, and energy. His main areas of emphasis within McKinsey are corporate transformation, strategy, mergers and acquisitions, and post-merger management. He holds a Dipl.-Ing.-Agr. as well as a PhD in agricultural economics from the University of Bonn.

Dr. Heiner Frankemölle, McKinsey & Company, Inc., Magnusstr. 11, 50672 Cologne, Germany

Christian Günther is a consultant in McKinsey's Dusseldorf office. He is a member of the global energy and materials practice and has worked in several process industries: petrochemicals, specialty chemicals, mining, and energy. Functional focal points of his work have been operational improvements, organizational effectiveness, and strategy. He holds a diploma and a PhD in chemistry from the University of Würzburg.

Dr. Christian Günther, McKinsey & Company, Inc., Königsallee 60C, 40027 Dusseldorf, Germany

Jürgen Hambrecht is chairman of the board of executive directors of BASF Aktiengesellschaft, a company he joined in 1976. He coordinates board activities and is responsible for legal, taxes, and insurance; strategic planning and controlling; executive management and development; communications for the BASF group; and investor relations. Since 2003 he has been president of the German Chemical Industry Association and vice-president of the Federation of German Industries. He studied chemistry in Tübingen and was awarded his doctorate in organic chemistry in 1975.

Dr. Jürgen Hambrecht, BASF AG, Carl-Bosch-Str. 38, 67056 Ludwigshafen, Germany

Bernd Heinemann is a principal in the Munich office of McKinsey. He is a member of the chemicals practice and co-leader of the German corporate finance practice. Bernd Heinemann has served many chemical clients on strategic and operational issues, with a focus on corporate finance and manufacturing. He holds a diploma in physics from RWTH Aachen and an MBA from Sloan School, MIT.

Bernd Heinemann, McKinsey & Company, Inc., Prinzregentenstr. 22, 80538 Munich, Germany

Karsten Hofmann is the global practice manager of McKinsey's chemicals practice, based in Frankfurt. Since joining the firm in 1995, he has worked as a consultant on numerous engagements in the process industries, focusing on chemicals and energy. In 2002, Karsten Hofmann assumed additional responsibility as manager of the global chemicals practice and as head of the worldwide chemicals research and information group. He holds a PhD in organizational psychology from the University of Mannheim, and a master's degree from Portland State University.

Dr. Karsten Hofmann, McKinsey & Company, Inc., Taunustor 2, 60311 Frankfurt/M., Germany

Baron Georges Jacobs is chairman of the board of directors of UCB Group, a company he joined in 1970 and led as president and chief executive officer from 1987 until the end of 2004. Baron Jacobs also serves on the board of directors of Delhaize-Group (as chairman), Bekaert, Belgacom, and SN Brussels Airlines. He is the former chairman of FEB (the Federation of Belgian Enterprises), UNICE (the Federation of European Enterprises), and Fedichem (the Federation of Belgian Chemical Industries). He is a doctor in law as well as holding a master's degree in economics from the Catholic Univerity of Louvain and an MA in economics from UC Berkeley.

Thomas Jetter is a partner in the Frankfurt office of Permira where he leads the chemicals team. He has worked on numerous transactions including Cognis, ESW-Extel Systems, Hofmann-Menü, and Sirona Dental Systems. Before joining Permira in 1995, Thomas Jetter worked for McKinsey in Germany and Brazil for six years, focusing on the engineering, paper, and consumer goods industries. He also spent three years with JP Morgan in Frankfurt and New York. He holds a master's degree and a PhD in business administration from Saarbrücken University.

Dr. Thomas Jetter, Permira Beteiligungsberatung GmbH, Clemensstr. 9/Falkstr. 5, 60487 Frankfurt/M., Germany

Helge Jordan is a consultant based in Leverkusen. Before going freelance, he was an associate principal/expert in McKinsey's Boston office and a member of the firm's purchasing and supply management practice. Helge Jordan worked in various fields of the process industry sector, with a particular emphasis on specialty chemicals and basic materials. His main focal areas within McKinsey were purchasing performance transformations and operational performance improvement efforts. He holds a master's degree in chemistry from the University of Cologne as well as a PhD in chemistry from the Swiss Federal Institute of Technology.

Dr. Helge Jordan, Jordan Consulting, P.O. Box 10 06 47, 51306 Leverkusen, Germany

Tomas Koch is a principal in McKinsey's Seoul office where he heads the Asian chemicals practice. He joined the Firm in 1992 and has led a variety of strategic, operational, and organizational engagements in the chemical industry. Tomas Koch studied economics and physics and has a PhD in polymer physics from the University of Freiburg.

Dr. Tomas Koch, McKinsey & Company, Inc., 27fl Seoul Finance Center, 84 Taepyungro 1-ga, Jung-gu, Seoul 100-768, South Korea

Birgit König is a principal at McKinsey, based in Berlin. She leads the German strategy practice and focuses on strategy and innovation in specialty chemicals, agrochemicals, and pharmaceuticals, predominantly working with research and development organizations. Birgit König holds a doctorate in molecular biology. Before joining McKinsey, she worked at the German Cancer Research Center on tumor promotion and with the Max Planck Society and Hanover Medical School on immune regulation.

Dr. Birgit König, McKinsey & Company, Inc., Kurfürstendamm 185, 10707 Berlin, Germany

Colja Laane is corporate science manager in the Life Science Products cluster at DSM. He is (co)author of over 120 papers and about 20 patents, mostly in the area of industrial biotechnology. He is a member or chairman of several committees involved in chemistry education and setting up life sciences startup companies in the Netherlands. He also chairs the Industrial Biotechnology section of the European Platform on Sustainable Chemistry. He graduated in biochemistry from the University of Groningen and holds a PhD from Wageningen.

Dr. Colja Laane, DSM – Food Specialties, P.O. Box 1, 2600 MA Delft, the Netherlands

Martin Lösch is a principal in McKinsey's Stuttgart office. He leads the European supply chain management practice and is a member of the leadership teams of the global pharma-healthcare, process, and operations practices. Martin Lösch has worked in various fields with a particular focus on agrochemicals, pharmaceuticals, and specialty chemicals. His main areas of emphasis are supply chain management, corporate strategy, operations, pricing, and business-building. Martin Lösch has a PhD in engineering from the University of Hohenheim.

Dr. Martin Lösch, McKinsey & Company, Inc., Birkenwaldstr. 149, 70191 Stuttgart, Germany

Christophe de Mahieu is the managing partner of McKinsey's Asia Pacific and Middle East energy and materials practice, based in Dubai. He has served chemical clients across Asia, the Middle East, and Europe on corporate strategy, growth, M&A, and operational excellence for the last twelve years. He has advised both multinationals and local chemical corporations in the petrochemical, fine chemicals, and agrochemical sectors. Before McKinsey, Christophe de Mahieu worked for Exxon. He holds master's degrees in economics and engineering and in corporate tax management, both from the Solvay Business School, Belgium.

Christophe de Mahieu, McKinsey & Company, Inc., 2903 Al Moosa Tower II, Sheikh Zayed Road, P.O. Box 335 38, Dubai, United Arab Emirates

Ralph Marquardt is an expert in McKinsey's Frankfurt office. He is a core member of the firm's global chemicals practice, focusing on specialty chemicals, agrochemicals, life-science products, and petrochemicals. His main focus is on commercial transformation programs, sales force effectiveness, pricing, strategy, mergers and acquisitions, and innovation management. Before joining McKinsey he worked as an R&D project manager at Degussa. He holds a master's degree in chemistry from TU Braunschweig and a PhD in organic chemistry from the Ruhr-University Bochum.

Dr. Ralph Marquardt, McKinsey & Company, Inc., Taunustor 2, 60311 Frankfurt/M., Germany

Gregory Nelson is group vice-president and president, service, equipment, and alternate sales channels at Nalco. Previously he worked at Sun Chemical, Union Carbide, and Dow Chemical in senior procurement and supply chain positions, and for 20 years at Exxon in a range of roles. Gregory Nelson holds a BS in mechanical engineering from Cornell University.

Gregory Nelson, Nalco Company, 1601 W. Diehl Rd, Naperville, Il 60563, USA

Karsten Neuffer is an associate principal at McKinsey and is based in the Berlin Office. He is a member of the global chemicals practice and the Asia House Frankfurt and co-leads the knowledge initiative on 'Chemicals in China'. He has worked in various fields in the chemicals and non-ferrous metals industries, with a particular focus on corporate strategy, organization, corporate transformation, as well as China. He holds a master of science in industrial engineering and an MBA from the University of Chicago.

Karsten Neuffer, McKinsey & Company, Inc., Kurfürstendamm 185, 10707 Berlin, Germany

Uwe Nickel is head of the Global Pigments & Additives division and a member of the board of directors of Clariant International. He has wide experience in operations, change management, and innovation management. Before his current assignment he held senior management posts in several functions at Clariant, Hoechst, and Cassella. He holds a PhD in organic chemistry from the University of Frankfurt am Main.

Dr. Uwe Nickel, Clariant International Ltd., Head of Division PA, Rothausstr. 61, 4132 Muttenz, Switzerland

Peter Peters is an associate principal in McKinsey's Business Technology office, based in Frankfurt. He leads the firm's global process architecture initiative, which combines process improvement measures such as lean, outsourcing, or offshoring to improve global operations, and is also a member of the leadership group of the global ERP initiative. His client experience includes various ERP projects in the high-tech and chemical industries, as well as IT and business restructuring engagements across various industries that leveraged outsourcing, offshoring, and shared services. He holds a diploma in computer science from the University of Dortmund and a PhD in computer science from RWTH Aachen.

Dr. Peter Peters, McKinsey & Company, Inc., Taunustor 2, 60311 Frankfurt/M., Germany

Michael Pragnell is chief executive officer and member of the board of Syngenta. Previously he was a director of AstraZeneca and of Zeneca Group, having also worked for 20 years at Courtaulds in a number of senior positions. He has a degree in modern languages from the University of Oxford and an MBA from INSEAD.

Michael Pragnell, Syngenta International AG, Schwarzwaldallee 215, 4002 Basel, Switzerland

Nicolas Reinecke is a principal in McKinsey's Hamburg office, with experience in the firm's New York/New Jersey and Silicon Valley offices. He leads the purchasing and supply management practice, and is a member of the leadership group of the European operations strategy and effectiveness practice and the European high-tech and chemicals practices. Since joining McKinsey, he has focused on operations engagements and in particular on manufacturing and procurement. He holds a PhD in energy and chemical engineering and a Dipl.-Ing. in mechanical engineering from the University of Hanover.

Dr. Nicolas Reinecke, McKinsey & Company, Inc., Am Sandtorkai 77, 20457 Hamburg, Germany

Jens Riese is a principal at McKinsey, based in Munich from where he co-leads the firm's innovation and biotech initiatives. He serves petrochemical, chemical, and biotech companies around the world on topics related to innovation, strategy, and operations. Jens Riese has worked with, among others, BIO, EuropaBio, and six leading companies on a perspective on the emerging industrial biotech sector. He holds a PhD in biochemistry from the University of Cambridge, an MSc in biology from Heidelberg, and a diploma in environmental management from Dresden.

Dr. Jens Riese, McKinsey & Company, Inc., Prinzregentenstrasse 22, 80538 Munich, Germany

Eric Roegner is a principal in the Cleveland office of McKinsey. Since joining the Firm in 1994, he has served primarily business-to-business clients in North America and Europe on a wide range of go-to-market, strategy, and operational topics. He is the leader of McKinsey's North American industrial marketing practice as well as the global coordinator of the Industrial Pricing Center. Eric has written on an array of marketing/operational improvement topics, including co-authoring 'The Price Advantage', a comprehensive business text on pricing management. Eric holds an MBA from Case Western Reserve University and a BS in mechanical and aerospace engineering from Princeton University.

Eric Roegner, McKinsey & Company, Inc., BP Tower, 200 Public Square Suite 3900, Cleveland OH 44114, USA

Thomas Röthel is a principal in McKinsey's Cologne office. He leads the German utilities practice and is a member of the global chemicals sector. He has worked in various fields of the process industry with a particular focus on power and gas utilities, petrochemicals, and energy. His main areas of emphasis are strategy, regulatory issues, mergers and acquisitions, and post-merger management, as well as operational performance and continuous improvement. Thomas spent two years at the Houston office and has a PhD in technical chemistry from RWTH Aachen.

Dr. Thomas Röthel, McKinsey & Company, Inc., Magnusstr. 11, 50672 Cologne, Germany

Gabriele Roolfs-Broihan is vice-president corporate development of Brenntag Holding. She is responsible for strategy evaluation, market analyses and research, as well as internal project consultancy and the program management of Brenntag's top projects. Before joining Brenntag she headed the investor relations department of Stinnes. She has a PhD in business economics from the Georg-August-University in Göttingen.

Dr. Gabriele Roolfs-Broihan, Brenntag Holding GmbH & Co. KG, Corporate Development, Stinnes-Platz 1, 45472 Mülheim a.d. Ruhr, Germany

Detlev Ruland is a principal at McKinsey based in Dusseldorf. He is a member of the Business Technology office and co-leads its global IT performance management practice. He has worked in various fields of IT in the process, pharmaceuticals, healthcare, and energy industries, focusing on IT-enabled operations and performance transformations. He studied mathematics and computer science at RWTH Aachen and holds a PhD in natural sciences.

Dr. Detlev Ruland, McKinsey & Company, Inc., Königsallee 60C, 40027 Dusseldorf, Germany

Wilfried Sahm sits on the board of the European Chemical Industry Council and is director-general of the German Chemical Industry Association, where he also sits on the Presidential Council. Wilfried Sahm worked at Hoechst for 20 years, ultimately as head of agrochemical production and deputy head of the company's agrochemical sector. He studied chemistry in Münster, receiving his doctorate in 1969. From 1992 to 1998 he was an expert member of the Enquete-Kommission set up by the German federal parliament to investigate the protection of humanity and the environment.

Dr. Wilfried Sahm, VCI, sahm@vci.de

Sunil Sanghvi is a director of McKinsey and manages the firm's Chicago office. He is a leader in the firm's chemicals practice and has served clients on a variety of strategic, organizational, and operational issues, including in the areas of commercial excellence, portfolio reshaping, and multiyear growth. Sunil Sanghvi joined McKinsey in 1986 after working as a process engineer for Chevron Corporation. He has an MBA from Stanford Business School and bachelor's and master's degrees in chemical engineering from MIT.

Sunil Sanghvi, McKinsey & Company, Inc., 21 South Clark Street, Suite 2900, Chicago IL 60603-2900, USA

Christoph Schmitz is a consultant in McKinsey's Frankfurt office. He is a member of the global energy and materials practice and has worked in various fields of the process sector with a particular focus on the chemical and pharmaceutical industries. His main areas of emphasis are supply chain management, total operational excellence, and sourcing. He holds a Dipl.-Chem. and a PhD in chemistry from the University of Bayreuth.

Dr. Christoph Schmitz, McKinsey & Company, Inc., Taunustor 2, 60311 Frankfurt/M., Germany

Thomas Schreckenbach is a general partner and member of the Executive Board of Merck KGaA in Darmstadt, Germany. He is responsible for the Chemicals business sector, some Asian countries, and the Chemicals business in North America. He graduated from the University of Munich with a doctoral thesis in biochemistry. Since 1992 he has held a professorship in biochemistry at the Technical University of Darmstadt, and he has an honorary doctorate from the University of Southampton.

Prof. Dr. Thomas Schreckenbach, Merck KGaA, Frankfurter Str. 250, 64293 Darmstadt, Germany

Lennart Selander was Vice-Chairman of the Operational Board of Linde Gas and Engineering in Munich from 2000 until his retirement on 30 April 2005. He worked in the gases industry for more than thirty years and held various senior positions including President and CEO of the former AGA in Sweden. He has an MEng. degree from the Royal Institute of Technology in Stockholm.

Lennart Selander, Linde AG, Gas & Engineering, Seitnerstr. 70, 82049 Höllriegelskreuth, Germany

Feike Sijbesma has been a member of the managing board of directors at DSM since 2000. In 1987, he joined the Industrial Pharmaceuticals division of Gist-brocades, with responsibility for strategic planning and business development. After several jobs within the company he joined its executive committee in 1995. Gist-brocades became part of DSM in 1998. Feike Sijbesma studied medical biology at the University of Utrecht and business administration at Erasmus University in Rotterdam.

Feike Sijbesma, Managing Board of Directors DSM NV, P.O. Box 6500, 6401 JH Heerlen, the Netherlands

John Warner is a director in McKinsey's Cleveland office, and is a leader of the firm's chemicals practice. Since joining McKinsey in 1993, he has served a wide variety of clients in the chemical industry. John has worked on specific programs in areas such as growth, revenue management, corporate strategy, supply chain optimization, transformation, people development, and sourcing. He has an MBA from the Harvard Business School, a BA in economics from the College of William and Mary, and a General Course degree from the London School of Economics.

John Warner, McKinsey & Company, Inc., BP Tower, 200 Public Square Suite 3900, Cleveland OH 44114, USA

Thomas Weskamp is a consultant with McKinsey, based in Cologne. He is a member of the global chemicals practice and has worked in various fields of the chemical industry, focusing in particular on strategy and innovation-related topics. Before joining McKinsey he worked for Symyx Technologies in Santa Clara, California, a startup company applying high-throughput screening for the discovery of new materials. He holds a PhD in chemistry from the Technical University of Munich.

Dr. Thomas Weskamp, McKinsey & Company, Inc., Magnusstr. 11, 50672 Cologne, Germany

1
Today's Chemical Industry: Which Way Is Up?

Karsten Hofmann and Florian Budde

The chemical industry has emerged from a long struggle to sustain profitability, become leaner and more competitive, and use more sophisticated management approaches. Several eras in the industry's development have seen it change from exciting and innovative beginnings as the first ever science-based industry to its current maturity. Now a new era has begun which is once more about to reshape the world of chemicals: the era of shareholder value.

1.1
The Chemical Industry Today – A Snapshot

Today, the chemical industry is one of the largest and most diversified in the world. The total value of chemical products sold in 2003 was about USD 1.24 trillion (excluding pharmaceutical and consumer product sales of approximately USD 660 billion). Western Europe accounted for 35 percent of this, the United States for 23 percent, and Japan for 11 percent (Fig. 1.1). Within the OECD member states, chemicals and petroleum products make a larger contribution to GDP than any other manufacturing industry. In Western Europe, chemicals account for approximately 1.3 percent of the total economy.

The chemical industry consists of hundreds of segments with an estimated 70,000 different product lines manufactured by more than 1,000 large and mid-sized companies, plus countless very small ones. The sector can be characterized as consisting of many "mini industries" of varying sizes, and it counts virtually every other industry among its customers – from agriculture to construction and electronics (Fig. 1.2). The automotive industry, for example, relies on several different chemicals in the production of tires, seats, dashboards, and coatings, to name only a few major components.

Value Creation: Strategies for the Chemical Industry. 2nd Edition. F. Budde, U.-H. Felcht, H. Frankemölle (Eds.)

ISBN: 3-527-31266-8

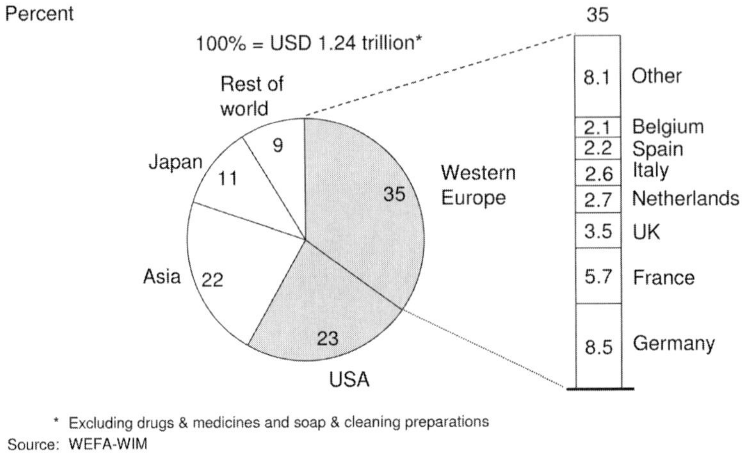

Fig. 1.1 Total chemical output by region, 2003.

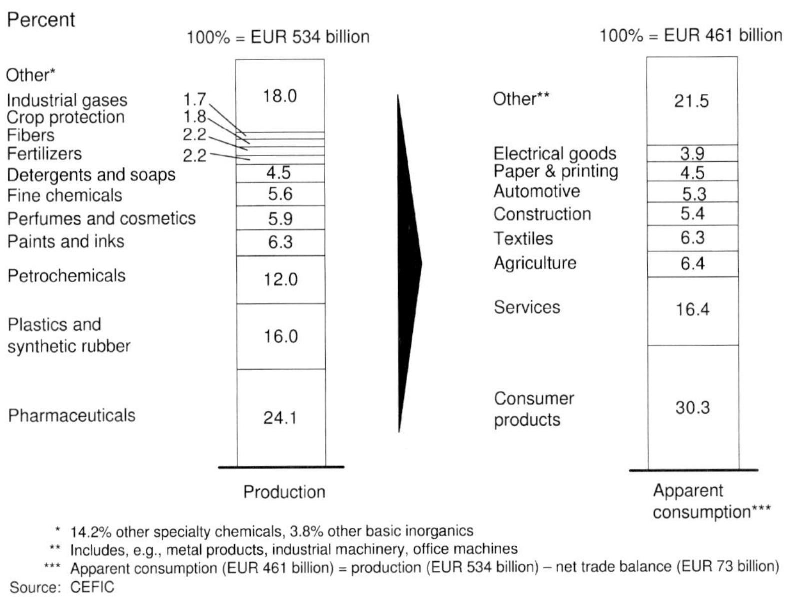

Fig. 1.2 Production and consumption of chemicals – Western Europe, 2003.

Chemical products can be roughly segmented into basic chemicals, polymers, specialty and performance products, industrial gases, and agrochemicals. However, for practical purposes they are usually classified as "commodities" and "specialties", the latter having higher value added and distinctive key success factors that allow for some additional differentiation between competing products. Albeit this distinction is commonplace in the industry, it should be kept in mind that chemical commodities are not nearly as commoditized as true commodities such as copper or crude oil, and therefore still offer much more opportunity for differentiation than these. Since chemical businesses are very heterogeneous, the range from “commodities” to “specialties” is a continuous spectrum.

Growth rates in the mature markets of Western Europe and the United States tend to be close to that of GDP – the minimum rate to maintain significance as an industry in the general economy over time. However, there is significant growth in the emerging markets. In China and South East Asia in particular, the industry’s growth rates have been impressive and are also forecasted to remain higher than the growth of GDP (e.g., 9.8% per annum compared with 7.3% per annum in the period from 2003 to 2010 in China, 7.0% per annum compared with 5.3% per annum in Malaysia).

For the period until 2010, annual double-digit growth of chemicals output is projected for China and for Asia excluding China and Japan (Fig. 1.3). Since China is expected to remain a net importer of many chemicals far beyond 2010, this demand will also drive output growth in other regions where domestic demand is only increasing slowly.

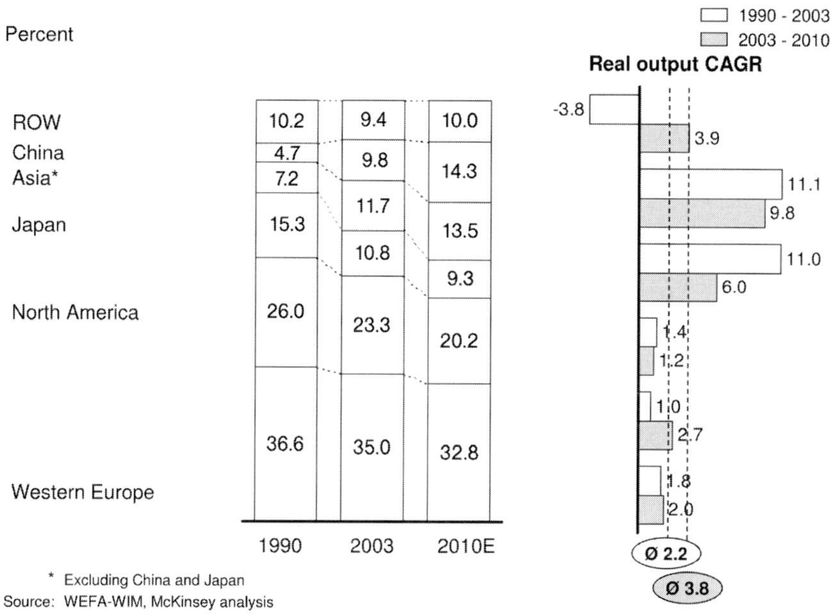

Fig. 1.3 Global chemicals output by region, 1990–2010 (nominal gross output, excluding drugs & medicines and soap & cleaning preparations).

The top players in the chemical industry are as global as any of their counterparts in other industries. Almost 40 percent of the sales of the top ten chemical companies originate from overseas countries. The comparable number for the top ten automotive and electronics firms is roughly 30 percent, and for the top pharmaceutical and oil companies around 50 percent. The industry has a distinctly regional character, however, in the sense that there are only limited trade flows between the three main manufacturing regions (North America, Europe, and Asia including Japan). Only 11.2 percent of total output in 2003 was shipped between the three regions (Fig. 1.4).

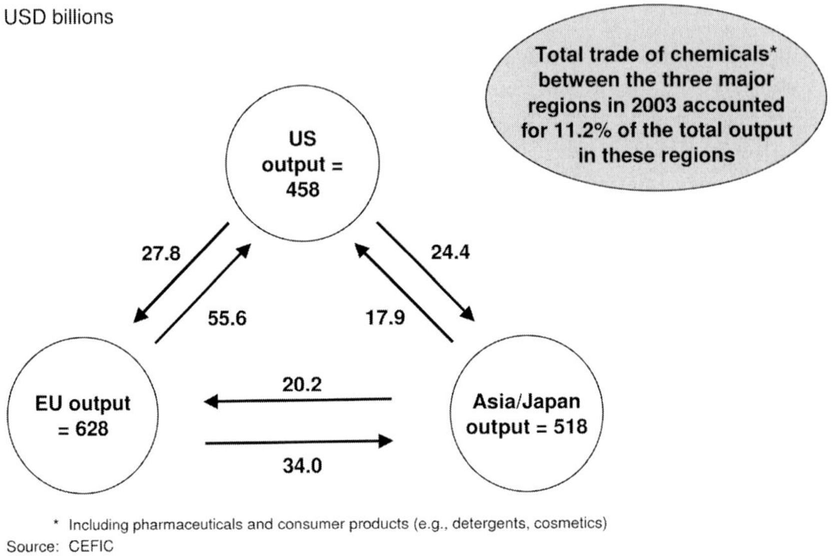

Fig. 1.4 Inter-regional chemicals trade, 2003.

Not surprisingly, inter-regional trade is particularly limited for volume products, which are relatively expensive to transport. Nevertheless, the limited inter-regional trade is sufficient to couple prices and industry cycles in the different regions worldwide. The prices of basic commodity plastics, for example, moved in remarkably close harmony in all three regions for the last 25 years.

In most places, however, trade within the regions is very strong. For instance, 55 percent of the European Union's total chemicals output in 2003 was exported within the region (including Central and Eastern Europe); 20 percent of it was exported outside Europe; and only 25 percent was consumed domestically.

Germany and the United States are the world's largest exporters of chemical products. In 2003, Germany exported USD 99.0 billion's worth and the United States USD 91.4 billion's worth. However, the United States is a net importer of chemical products overall while Germany is still the largest net exporter (Fig. 1.5).

Trade patterns for chemical products differ markedly from those of overall trade. Countries like China and Taiwan, for example, which have large surpluses in their overall trade accounts, have sizeable deficits in their trade in chemicals. The United States and Britain, on the other hand, have large trade deficits overall, but trade surpluses, or at least somewhat balanced trade flows in chemicals.

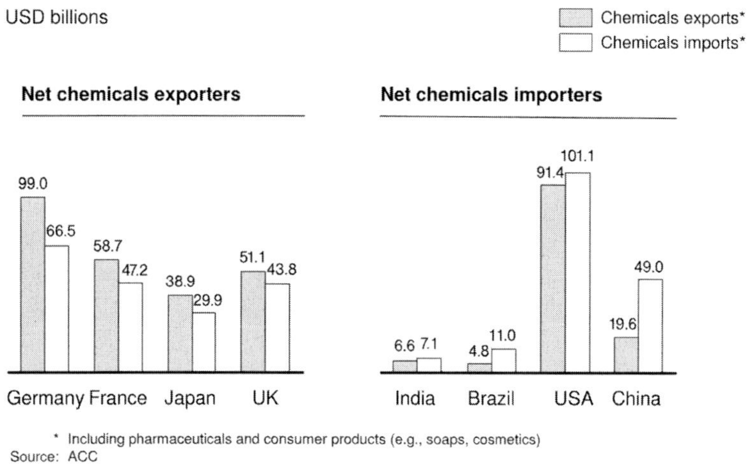

Fig. 1.5 Chemicals import/export balance of major economies, 2003.

As mentioned above, one interesting aspect of the chemical sector is that the industry as a whole is highly fragmented. The top ten companies in chemicals (excluding pharmaceuticals) account for only 18 percent of the total market, well below comparable values in other industries, such as automobiles, where the top ten firms account for 85 percent of sales, or pharmaceuticals, where the top firms account for more than half of all sales (Fig. 1.6). While the overall consolidation level is low, several product segments have already consolidated. The top ten manufacturers of acrylic acid, for instance, account for 85 percent of their market. The top ten manufacturers of crop protection products account for 87 percent of their market, and the top ten in paints and coatings for 42 percent.

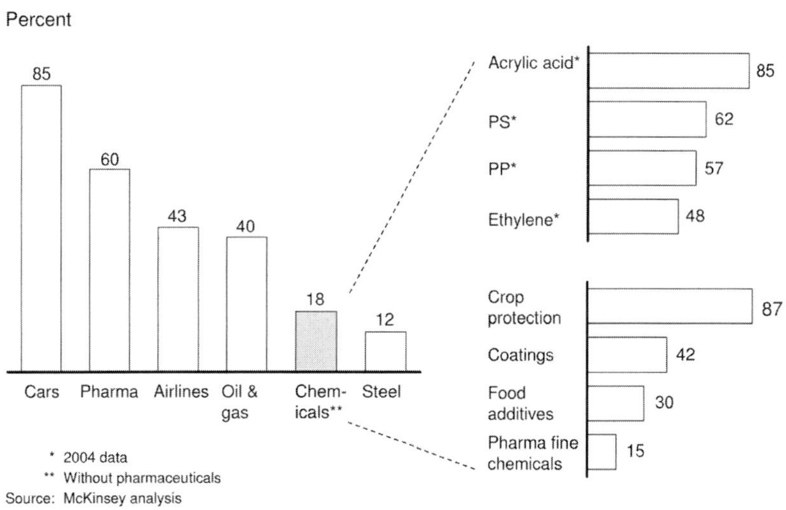

Fig. 1.6 Concentration level of the chemical industry – top ten worldwide market share, 2003.

The level of concentration in the industry also varies by region, with North America in general showing the highest concentration and Asia the lowest. For acrylic acid, for example, the top four manufacturers in the United States account for the whole market there. In Western Europe the top four account for 98 percent, but in Asia (excluding Japan) they account for only 63 percent of the market. The comparable figures for the top four producers of polystyrene in North America, Western Europe, and Asia (excluding Japan) are 84 percent, 72 percent, and 37 percent respectively.

There are plenty of other examples (e.g., polypropylene and HDPE) where the Herfindahl Index, a measure of industry concentration frequently used by antitrust authorities, is rather high in North America and Western Europe. This necessarily limits the opportunities for Western players to grow by means of mergers and acquisitions in their own domestic markets.

The large number of Asian players in several industry segments often results in sub-scale plants with inefficient operations. The average plant size of an acrylic acid manufacturer in Asia (excluding Japan), for instance, is less than one-fifth of the average size of a corresponding plant in North America. It should be noted, however, that since Asia is the major growth region for the chemical industry, the Western players' scale advantage will quickly diminish as new world-scale plants come on-stream in Asia over the next few years. In ethylene or polypropylene, for example, the five largest plants in East Asia will measure up to their North American and Western European counterparts or even surpass them in terms of average size by 2008.

Nine of the top ten firms in the industry are Western, and for the best part of the last two decades of the twentieth century there was little change in the names appearing in annual lists of the top ten. Eight of those in the 1980 listing, for example, also appeared in 1998 (Fig. 1.7).

Changes in line-up of chemical players*

1980	1998	2003
1. Hoechst	BASF	Dow
2. Bayer	DuPont	BASF
3. BASF	Bayer	DuPont
4. DuPont	Dow	Bayer
5. ICI	ICI	Total
6. Dow	Hoechst	ExxonMobil
7. Union Carbide	Shell	BP
8. Shell	Akzo Nobel	Shell
9. Exxon	Exxon	Mitsubishi
10. Montedison	Elf Aquitaine	Degussa

Major MA&A changes since 1998

- Hoechst and Rhône-Poulenc spun off their chemicals businesses (Celanese and Rhodia) and merged the life science parts to a new entity (Aventis)
- Mergers of oil companies
  - Exxon and Mobil
  - Total, Petrofina, and Elf
  - BP and Amoco
- Mergers of chemical companies
  - Dow and Union Carbide
  - Degussa-Hüls and SKW

* Chemical sales only
Source: C&EN, McKinsey analysis

Fig. 1.7 Changes in top ten chemical companies.

However, between 1998 and 2004, some dramatic shifts occurred as the result of a number of large mergers and acquisitions. On the one hand, Hoechst and Rhône-Poulenc spun off their traditional chemicals businesses (to become Celanese and Rhodia) and merged their life science divisions into a new entity named Aventis, now part of Sanofi-Aventis. On the other hand, several other companies – mainly oil-based – merged their businesses in order to gain economies of scale. The result was that several new names appeared around the turn of the century in the top ten: BP Amoco, ExxonMobil, and TotalFinaElf. The top ten line-up continues to see major changes as Total, BP, and Bayer are spinning off major parts of their chemical businesses, i.e., Arkema, Innovene, and Lanxess, respectively. The acquisition of Millennium by Lyondell will propel another North American player into the global top ten list.

Despite the regional structure discussed earlier, all the top ten giants are truly global companies with large parts of their business in each of the three main regions. In addition, the portfolios of products manufactured by these giants are very diverse, with the major conglomerates ranging from primarily commodity players to hybrids which operate at all levels of the industry, with Degussa the only one of the group to focus entirely on specialty chemicals.

The management challenges in each of these major categories differ considerably. In Chapter 3 we will return to these and describe the potential strategic choices for the chemical industry.

1.2
Eras of the Chemical Industry

Early chemical companies can be compared to the upcoming software and computer giants of the 1990s: new, innovative, and completely revolutionary. The industry remained exciting and attractive until as recently as the last twenty to twenty-five years, but with increasing maturity and declining innovative potential it has come to be regarded as somewhat dull, a less attractive investment or employment prospect. What, then, does the future now hold?

It may be useful to set the scene for our analysis of the chemical industry's current situation and future challenges by taking a brief look at its evolution, and, in particular, by outlining our understanding of the changes in general industry dynamics in the various chemical businesses over the past decades and up to the present day.

Industrial historians trace the origins of today's chemical industry back to the discovery in 1856 of a synthetic process for manufacturing mauve dye. Before that date, chemical production was largely confined to the unsophisticated processing of inorganic compounds found in the ground, and was little more than an offshoot of mining. The discovery of this organic dye, by the Englishman William Henry Perkin, marked the beginning of the world's first science-based industry.

From the moment of Perkin's discovery until today the development of the chemical industry falls into six major eras, during which it has been transformed into the world's largest manufacturing industry, now indispensable to almost every manufacturing process (Fig. 1.8).

	Foundation	Develop-ment	Expansion	Diversifi-cation	Maturity	Shareholder value
Scope	• Scientific discoveries • New mole-cules	• Upscale from laboratory to production • Emergence of polymers	• Substitution • Internation-alization	• Broadening of product range	• Global competition • Restructuring • Price-cost squeeze	• Focus on financial returns • Ongoing price-cost squeeze
Key fac-tors of success	• R&D • Access to raw material	• Process development • Engineering • R&D	• Process technology • Integration • International presence • R&D	• Sales • Marketing • Process and application technology	• Operational excellence • Economies of scale • Market conduct	• Operational excellence • M&A capabilities • New business generation
Growth relative to GDP	• Low	• Medium/ high	• High	• Medium/ high	• Medium	• Medium
Industry structure	• National, fragmented	• National, multidivision companies	• Appearance of multi-nationals	• Move to con-glomerates	• Consolidated multibusiness chemical con-glomerates	• ?

Timeline —— 1920 —— 1950 —— 1970 —— 1985 —— 1995 ——→

Source: McKinsey

Fig. 1.8 Era analysis chemical industry.

Foundation: In this era the chemical industry was putting down its roots. As the industrial revolution had created demand for many chemical products such as dyestuffs, this period was characterized by the discovery of new molecules, based for the most part on different types of hydrocarbons. In addition to R&D, access to raw materials was the key success factor. With the development of the coal industry in the nineteenth century discoveries were plentiful. This was the starting point of many national chemical companies which still exist today as global players. Production during this period, however, was small-scale and fragmented.

Development: The industry then moved out of the laboratory and into the factory. The emphasis shifted from discovery to large-scale organized production. Engineering and process development became as important as chemical R&D. The Haber-Bosch ammonia process was the leading example of a catalytic process operated on a large scale, requiring important advances in process design and engineering. Companies in the industry began to develop more sophisticated multidivisional structures, and growth began to accelerate. One important reason for that was the emergence of polymers: most of the modern mass polymers like polystyrene, PVC, polyethylene, and polyurethanes were discovered between 1930 and 1940.

Expansion: Chemical companies grew rapidly as a result of the increasing substitution of synthetic products for natural materials. Plastics, in particular, replaced natural products such as paper, wood, or cotton in many applications, as they were cheaper and easier to process. Based on the importance of crude oil as a raw material for nearly all organic chemicals, large integrated chemical complexes like those in the North American Gulf Coast region or Rotterdam and Antwerp in Europe were established, and economies of both scale and scope became important. Another major trend was that the chemical companies began to spread outside their home markets. True multinational organizations began to emerge, as international presence became as important to success as scientific discovery and process development.

Diversification: Companies broadened their product ranges even more widely. The key to success here switched from R&D to processes and application technology as well as customer service. This change resulted in the build-up of strong customer service and technical support departments. This was the stage when many players were transformed into chemical conglomerates, with highly diversified product portfolios ranging from commodities through specialty chemicals to pharmaceuticals. The industry's growth rates – still well above GDP – decreased, however, after the oil crises (1973 and 1979).

Maturity: The creation of new products slowed down and the industry became more focused on global competition between firms. It was harder for companies to grow faster than GDP, and they were forced to achieve operational excellence through aggressive restructuring and cost control. The number of mergers and acquisitions began to accelerate in order to consolidate the industry and to achieve further cost synergies through economies of scale.

Shareholder value: In this most recent phase the main focus is on financial returns, that is, on the value created for investors. M&A activity in the industry continues to be on the surge. It is mainly driven by the need of established players to focus on the most promising parts of their portfolios to create value for their shareholders. This translates at the same time into disinvestment of non-core assets and into bolt-on acquisitions to strengthen existing core platforms. Private equity players, who are responsible for a good share of chemical M&A, are playing an important role in demonstrating that significant hidden value is still waiting to be unlocked.

Obviously, these six eras have occurred at different times in different parts of the world. In addition, the degree of participation has varied. In the early years, Europe led the way. Britain and Germany, with the USA joining in from the 1920s onwards, mainly drove the foundation and development phases of the industry. The expansion, diversification, and maturity eras took place in Europe and the United States at more or less the same time. They occurred in Asia rather later and somewhat less intensely. The era of shareholder value, however, came first and fastest to the United States. European firms have followed their US counterparts into this phase in the last ten years, and the same is currently happening at many Asian companies.

To understand this recent evolution and the perspectives for the chemical industry, it is essential to understand the capital market point of view, which drives the strategies of most global chemical players today. The importance of the shareholder value era for the chemical industry is discussed in the following two chapters.

1.3 Summary

In the early years of the shareholder value era, the current status and past development of the chemical industry are outlined:

- The chemical industry is one of the world's largest and most diversified, consisting of many "mini industries" of varying sizes.
- Its growth is close to GDP, the minimum rate to maintain significance as an industry over time. However, there is significant growth potential in the emerging markets, particularly China and South East Asia.
- The top players are global companies, although the industry has a distinctly regional character.
- Though some major mergers and acquisitions have occurred, the industry is still very fragmented in most segments
- The authors define six major eras of the chemical industry from its foundation through to maturity, ending with the most recent phase in which the focus is on the value created for investors – shareholder value.

2
Shareholder Value Orientation: Not a Question of Whether, but How

Bernd Heinemann and Thomas Augat

After enthusiastically embracing the shareholder value orientation in the 1990s, many companies view the concept today with much more sober eyes. There is considerable doubt as to whether the stock market is really a solid yardstick of long-term corporate performance rather than favoring short-term earnings effects and management fads, effectively inclining companies to make wrong decisions. Nevertheless, we demonstrate here that the foundation for shareholder value orientation is as valid as ever, although the way it is actually practiced needs to be significantly enhanced. There is conclusive evidence that capital markets do generally value companies in line with their long-term value creation potential – but also some evidence of situations where capital markets temporarily deviate from a substantiated assessment of fundamental value creation. The concept of an advanced shareholder value orientation proposed here consists of two elements: first of all, executives should be aware of the fact that value orientation is about creating long-term fundamental value rather than maximizing stock prices in the short term. Second, companies should free themselves from a too naïve faith in capital markets, and take a more sophisticated approach to analyzing capital market valuations and managing their implications for long-term value creation in terms of financing choices and communication to capital markets. This advanced shareholder value orientation can guide chemical companies towards better shareholder returns.

2.1
Capital Market Deviations as the Key Challenge for Shareholder Value Orientation

Shareholder value orientation (SHV) is by no means a new concept. Maximizing fundamental value to shareholders has been the dominant paradigm in finance and economics for decades and, after all, running a corporation in the primary interest of its shareholders, i.e., its owners, appears to be a straightforward concept in a market economy.

Over the last decade in particular, shareholder value orientation has established itself as a guiding principle for corporate management, including the vast major-

Value Creation: Strategies for the Chemical Industry. 2nd Edition. F. Budde, U.-H. Felcht, H. Frankemölle (Eds.)

ISBN: 3-527-31266-8

ity of chemical companies. Almost every chemical company states that providing a good return to shareholders is at least one key corporate objective, if not the dominant one. Many chemical companies have adopted value creation-based metrics such as return on capital or economic profit and are actively communicating them to the capital markets. For example, BASF set itself a target of covering a return on capital of 10% over a cycle, and Degussa is shooting for a two percentage point spread on its cost of capital.

Today, though, many enthusiastic adopters take a very much more sober view of an SHV tainted by its association with the rising stock markets of the late 1990s. The bursting of the technology bubble, some cases of excessive management compensation through stock options, and several prominent examples of plain accounting fraud have raised considerable doubts as to whether a primary orientation towards capital markets is a sustainable approach to corporate management. Many feel that SHV inclines companies to focus on next quarter's earnings expectations instead of on a sound strategy based on reasonable long-term objectives.

In actual fact, however, shareholder value orientation is essentially about management decisions that create fundamental value. Fundamental value, defined as the sum of expected discounted cash flows available to shareholders, is by that definition a long-term concept. In theory, a company could pursue SHV without even caring about its share price (Fig. 2.1).

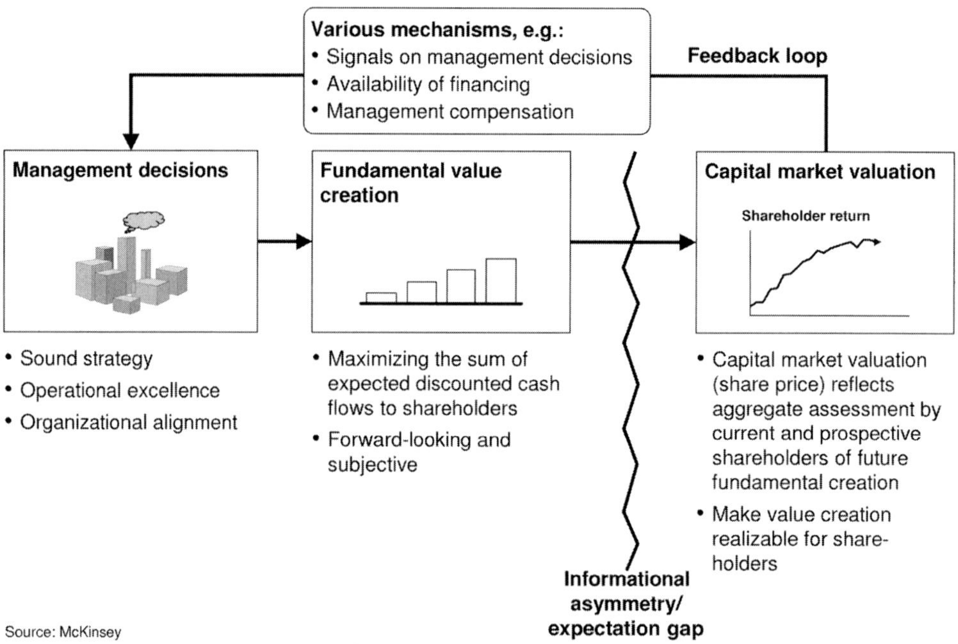

Fig. 2.1 Conceptual framework for shareholder value orientation.

Nevertheless, to disregard the concerns mentioned above would be to ignore important issues and be simply impractical. Stock prices make fundamental value creation tangible: in order to realize the value creation potential, shareholders need to sell their stocks to somebody else at a current market price.

Furthermore, the capital market valuation of a company directly affects fundamental value creation by several feedback mechanisms: fundamental value is a subjective concept, since it is by definition about expectations rather than observable facts. And there is always informational asymmetry between shareholders and management – managers on the spot will always know more. On the other hand, stock market prices are observable and reflect the aggregate opinion of current and prospective shareholders on the company's value. Furthermore, a significant share of management compensation is often tied to share price development, to align the interests of management and shareholders. Hence, the signals stock market reactions send to management and shareholders are essential.

Additionally, if there were significant discrepancies between fundamental value creation and its capital market assessment as reflected in stock prices, it would be impossible in many cases to actually initiate or sustain a strategy perceived to increase fundamental value, because, for example, the necessary financing would not be available.

Thus, doubts as to whether capital markets reflect a reasonable representation of the fundamental value of a company, and the significance and persistence of potential deviations, pose a significant conceptual and practical challenge to shareholder value orientation.

2.2 How Capital Markets Reflect Fundamental Value Creation

It is illuminating to examine the evidence on how capital markets reflect the chemical players' fundamental value creation. The analysis below employs a variety of new analytic approaches and makes use of an exhaustive proprietary performance database for the chemical industry.

2.2.1 Capital Markets and Effects of Earnings Announcements

Many perceive capital markets as mainly reacting to short-term earnings fluctuations, effectively reducing their assessment of companies to the question of how earnings per share changed from one year or even quarter to the next in relation to expectations. The resulting pressure to meet or exceed earnings per share expectations would indeed force executives to focus on the short term instead of on long-term value creation, because many key variables for long-term value creation, such as the R&D budget or investments, are likely to depress earnings in the short term. Furthermore, the emphasis would shift towards accounting metrics rather than cash flows, inclining executives to think about creative accounting (in its legal or even borderline variety) rather than about how to improve their business.

Fortunately, these concerns are not in line with empirical evidence. Analysis reveals that chemical company values do not show a systematic relationship between discrete earnings surprises relative to the consensus expectations of analysts on excess shareholder returns, either a fairly long time after the announcement, or, even more interestingly, very close to it. Figure 2.2 shows what happens when earnings per share either disappoint or delight unexpectedly. The "surprise factor" is based on the difference between the reported and the forecast EPS for the year. Looking on either side of the baseline, it is evident that the number of companies with returns to shareholders above or below FTSE Chemical Europe performance is fairly evenly distributed. If the capital markets followed short-term fluctuations, we would expect to find all or most of the "disappointing" companies below zero on excess returns, and the "delightful" ones above (see also Koller, T. and Rajan, S. R., 2002).

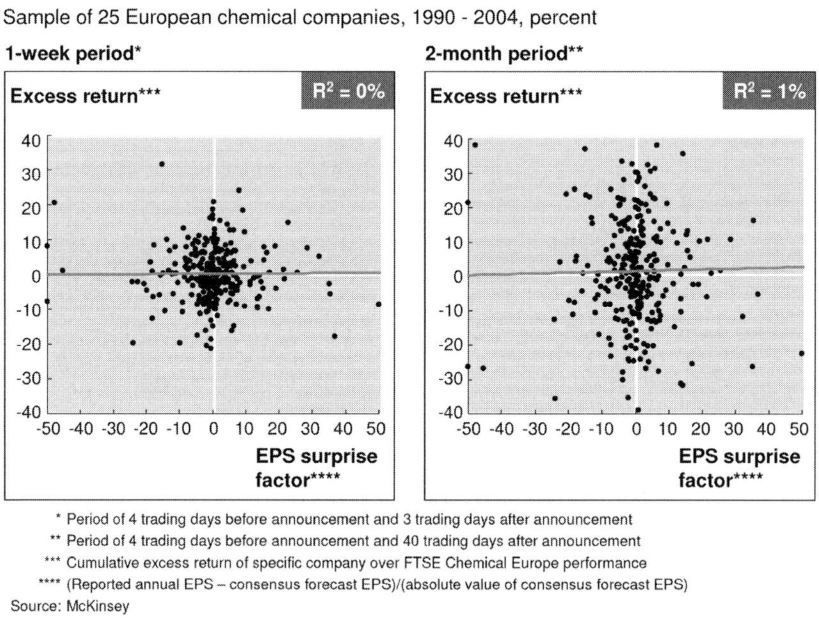

Fig. 2.2 Effect of earnings surprises on shareholder returns.

This means that capital markets appear to see through short-term earnings surprises caused by accounting issues or short-term deviations from a longer trend. Naturally, this does not mean that earnings announcements are irrelevant. In many cases, they actually convey important fundamental information to capital markets. For example, it is obvious that the capital market will not develop any confidence in a chemical company's growth story if this growth is not reflected in the reported profits over a prolonged period.

Short-term profit volatility and analysts' forecasts, then, have much less impact on a company's share price than is often assumed. Capital markets have a longer-term perspective. Furthermore, in an industry like chemicals, even if we assume a continuous improvement in fundamental performance, the attempt to generate an equally continuous, volatility-free growth in profits will be virtually impossible because of structural factors like feedstock volatility, currency exposure, and seasonal effects. Chemical companies should learn to live with earnings surprises and concentrate instead on communicating a concrete long-term strategy – and of course deliver appropriately on capital market expectations, which will be correctly set on this basis.

2.2.2
Capital Markets and Expectations on Fundamental Value Creation

The technology and Internet bubble passed by the chemical industry and other traditional industries. Additionally, we have demonstrated above that capital markets pay less attention to short-term earnings than many assume. But how, then, does the capital market value the chemical industry?

The answer is: on average, very much in line with the principles of fundamental value creation.[1] This can be demonstrated using a simple valuation model which maps the key value drivers of return on capital, long-term growth, and cost of capital to a predicted fundamental valuation level for the industry (Fig. 2.3). To generate the input data for this valuation model, a proxy for the capital market expectations on the future development of these value drivers has to be identified. We used a two-stage model where return on equity and growth at a given point in time were assumed to converge to long-term industry averages over a period of 4 years.[2] Although this 'mean-reverting' approach is a rather crude, but conservative, assumption on capital market expectations, it is evident that the actual price-earnings ratios track the fundamental line based on long-term averages pretty closely on average.[3] For example, the valuation level of the industry in 2003 can

1) See for a more general discussion in the context of behavioral finance theory Goedhart, M. et al., Spring 2005.

2) To allow for maximum consistent data over a long time frame, we used equity related metrics for this analysis. The basic approach is that the valuation level of the industry (its price-earnings ratio) can be linked to the fundamental value drivers as follows:

$$P/E = (1-g/RoNE)/(CoE - g)$$

with

P/E	Price-earnings ratio
RoE	Return on equity
g	Growth in equity
RoNE	Return on new invested equity
CoE	Cost of equity

As it proved to provide more stable results, growth and cost of equity are modeled in real instead of nominal terms. The actual formula used is more complex to allow for a linear reversion of current values to long-term averages.

3) The fact that our fundamental valuation model appears to generate values which are, although significantly correlated to actual valuation levels, generally somehow lower than the actual ones might be attributed to a certain misspecification of the model. However, given the factual TRS of US chemical companies, it might also partially reflect a tendency of overoptimistic expectations on specialty chemical companies' performance, see Goedhart, M. et al., Autumn 2001.

be reconciled with the currently rather low forward-looking real cost of equity for the industry of approximately 6.5 percent, a gradual improvement of the industry's return on equity to its long-term average of around 14 percent (from approximately ten percent), and a gradual decline of the strong growth rates back to their long-term level of around one percent per annum in real terms. Indeed, given the actual capital market expectations of a strong cyclical recovery, this model explains the actual valuation level of the industry well.

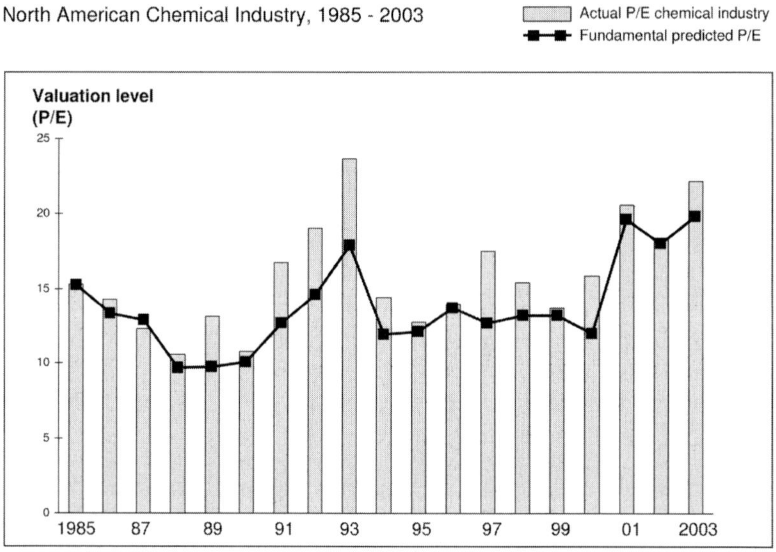

Source: McKinsey

Fig. 2.3 Actual vs. 'predicted fundamental' valuation level.

However, this does not rule out the possibility of exaggerations and temporary wrong valuations. The fine chemicals bubble of the late '90s is a case in point. At this time, fine chemicals was seen as a sector with enormous growth prospects, at least for those companies with scale advantages, with returns comparable to the pharmaceuticals business. In the eyes of the capital markets, a significant share of fine chemicals business was a decisive value driver even in large, diversified chemical companies, and was rewarded with valuation premiums. Companies with a focus on fine chemicals, such as Lonza in Europe or Cambrex in the USA, were actually valued at the same level as large pharmaceutical enterprises with their patent-protected product portfolios. As we now know, these highly optimistic estimations proved false, and many fine chemical companies have been in a trough since 2002.

Of course, the very optimistic expectations were not a phenomenon restricted to capital markets, nor were they necessarily irrational at that time – there may have been good reasons for a very optimistic outlook, just as there were good reasons to believe in sustained above-average growth in the new economy as a result of the productivity boost provided by the new communication technologies. But one feature was to be observed where capital market expectations and price reactions undoubtedly had a major role to play: the behavior was highly homogeneous, everyone wanted to take part in the boom. Many companies announced huge investments in their fine chemicals businesses, enormous sums were paid for the acquisition even of fairly small companies – even measured against the flourishing profits of the time. After the high expectations turned out to be wrong, many of the investments had to be written off completely, which caused a number of companies considerable balance sheet problems, not to mention the destruction of shareholder value. Hence, capital markets can from time to time send the wrong signals, leading companies into ventures that would better have been avoided.

Let us look at another example. Commodity chemicals, too, suffer from potential overreactions – not, in this case, because of wrong estimations of growth opportunities, but because capital markets are fooled to some extent by the industry's marked cyclicality.

Perfect capital markets should "see through" the cycle and the reaction of valuations to purely cyclical fluctuations in the industry return on capital should be insignificant. To examine this, we again compared the actual valuation level to a fundamental predicted valuation level for commodity chemical companies. To pinpoint the effects of cyclicality, however, we used a slightly different approach. As we wanted to evaluate capital market expectations, we compared the actual valuation level with a fundamental valuation level based on 'perfect foresight', i.e., assuming that the capital market knew the actual development of the key value drivers for the next 8 years and evaluated this information in line with its implied fundamental value creation.[4] Furthermore, we used a capital structure-adjusted market-to-book ratio instead of an earnings-based metric for the valuation level.

It can be demonstrated that the capital market overreacts – both upwards and, apparently to a lesser extent, also downwards – to purely cyclical fluctuations in industry performance.[5] Our chart shows that the indexed market-to-book ratio roughly follows the normalized ROIC over time, rather than staying close to the 100 percent valuation level which is perfectly in line with future value creation (Fig. 2.4).

In this case the key problem is that financing might be very hard to obtain in trough times, which is the only time commodity chemical companies should consider adding capacity, in theory at least. Conversely, in an upturn, financing is available more easily and there might even be some pressure from capital markets to invest in new capacity – in fact preparing the overcapacity that will cause the next downturn (see also Chapter 6).

4) And assuming a performance at industry average for the years thereafter.

5) It appears that capital markets react more strongly to peaks than to troughs in industry performance. However, it should be noted that this is potentially not a reflection of a specific pattern of overreactions to cyclicality but is rather driven by the declining trend in industry ROIC over the last years.

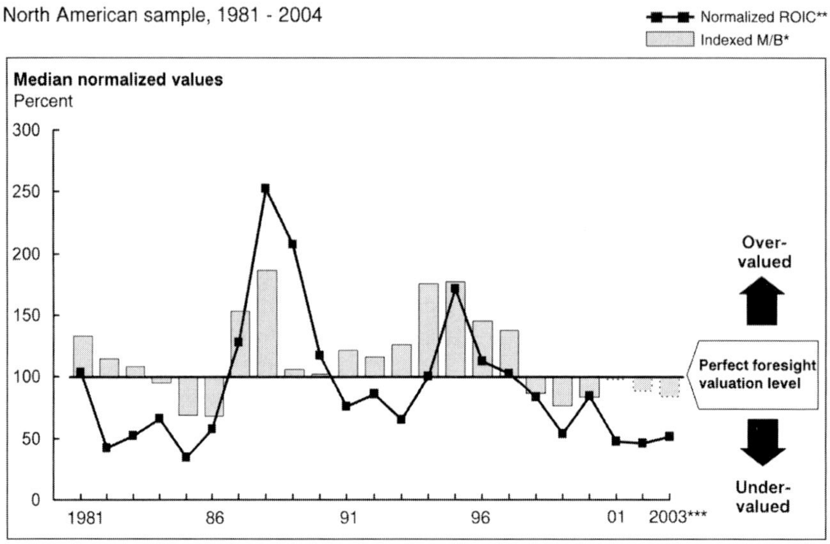

* Indexed M/B = actual median M/B divided by M/B with perfect foresight
** Normalized ROIC = industry median ROIC divided by long-term average median ROIC of 10.4%
*** Perfect foresight approximated by using data from last observed cycle
Source: McKinsey

Fig. 2.4 Overreaction to cyclicality in commodity chemicals.

Many experienced investors might even be aware of this mechanism, at least qualitatively. However, it is very difficult to exploit it under the typical institutional constraints for most investors. Assuming that it were possible to pinpoint a cyclical low precisely, the stock would have to be held for an average of four years. The typical fund manager, who has to measure his investment results against the benchmarks annually or even every six months, would have to summon up a great deal of courage to place a significant bet on an investment opportunity of this nature. In addition, even where the signs indicate that the historical pattern will basically continue in the future, there is still a considerable risk that it will change in terms of timing and magnitude. For example, in the late 1990s many expected to see the next cyclical peak for commodity chemicals in 2001 or 2002 – but then September 11 and other events sent the industry into a prolonged trough.

On average, then, capital markets follow the principles of fundamental value creation remarkably closely. Nevertheless, there are deviations, which are likely to be persistent or recurring because they are difficult to exploit. These deviations can put chemical companies at considerable risk in pursuing a strategy focused on fundamental value creation, but as we see below there are tools to help manage this risk.

2.3
New Tools for an Advanced Shareholder Value Orientation

In light of the results above, the rationale for SHV seems as valid as ever. However, to address their skepticism and try to manage the risks from capital market deviations, many companies need to develop a more sophisticated understanding of the true meaning of SHV – an advanced shareholder value orientation.

Making business decisions that create fundamental value is the only way to increase a company's market value in a sustainable way. All other measures often associated with SHV, such as stock options, switching to international accounting standards, dual listings, or the like, are auxiliary. Shareholder value results from a superior strategy and operational excellence. It is not a substitute for insight, good management decisions, and organizational alignment around these decisions, but rather the outcome of these factors. Hence, the first element of an advanced SHV orientation is for a company to rejuvenate (if necessary) the conceptual foundations of focusing on long-term value creation.

The fact that capital markets do sometimes deviate from fundamental value creation means that a pure focus on it is not enough. As mentioned above, these deviations represent a significant risk to the pursuit of a strategy centered around long-term value, yet they may also from time to time provide opportunities for further value creation. To address the challenges of these opposing phenomena, an advanced shareholder value orientation has to focus on fundamental value creation while liberating itself from a naïve faith in capital markets. Companies need to develop a superior understanding of the drivers of their capital market valuation and potential gaps to their own best estimate of their fundamental value. Integrating financing choices with strategy helps in managing downside risks arising from the times when capital market valuations deviate from fundamental value. And finally, a new approach to investor relations might help to close gaps between fundamental value and the current capital market valuation. Some useful approaches and tools are discussed below.

2.3.1
Capital Market Diagnostic: Gain an In-depth Understanding of Capital Market Signals

The starting point for an advanced SHV orientation is a clear understanding of the company's capital market valuation. This may appear an obvious point, but in our experience many companies rarely go beyond tracking TRS in comparison with competitors and talking regularly to analysts and I-banks. There is more to it than that: chemical companies need to develop a superior understanding of the drivers of their capital market valuation and potential gaps to their own best estimate of fundamental value.

The first step is to conduct a valuation of the current business plan, using multiple scenarios if possible. A continuous, internal company valuation is relatively easy to implement, and creates crucial transparency of the value creation potential of the current business plan and its drivers.

Second, the capital market's valuation should be analyzed using a broad set of analyses that will enable the company to decipher the information embedded in stock prices:

- *Historical capital market performance:* at first glance, historical shareholder returns often appear rather erratic. However, it takes only a little math to create transparency for the key drivers: historical shareholder returns can be decomposed into components such as peer group performance and company-specific (excess) performance. This company-specific element can again be decomposed into dividend yield, changes in realized performance (earnings growth), and changes in expected performance (changes in valuation level). This type of analysis often provides interesting insights into the longer-term drivers of past capital market performance.
- *Explicit formation of expectations by analysts:* although it is arguable whether sell side analyst opinions are truly representative of the market, an analysis of research reports provides further important information. First of all, this means making a systematic analysis of the analysts' forecasts for the coming years and the important factors of influence from their point of view. Quite often, significant differences are found here from one analyst to another, but in particular differences will be found between the analysts' forecasts and the company's own corporate planning. Second, companies should develop a profound understanding of analysts' approaches to valuing the company in the context of the industry and their assessment of structural trends. Furthermore, because highly technical details of valuation methods often have a substantial impact (e.g., cost of capital, terminal value, definition of net debt, etc.), an in-depth understanding of the analysts' approaches to this area should also be obtained.
- *Implicit capital market expectations:* it is meaningful to translate observable – and normally fundamentally correct – capital market valuations into expectations about fundamental performance. Companies can use a standard valuation model to reengineer the current market value by entering assumptions on growth and profitability in such a way that the resulting fundamental DCF value equals the current market value. Frequently, however, only very vague conclusions can be reached because of the large number of conceivable constellations of the various parameters. It is therefore often more useful to make a rough breakdown of the present market value, supported by a comparison with competitors, into the value of present performance and growth expectations.

Third, a number of important insights can be gained by comparing the internal company valuation with the results of the analysis of the capital market valuation:

- A comparison with capital market expectations is a valuable touchstone for in-house corporate planning. Currently, capital market expectations will fairly frequently be much lower than internal plans, indicating a possibly over-optimistic estimation (the well-known "hockey stick").[6]
- Explicit translation of the company's own corporate plans into a fundamental value and, conversely, of the valuation into implicit assumptions also helps to identify wrong perceptions in the capital market, and to protect the company at the same time, for example, from excessively expensive acquisitions.
- Finally, the company's own estimation of fundamental value will frequently be more or less in line with the capital market valuation. This would mean that if the corporate plans are realized in the succeeding years, a shareholder return can be expected of exactly the level of the cost of equity, i.e., roughly the industry average. Here, the valuation logic can be used to set more ambitious objectives in relation to capital market performance, and to translate these into operational targets for profit improvement.

An example of how the actual valuation level, the company's value creation targets, and capital market expectations can be integrated into a framework to assess and derive value creation aspirations is given in Figure 2.5.

6) The opposite case of a significant overvaluation if compared with business plans, although currently rather unlikely, might be a situation requiring thorough analysis as well (Jensen, 2004).

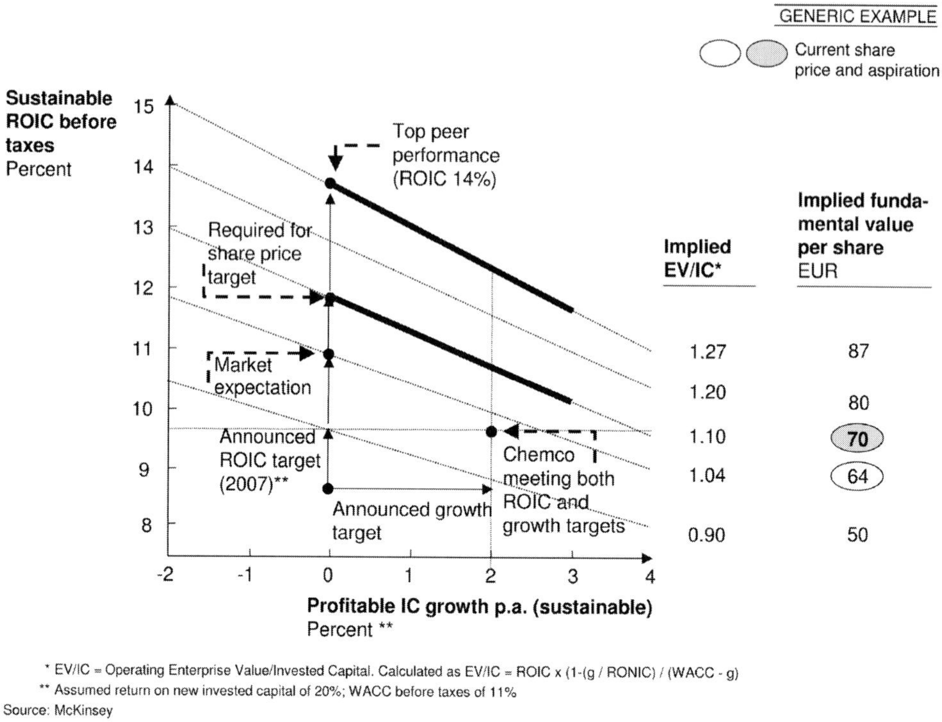

Fig. 2.5 Assessing value creation aspirations.

2.3.2
Financing Policy: Harmonizing Financing with Strategy

A profound understanding of the company's fundamental value and actual company market valuation is also a prerequisite for making more advanced financing choices. Dealing with capital market deviations requires informed decisions on the trade-off between the benefit and cost of financial flexibility.

In many companies, the choice of financing is pretty much disconnected from overall strategy, and mainly optimized with regard to lowest interest cost while maintaining an appropriate level of short-term refinancing risk. This is in some sense a reflection of the notion of perfect capital markets – the assumption is that value-increasing strategies will always find financial backing in this world, and other strategies will not. However, this assumption can put companies at considerable risk in pursuing a long-term value creation strategy, because, as demonstrated above, equity markets (and debt markets as well) may temporarily deviate from this notion. Although this might provide some opportunities to exploit market overvaluations, it is primarily relevant for coping with situations where there is or could be a significant undervaluation from a management point of view.

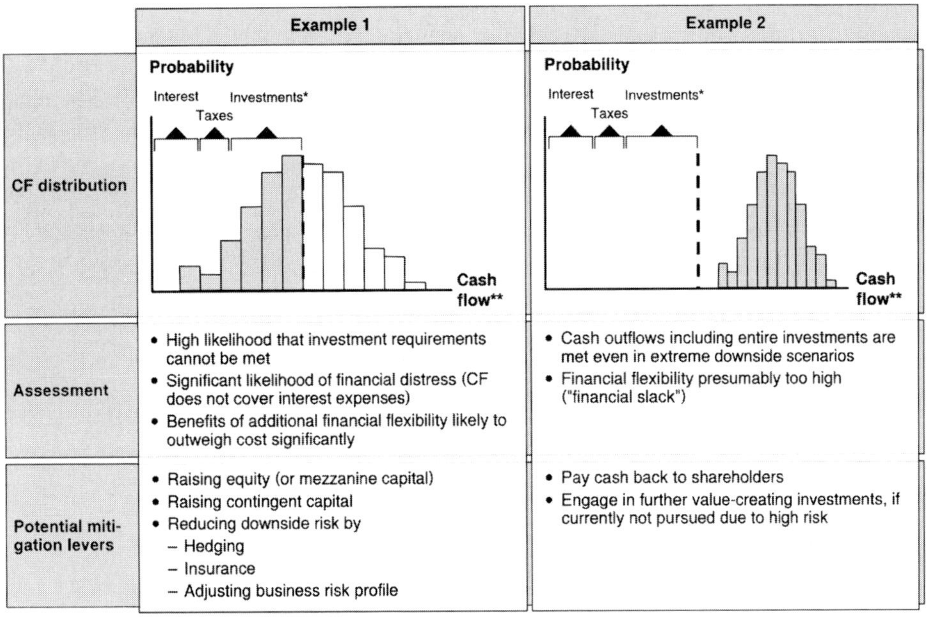

* Including required net working capital, CapEx, acquisitions
** Cumulative cash flow for planning horizon including available cash reserves
Source: McKinsey

Fig. 2.6 Assessing financial flexibility.

For example, in the case of commodity chemical companies, it seems a very reasonable idea to secure financial flexibility for cyclical troughs, when operating cash flows are low, financing is hard or impossible to obtain, but low asset prices provide good investment opportunities. Another set of cases is related to the downside risks of a given strategy. It is self-evident that in case of financial difficulties (e.g., breach of debt covenants, significant rating downgrades, and eventually insolvency) companies are forced by debtholders and legislation to follow actions which might be detrimental to long-term value creation for their shareholders, e.g., cutting NPV positive investments and R&D and carrying out enforced fire sales of assets to free up cash in the short term. Hence, it is essential to protect a risky, but value-creating strategy, for example a big acquisition, against the effects of potential financial distress by securing enough financial flexibility to cover a downside risk scenario.

However, establishing financial flexibility, e.g., by using equity instead of debt financing, holding significant cash reserves, or engaging in contingent capital contracts, also comes at a significant cost: directly, because shareholders might be able to invest this money more effectively (in terms of both return and tax effects), and many contingent capital arrangements are fairly costly; and indirectly, because this financial flexibility might easily turn into "financial slack", tempting companies, for example, to refrain from necessary restructuring measures by cross-subsidizing uncompetitive business units.

An advanced way to provide an analytical base for informed decisions on the trade-off between the benefits and costs of financial flexibility is to establish an integrated view of the risk profile of a given strategy and the corresponding financing decisions in a Cash-Flow-at-Risk model. The projected outcome of the current strategy has to be available not only as a deterministic plan or a variety of different scenarios, but as information on a probable distribution of outcomes. This might seem impractical, but such models can actually be set up using standard spreadsheet-based simulation software. In many cases, it is sufficient to assess inputs for a few key value drivers (e.g., feedstock prices) for which the required information is available or can be estimated with a sufficient degree of reliability. A generic example for the insights such an approach could generate is given in Figure 2.6.

Hence, by carefully assessing financing choices in the context of strategy, chemical companies can secure one key foundation of a strategy focused on long-term value creation.

2.3.3
Investor Relations: from a PR Approach to Targeted Marketing

As a further consequence of an advanced shareholder value orientation, a more targeted approach can be taken to investor relations. We feel that investor relations in many chemical companies today is predominantly aimed at providing timely, accurate, and exhaustive information to capital markets. Again, the assumption is that capital markets are perfectly efficient and will always evaluate this information correctly; again, this does not match up with the real world. Behind the capital markets are investors with a wide variety of expectations and important specific investment preferences and institutional restrictions. In addition, it can be demonstrated that only a small number of investors with significant trading volume mainly influence stock prices in the short and medium term (Coyne, K. P. and Witter, J. W.).

Investor relations should therefore actively sell its product to its divergent sets of key customers – which requires a much more systematic approach, incorporating insights from product marketing. We do not claim that this approach boosts a company's share price above the fundamentally justified level, but it is the right approach to mitigate temporary deviations and ensure that the fundamental value is fully reflected in its capital market valuation.

Particularly in a critical situation where the capital market's valuation deviates substantially from management's assessment or might be in danger of doing so, it is important to create a much more active dialogue with investors, trying to market the companies' equity story to those investors with the investment preferences that best fit its strategic posture. For example, in order to stabilize cash flow over the cycle, it might be a sound strategy for commodity chemical companies to set up exhaustive hedging programs in order to insulate them from fluctuations in feedstock and product prices. Typically, however, the investment style of many investors actively buying into commodity chemical stocks is to actually seek these commodity exposures. A strategy focused on hedging commodity risks therefore needs to be actively communicated to the market and a different set of investors needs to be attracted in order to avoid undue pressure on stock prices.

The typical approach to investor relations therefore needs to be developed along three key dimensions:

- *Know your investors.* Rather than focusing on sell side analysts, companies should establish direct communication with and a deep understanding of their most important investors. First of all, this requires taking an inventory of the investors, a process which will vary in difficulty depending on jurisdictions and types of shares issued. Once the list is drawn up (at least for the most prevalent institutional investors), a perspective needs to be developed of why these investors bought the stock, how long they are likely to keep their holdings, and what specifically would them induce to sell them – decisions driven by factors such as investment style, constraints, overriding investment themes, specific views on strategy, and future earnings development.
- *Conduct an investor segmentation.* Once a detailed knowledge of investors is available, they should be segmented by their propensity to trade and the size of their holdings. In very straightforward terms, the most important "customers" in the short term are those with a high propensity to trade and holdings of substantial size. As mentioned above, there is likely to be a very small number of investors in these segments driving short-term share price movements, and these should be given particular attention accordingly. Eventually, this could result in the ability to structurally predict these investors' trading behavior and their impact on stock prices when important announcements are to be made to the market
- *Tailor communications to important investors.* The third step is to tailor communications to the key investors' specific needs. This does not necessarily require more disclosure, but in many cases a stronger focus on forward-looking information on strategic targets and the measures being taken to achieve them. The results of investor segmentation often yield valuable insights on how to improve investor communications.

2.4 Summary

- Shareholder value orientation has recently been scrutinized. Many believe that it fosters short-term thinking and might be even detrimental to a company's long-term health, because capital markets are inherently myopic.
- It is demonstrated here that capital markets are rational and do value companies at close to their fundamental value, in the chemical industry as in others. However, there is also evidence that they sometimes deviate from that value in the short term, which might put a strategy focused on long-term value creation at significant risk.
- The conclusion from that is not to abandon the shareholder value orientation, but to enhance the way it is practiced. Such an advanced shareholder value orientation consists of two key elements:
  - First, executives should be aware of the fact that value orientation is about long-term fundamental value creation rather than stock price maximization, and that they should concentrate on sound management decisions rather than stock market values.
  - Second, they should take a more sophisticated approach towards managing the implications of capital market deviations for long-term value creation, by gaining an in-depth understanding of capital market signals, harmonizing financing policies with strategy, and taking a targeted marketing approach to investor relations.

References

Coyne, K. P., Witter, J. W., What Makes your Stock Price Go Up and Down. In: McKinsey Quarterly 2/2002.

Goedhart, M. et al., Do Fundamentals – or Emotions – Drive the Stock Market? In: McKinsey on Finance, Spring 2005.

Goedhart, M. et al., Prophets and Profits. In: McKinsey on Finance, Autumn 2001.

Jensen, M., Agency Costs of Overvalued Equity. Harvard Working Paper 2004.

Koller, T. and Rajan, S. R., Who is Afraid of Variable Earnings? In: McKinsey on Finance, Summer 2002.

3
Structural Drivers of Value Creation in the Chemical Industry

Eric Bartels, Thomas Augat, and Florian Budde

A long-term analysis of the chemical industry has revealed that many factors that are thought to increase value actually do not, and delivered interesting insights into the true drivers of value creation. The study shows that the primary driver of value creation in chemical companies is in actual fact the composition of their product portfolios. Strategies to improve return on invested capital (ROIC) as the primary value driver in each of the product segments should focus on capital efficiency as well as margins and take a cautious stance on trying to outgrow performance problems. On a corporate level, a rigorous approach to portfolio management can be a significant factor in value creation, if centered around improving long-term ROIC rather than on size and growth targets, and if it ensures that the different parts of the portfolio are managed in accordance with the type of business.

3.1
Introduction to the Study

With more than 70,000 product lines and dozens of geographic markets, the chemical industry of today is deeply complex. So many strategies – or combinations of strategies – have been developed over the years for so many markets that industry analysts and executives struggle to understand what really creates shareholder value.

While previous performance is of course no guarantee of future returns, we felt for several reasons that an examination of the historical patterns of value creation might deliver insights to support forward-looking strategy formulation. The industry reached maturity about 20 years ago, the evolution of supply and demand has become, on average, more or less predictable. Moreover, unlike pharmaceuticals or telecommunications, the industry is not likely to be transformed by technologies or regulations in the near future. Hence, historical trends are fairly likely to reflect the future development of the industry as well. In order to establish a sound basis, we compiled more than 40 years of financial and stock market data on 130 publicly traded chemical companies in Europe and the United

Value Creation: Strategies for the Chemical Industry. 2nd Edition. F. Budde, U.-H. Felcht, H. Frankemölle (Eds.)

ISBN: 3-527-31266-8

States[1], employing various performance metrics and detailed refinements to make the data as comparable as possible over geographies and time.

Since strategies are hard to classify, we chose to examine performance relative to some easily measurable dimensions. These dimensions, however, directly relate to important strategic choices for individual industry players. Our research suggests that, on the industry level, none of the factors commonly regarded as strategic levers for value creation in the industry – scale, geography, market position, or degree of focus – makes much difference. The performance of a chemical company correlates strongly only with its product portfolio – commodity, specialty, or diversified chemicals[2]. Within each of these segments, distinct trends in the factors can be identified which yield valuable insights for chemical companies.

3.2
Mobility in a Mature Industry

The chemical industry is often regarded as sluggish and slow-growing – a reputation due partly to its maturity and partly to comparison between its shareholder returns and those of technology companies during the bubble of the late 1990s. However, the long-term data show that this reputation is largely unjustified.

To begin with, though the percentage of overall economic activity accounted for by the industry continues to shrink – in the United States, down from four percent to less than two percent over the past 25 years – shareholder returns were roughly on a par with those of the broad market indexes in the United States and Europe over the same period. In the United States, for example, both the chemical industry and the broad market returned roughly thirteen percent a year to shareholders,[3] a more robust rate than that of other asset-heavy mature industries, such as steel, automotive, and pulp and paper. This is certainly quite an accomplishment.

Furthermore, the chemical industry stands apart from other commodity industries in that a wide gap separates the top and bottom performers. In November 2004, for example, the top quartile of chemical companies had a market-to-book ratio 3.1 times the bottom quartile's – a far wider spread than that of other asset-heavy industries, such as automotive (1.9), pulp and paper (1.8), and steel (1.6). Companies in those industries operate in more transparent and global commodity markets with fewer strategic options at the corporate level. By contrast, the far more complex chemical markets are fragmented into thousands of submarkets

1) Our database includes companies that have ceased to exist as independent entities as well as companies listed throughout the period covered by the research. The data for the European companies extend from 1988 to 2003. At the time of writing, not all the 2004 data have been published.

2) Diversified companies sell both commodity and specialty products, with neither accounting for more than 70 percent of total revenues.

3) The exact values for the annualized Total Return to Shareholders are 12.9 percent for the chemical industry and 13.8 percent for the US market from 1980 until November 2004 (Source: Thomson Financial).

defined by product, geography, and customer industry. Although this complexity makes it difficult to identify industry-wide patterns of success, it offers a diverse range of opportunities to exploit niches and create shareholder value.

In addition, there is significant mobility between the upper and lower performance quartiles of the industry – demonstrating its potential for creating (and destroying) value and underscoring the fact that stock markets pay careful attention to the performance of individual companies. Fewer than half of today's top-quartile companies, for example, occupied the top quartile a decade ago. In fact, several of today's top-performing chemical companies – including Ecolab and Valspar – were formerly average or even below-average performers (Fig. 3.1). And as recently as a decade ago, 20 percent of today's bottom-quartile companies were above-average performers.[4]

This mobility must be understood in the context of the maturity of the chemical industry. No imminent disruptions will transform it as a whole, but myriad changes, in factor prices and end-user demands, for example, will allow companies to redefine their products and services dynamically in specific markets and geographies. Companies in other mature industries, such as pulp and paper or metals and mining, find it much harder to reposition themselves competitively.

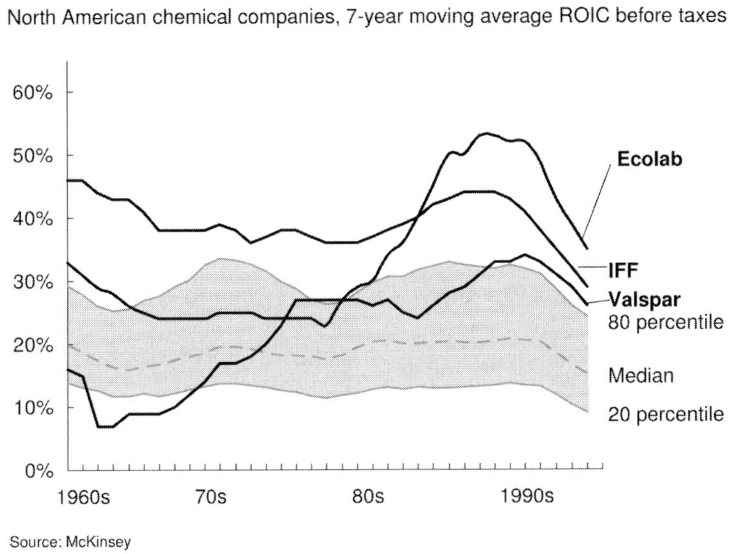

Fig. 3.1 Mobility in a mature industry.

4) This analysis is based on a subset of 60 European and North American companies which were listed throughout the period between 1992 and 2003.

3.3
What Drives Performance?

What strategy, if any, correlates with strong performance? Unfortunately, chemical companies' strategies are hard to classify, since most compete in a range of product and geographic markets requiring a variety of approaches: one unit might strive to produce at low cost, another to innovate. Moreover, a degree of subjectivity is involved in the classification of one company as, say, innovative and another as not innovative.

We therefore chose to examine the performance of chemical companies in relation to some easily measurable dimensions of their corporate activity – such as product portfolio, scale, geography, market position, and product focus. By classifying companies this way, we were able to test a number of hypotheses about what drives the creation of value, measured by total return to shareholders (TRS), market-to-book value, and pre-tax return on invested capital (ROIC). Data from the last full commodity cycle (1992 to 2003) generated surprising insights.[5)]

- *Scale:* In the chemical industry, bigger is thought to be better. Nevertheless, given the highly fragmented nature of its various markets, size is not a primary influence on performance. Economies of scale do, however, provide benefits in the commodity and diversified segments.
- *Geography:* Although North American chemical companies are widely thought to perform better than their European competitors, our research showed that the gap has significantly narrowed over the past decade. North American chemical companies still have a higher market-to-book value. However, as there are virtually no differences in return on invested capital anymore, we believe that this primarily expresses a general valuation difference between the capital markets rather than something specific to the chemical industry.
- *Market position:* Most industry analysts believe that companies with market-leading positions – those that derive more than 50 percent of their revenues from businesses ranking first or second in global or regional sales – should exhibit better performance than second-line players. Yet one of our more surprising findings was that companies with market-leading positions had no significant performance advantage
- *Focus:* Conventional wisdom also holds that tightly focused companies – those deriving more than 80 percent of their revenues

5) These metrics were selected in order to capture different aspects of value creation: TRS measures the realized value creation for shareholders. The market-to-book value is a proxy for the capital market's expectations on future value creation. And ROIC is, in an industry without significant growth, the key driver of fundamental value creation. We are using pre-tax returns to allow for better comparability over time and between different countries.

from only two kinds of businesses – tend to have better performance. In reality, again, there are no significant performance differences between focused and unfocused companies on an industry level. What is more, large diversified companies have been able to generate the highest total return to shareholders.[6] However, after a long period of insignificant differences in performance, focused specialty players nowadays appear to be performing better in terms of current ROIC and also future expected performance (market-to-book) than unfocused specialty chemical companies. Of course, we can enter a caveat here: the definition of focus as "product focus" may be over-simplified and fail to do justice to the complexity of the markets. A focused company may, for example, apply the same core competence in all it does (e.g., technology/production, the same customers, the same business model or approach). In other senses, therefore, the correlation between focus and success may still be valid.

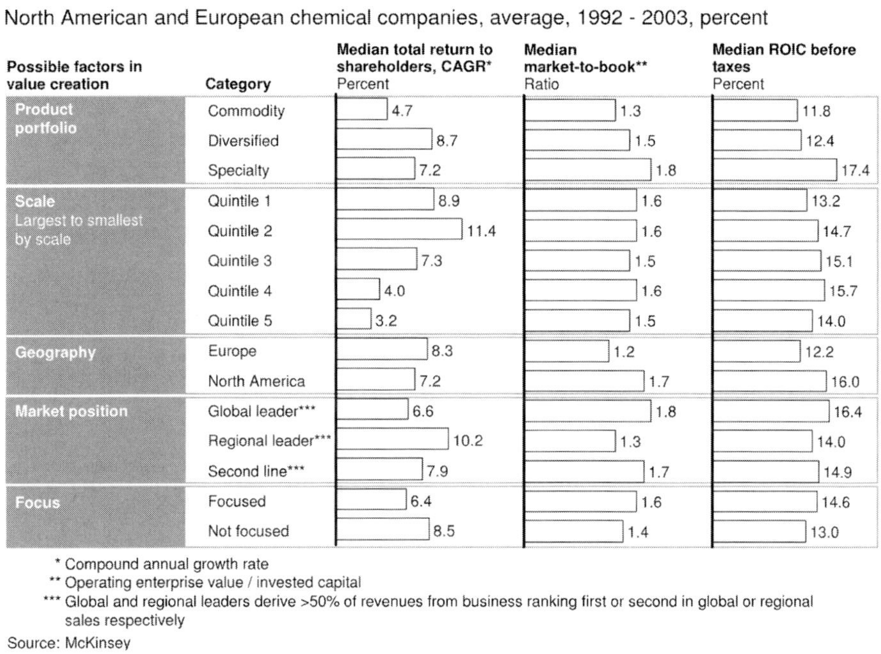

Fig. 3.2 Product portfolio matters most.

By the definitions applied, however, our analysis showed that the only statistically significant characteristic that correlated with the measures of performance –

6) Although this only means that they have been able to improve their performance better than other chemical companies.

total return to shareholders, market-to-book value, and pre-tax return on invested capital – was a company's product portfolio: commodity chemicals, specialty chemicals, or diversified (Fig. 3.2). In the period we studied (1992 to 2003), diversified companies generated higher total returns to shareholders (8.7%) than did specialty companies (7.2%), which in turn outperformed commodity companies (4.7%). As measured by return on invested capital and market-to-book value, specialty companies performed best. The diversified companies' higher TRS over this period can be explained by the expectations of shareholders, who thought, in an extrapolation of past performance, that specialty companies but not diversified ones would perform well financially. When diversified companies did better than expected, their shares rose considerably. Meanwhile, the specialty companies had to perform well merely to keep their total returns to shareholders stagnant.[7)]

Although the performance of companies in the chemical industry correlates well with their product segments, they should not rush to modify their portfolios, since some specialty companies perform much worse than some commodity players (Fig. 3.3). Besides, the wide range of performance within each segment proves that portfolio choices alone do not guarantee success. However, a diverse set of opportunities, some relatively overlooked, seems to be available for creating value in market niches.

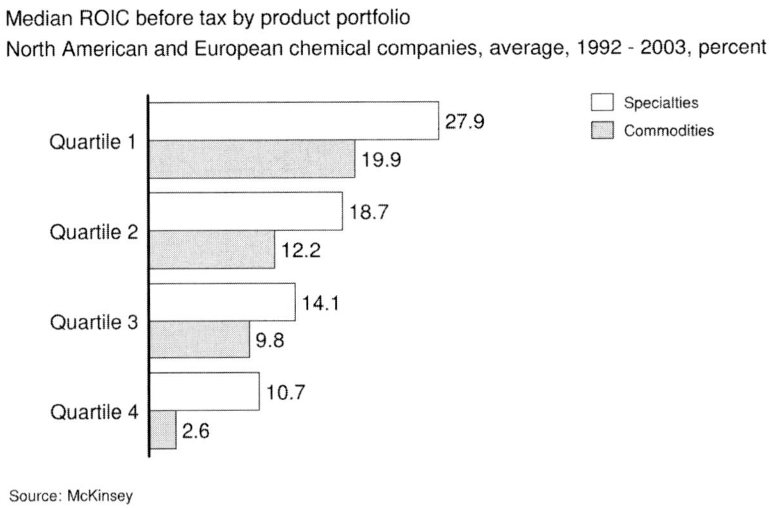

Fig. 3.3 Operating performance is relative.

7) The higher market expectations can be seen in the specialty segment's higher average market-to-book values.

3.4
A Closer Look at Value Creation in the Segments

The data sent some clear messages about how commodity and specialty companies could improve their performance. For diversified companies, the results were more ambiguous.

3.4.1
Commodity Companies

Although commodity companies[8] performed less well than their specialty and diversified counterparts from 1992 to 2003, plenty of evidence suggests that they can find the right strategy to create shareholder value – or at least to avoid destroying it. Many of our findings confirm accepted beliefs about what makes commodity companies successful, but few companies seem to heed them.

One significant phenomenon of the commodity segment is that this subindustry segment earns a margin above its cost of capital only on average over a full commodity cycle, usually a period of seven to eight years (see also Chapter 6).

The first finding for the commodity segment is that size does matter. Larger companies had less cyclical (though not higher) returns on invested capital, and the less cyclical returns correlated, albeit mildly, with higher total returns to shareholders. These companies were not only able to cut their production costs, but also had sufficiently diversified offerings – and therefore sufficiently stable cash flows – to weather cycles for individual products.

We also found that commodity players cannot simply grow into market appreciation, i.e., increase their valuations through strong top line growth, since sustainable growth is almost impossible for the markets to detect. The modest amount of nominal, volume-based growth we found was heavily masked by the industry's largely self-imposed pricing cycles, not by broader macroeconomic trends. As a result, investors have not rewarded companies that tried to grow, either organically or through acquisitions, at below-average profitability. In our sample, such companies had paltry median market-to-book ratios of around one.

There are two problems around growth in the commodity chemicals segment.

First, too many companies have sought to achieve scale without paying enough attention to the return on their invested capital. The clearest finding of our research reinforces this basic point: return on invested capital matters far more than revenue growth. Even those companies with above-average returns on invested capital had the same market-to-book ratios, no matter whether they were growing quickly or slowly. This is, as we believe, due to the commodity markets' inherent cyclicality, which makes it impossible for investors to tell real growth from temporary price driven fly-ups.

8) Our study did not include the integrated petrochemical activities of the large oil companies.

Second, our research also confirmed the idea that the timing of capital investments, rather than strong fluctuations in demand, is to blame for the industry's volatile cycles. An industry-wide herd instinct for capital investment decisions has created large supply-and-demand swings (see Nattermann, P.). Since companies make most of their capital investments during the cycle's upswings (Fig. 3.4), prices fall quickly as new supply floods the market about two years later.

Interestingly, the industry appears to be, in its current upswing, once again at one of these critical decision points. Executives with the courage to defy conventional wisdom and withstand pressure from their boards, investors, and bankers by investing in new capacity countercyclically (or at least independently of cycles) could generate substantial returns. Indeed, a McKinsey study estimates that companies might double their return on new capital investments by pursuing an independent approach (see Copeland, T. et al.). One might speculate that privately held commodity companies, which are free to buck the pressure to conform, stand a better chance than others of breaking out of the industry's self-destructive investment cycles.

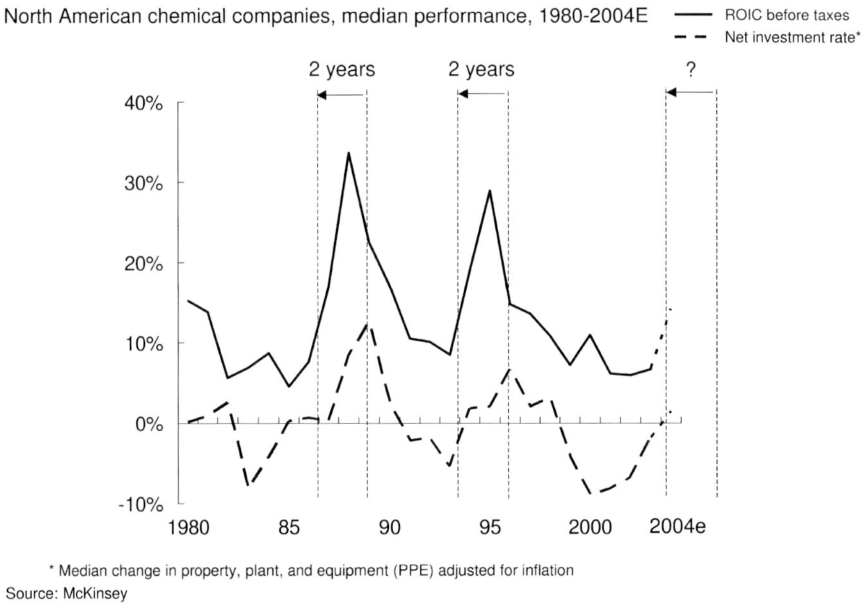

Fig. 3.4 ROIC and investment rate in commodity chemicals.

3.4.2
Specialty Companies

In the 1990s, there were two distinct periods for specialty companies: 1990 to 1997 was a time of good performance and significant expectations on value creation; after 1997, valuations fell sharply (Fig. 3.5). Despite continuing evidence of an industry slump, however, specialty companies have ample opportunities to create value in the thousands of product and geographic markets in which they compete.

During the boom years of the early to mid-1990s, specialty companies enjoyed real and sustained sales growth, higher returns on invested capital, and greater total returns to shareholders. Yet they did not deliver these strong results through higher margins – a somewhat counterintuitive finding, since specialties are thought to have higher average margins than commodities. In fact, notwithstanding higher prices and lower depreciation rates than commodity companies can boast, specialty companies turned out to have a higher cost base in R&D, marketing, technical support, and the like. Over full commodity cycles, their margins are thus roughly comparable to those of commodity companies.

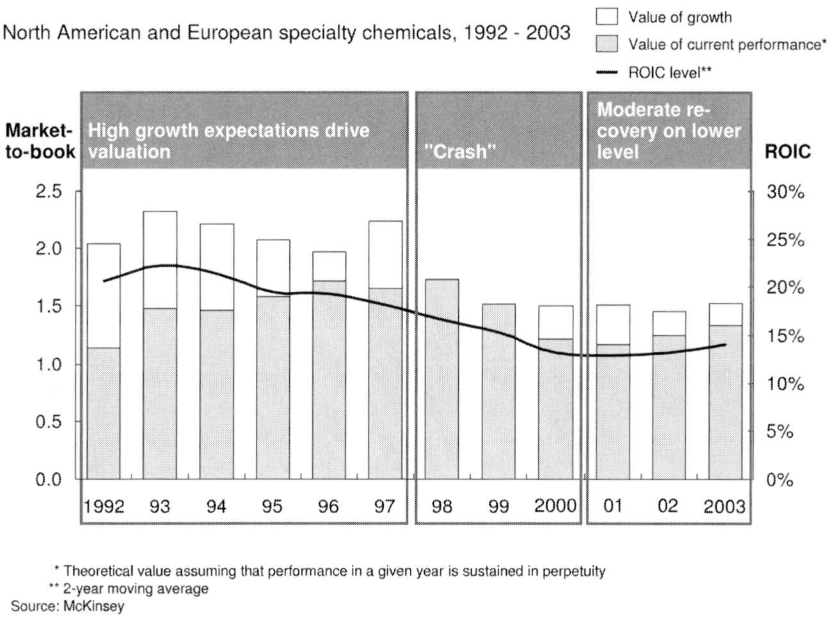

Fig. 3.5 Evolution of specialty chemicals' valuation.

Specialty companies have achieved higher returns than commodity companies through higher levels of capital productivity – the result of the smaller facilities needed to produce smaller quantities of chemicals that command higher prices. The specialty segment's higher levels of capital productivity have generated higher returns on invested capital, which (coupled with revenue growth) have created shareholder value. Since high capital productivity drives shareholder value in this segment, specialty players have to preserve it.

Thus, the massive decline in the specialty segment's capital productivity during the mid-1990s came as an unpleasant surprise. The decline was due in part to rising levels of invested capital, most notably from a string of acquisitions laden with goodwill, such as Clariant's purchase of BTP and ICI's acquisition of Unilever's specialty chemical businesses, attempts to gain scale in market segments deemed to be particularly attractive, i.e., life sciences, and a general expansion of capacity in the quest for growth which did not materialize. Moreover, as specialty products have come to face increased competition and commoditization from low-cost producers in China and India, prices have fallen, and the resulting decline in revenues has not helped. The fate of vitamin C is a good example: from 1990 to 2002 the global market share of Chinese producers rose from zero to 24 percent; meanwhile, prices per kilo dropped by 63 percent, to USD 4.7 from USD 12.7.

In the late 1990s, when investors noticed the combination of declining growth and declining capital productivity, many lost faith in the segment. While fears of repercussions from the Asian crisis proved unfounded, other structural factors such as Asian competition and the commoditization of many former specialty products took effect in their stead.

Currently, specialty chemical players too are benefiting from better industry conditions, significant cost improvement efforts by many players, and a very cautious stance on investments and acquisitions over the last few years. However, we predict that the pre-1997 glory days will never return. Rapid commoditization of parts of the portfolio and relentless price pressure will continue to endanger business models. For some players, a rigorous focus on cost will be the path to success, others – and these businesses will still earn a substantial premium on industry average profitability – will develop true customer solutions.

Some highly successful companies, Ecolab for instance, have already focused on organic growth through new business models and the extension of capital-light service lines, such as cleaning services for the food industry. For other companies with less favorable positions, tough choices on their portfolios of businesses will lie ahead. We are not necessarily advocating a focused business model, but many specialty chemicals companies with broad portfolios will need to explore every option to deal with the number of already-disadvantaged businesses, suffering from structural overcapacity and continuous price pressure. Companies that avoid overinvesting will have an advantage in maintaining their capital productivity rates and positioning themselves for stronger performance.

3.4.3 Diversified Companies

Diversified companies presented a much more puzzling picture during the last cycle. In terms of total return to shareholders, surprisingly, diversified chemical companies performed best among the industry subgroups from 1992 onwards.

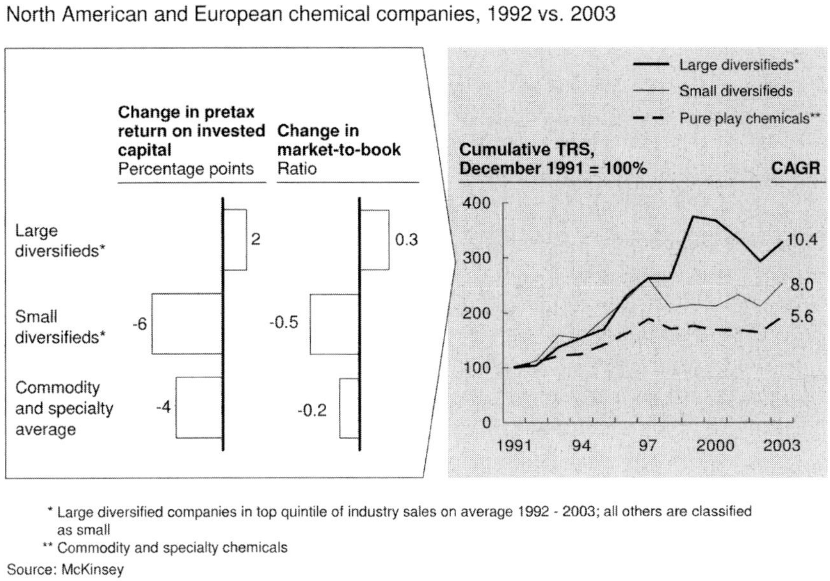

Fig. 3.6 Large diversifieds outperformed other chemical companies.

This seems hard to explain, because one would expect a performance between the specialty and commodity players, reflecting the diversified companies' business portfolios. Their high total returns to shareholders were driven, as we have seen, by their strong financial performance against low market expectations in the early 1990s. Although not all of them did well, from 1992 to 2003 the large ones (defined as those in the top quintile of sales) actually had significantly higher returns than either commodity or specialty companies (Fig. 3.6). In fact, large diversified companies also exhibit a trend in fundamental performance which has been different from the other subindustries. In the early 1990s their ROIC at best matched that of more focused commodity-based players. Diversified companies improved performance in the mid 1990s, and they have been able to sustain some of this performance improvement against the industry trend of declining return on capital. Consequently, their valuation levels have followed the improvements in fundamental performance which in turn led to exceptional returns to shareholders until the late 1990s. The smaller diversified companies, by contrast, have not recovered from the combined effects of an industry downturn and the 1997 Asian economic crisis, and our research did not identify clear performance drivers for this segment.

We feel there are several underlying fundamental factors in this development:

- Some diversified players (e.g., BASF, DSM) thrived as a result of rigorous portfolio rationalization which allowed them to concentrate on segments where they could be major players while avoiding overpaying to rush into specialties. At the same time, other

now-specialty chemical companies destroyed significant value and a single-segment focus for the sake of short-term capital market applause.
- Additionally, diversified players have often been able to build advantages of scale and scope in their businesses (in particular in commodities) and use globalization as an opportunity rather than a risk, while still coping successfully with the conflicts of running fundamentally different businesses within one organization.
- Although market position was not found to be a value driver for the industry in general, within the diversified segment it correlated moderately with higher returns. A focus on achieving market leadership by a well-targeted portfolio strategy may have helped transform these companies' fundamental financial performance over the past decade: DuPont, for example, almost doubled its pretax returns on invested capital from 1992 to 2003, driving up its capital market valuation by 40 percent over the same period.

Having improved their performance from a very low level to industry average, the large diversified companies now need to seek new ways to create value. Many could still pursue a much more rigorous portfolio strategy and performance management. While stabilizing returns against the industry trend is an achievement, rejuvenating the above-average shareholder returns will require diversified companies to make a further quantum leap in the continuous race for performance. It will be interesting to see how the large players here live up to the expectations underlying their higher valuations.

3.5 Summary

The highly fragmented and complex chemical industry is more dynamic than many people think. The performance difference between the top and bottom quartile is larger than in other asset-heavy, mature industries and mobility between the quartiles is significant over time. A thorough review of the past decade of the industry's value creation reveals that there are no generic textbook formulas or cookbook recipes for sustainable, value-creating strategies in the chemical industry. The industry's inherent complexity and obfuscating dynamics require tailor-made strategies for each situation and player which incorporate specific insight into these factors. The long proclaimed "death of strategy" certainly does not hold true for chemicals. However, a close examination of the interaction of fundamental value creation and capital market perception has created a clearer view of how commodity and specialty companies can and should manage capital productivity, growth, and portfolio structures to create shareholder value in the years ahead.

- On the business unit level, regardless of the type of business, chemical companies should focus on capital productivity as the key value driver of ROIC besides margins.
  - In addition to operational EBIT improvement levers, capital-light strategies and focus on the management of capital investments and net working capital are also means to this end.
  - For the extremely capital-intensive commodity segments, particular attention should be paid to the right timing of investments.
  - Even though the industry environment may now be geared towards more growth than in the past few years, chemical companies should in general take a cautious stance on trying to outgrow their performance problems. A strategy focused on growth will only be successful if based on a profitability clearly above the cost of capital by some margin in order to compensate for the considerable risk of growth strategies in asset-heavy industries.
- On the corporate level, there is evidence that portfolio management can be a significant driver of value creation for chemical companies, both for good and ill.
  - Portfolio transformations should be centered around improving long-term ROIC rather than size, growth targets, or a misperceived superiority of some business segments, such as the view of specialty chemicals in the past.
  - Furthermore, corporate centers which choose to stay diversified should recognize the substantially different nature of commodity and specialty chemical businesses. They hence need to manage them differently and in doing so bridge the chasm this will almost certainly create within their organizations.

References

Copeland, T., Koller, T., and Murrin, J., Valuation – Measuring and Managing the Value of Companies, third edition, New York: Wiley, 2000 p. 334.

Nattermann, P. M., Best practice ≠ Best strategy, The McKinsey Quarterly 2/2000, pp. 22–31.

4
Chemicals – Driving Innovation in Other Industries

Thomas Schreckenbach and Werner Becker

Chemistry – or chemicals – lies at the heart of everything: fibers for new textiles, catalysts for a clean environment, colors that gleam in the sun, polymer electrolyte diaphragms for fuel cells, chemicals for chip production, or fertilizers that help plants grow, ensuring global food production. The chemical industry develops the intermediate products that other industries use, and often develops new processes and products itself, or in conjunction with customers.

The industry invests, on average, about eight percent of its revenues in research and development (R&D). Such a research-intensive approach helps end customers immensely, and it is little wonder that chemicals are seen as the engine of innovation. According to a February 2003 study from the Center for European Economic Research, the chemical industry provides more industry-overlapping R&D transfers than any other – a fifth of the total, in fact. Thanks to the properties of new materials and components, other industries are able to develop new products and applications – and these innovations succeed because they boost companies' economics, improve the effectiveness of existing products and processes in other industries, and are less polluting (Fig. 4.1).

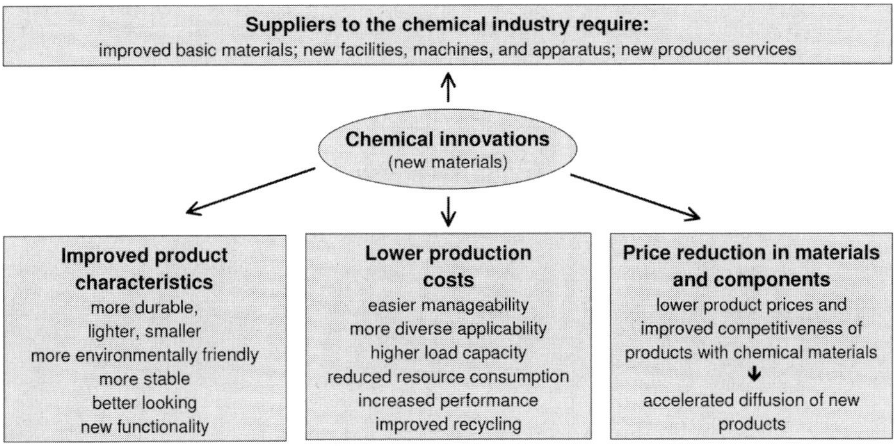

Fig. 4.1 Impact of chemical innovations beyond the industry.

Value Creation: Strategies for the Chemical Industry. 2nd Edition. F. Budde, U.-H. Felcht, H. Frankemölle (Eds.)

ISBN: 3-527-31266-8

To investigate how the chemical industry acts as an innovation engine, we have chosen to examine in detail something that has quite literally changed the face of communication in the last fifty years. We will look at the evolution of this particular material, how our company Merck and its customers have created value from it, how it has become a worldwide success, and how its development is continuing.

4.1 Liquid Crystals: Superlative, Not Superfluous

Modern communication operates independently of time of day or geographical location. It brings people closer, makes life easier, and has changed the working environment profoundly. At the heart of all this sits a product so common that most people do not even think about its presence. Liquid crystals (LCs) are hi-tech chemical materials with unique properties. Without them we would not have Liquid Crystal Displays (LCDs) – lightweight, flat displays that consume little energy.

Liquid Crystals/Liquid Crystal Displays

Liquid crystals are elongated, rod-like organic molecules with a dipole due to charge separation. Macroscopically they are, within a certain temperature range, more or less viscous liquids that behave physically like crystals. Just as in crystals, some of their properties are directional and thus they are anisotropic. This gives liquid crystals their two most important properties: they can modulate polarized light and they can change their orientation in an electric field in reaction to an applied voltage, i.e., they can act as optical switches.

These are the properties used in Liquid Crystal Displays (LCDs). The liquid crystal material lies in a very thin layer between two glass plates. If no voltage is applied, polarized light is transmitted without hindrance and the pixel appears bright. Under a low voltage, the molecules rotate from their original orientations, no light is transmitted and the pixel becomes dark. Combinations of such pixels may represent numbers, letters, and symbols. Very little current flows during switching, so LCDs consume tiny amounts of energy. Liquid crystal mixtures used in displays may contain up to 25 components, sometimes even more, and for each application a separate liquid crystal cocktail may be used (Fig. 4.2).

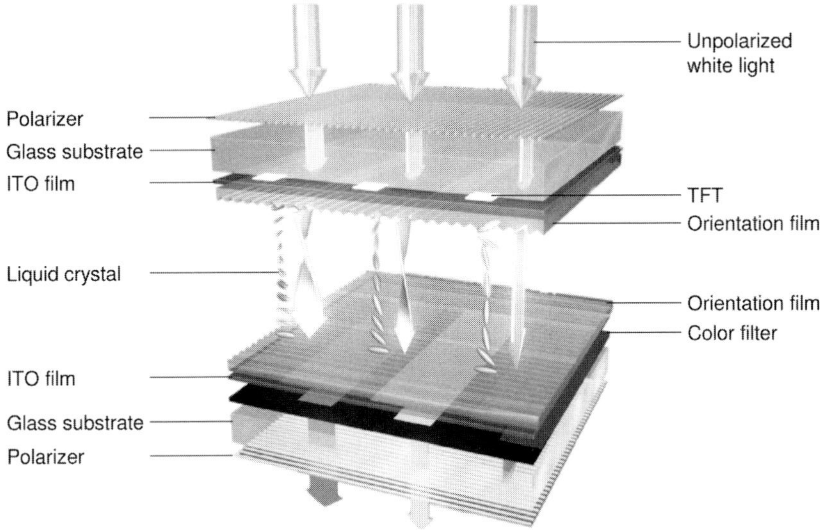

Fig. 4.2 Structure of the different layers of an LC display shown in cross-section of a single pixel (twisted nematic cell).

Today, neither household nor office is conceivable without LCDs. They are flat, lightweight, low in energy consumption, and transportable, and continue to displace older technologies such as cathode ray tubes (CRT). After all, a flat television that hangs on the wall like a painting is highly desirable – especially in the living rooms of East Asian cities where space is often in short supply. In fact, surely an LCD TV would be better than a bulky TV set in any room anywhere!

The predicted boom in LCD TVs has been made possible by a new generation of liquid crystals. Three of Merck's scientists are at the very forefront of this development: Melanie Klasen-Memmer, Kazuaki Tarumi, and Matthias Bremer.

As the mobile phone boom helped push LC consumption for displays, the cost of LCDs has fallen, which has helped cut costs for LCD TVs. Independent market research predicts a sharp increase in demand. In 2004, about nine million LCD televisions were sold worldwide; by 2008 that is expected to have reached 50 million per annum. This is also reflected in worldwide investments: the five or six largest LCD manufacturers are expected to invest about USD 30 billion in flat TV technology in the next three to five years.

Specialty chemicals are generally successful only if they are pushed into companies at the very cutting edge of each wave of technological advance. However, the developers also have to be able to create value from their investments in innovation, which requires both the scientific research to develop the chemicals in the first place, and also a business-focused mindset to ensure their longevity in the market.

For liquid crystals, the following factors have proved critical in Merck's success:

- Staying power
- An active patent strategy

- A customer-focused approach built on expertise in production and application technology
- Timing of internationalization and diversification
- The best intellects.

4.1.1
Staying Power

Over a century ago, Professor Otto Lehmann of the University of Karlsruhe – who is acknowledged as the pioneer of liquid crystal technology – disagreed with the scientific establishment about the existence of the liquid crystalline state of aggregation. Lehmann turned to Merck. He asked the company to supply him (and others) with liquid crystalline compounds so that he could clarify what no amount of theoretical debate could resolve. This was Merck's first foray into the LC business – although there was no actual business at this stage. Nevertheless, it was already well known that Merck supplied products of the highest purity, which allowed results to be replicated accurately.

However, it was not until 1967 that the technology was woken from its slumber by Americans James Fergason and George Heilmeier. These two developed the first technical applications of liquid crystals in thermography and the first liquid crystal displays; Merck resumed its interest in liquid crystals, and the business began.

Many at Merck were skeptical of this technology and sarcastic comments floated around, generally referring to 'superfluous crystals' (in German, *flüssig* means liquid, but *überflüssig* means superfluous); but despite this internal resistance, our liquid crystal pioneers remained resolutely positive.

> *"Luck, perseverance, and ignorance" are, according to a Japanese proverb, the basic conditions for a successful scientist's existence. Kazuaki Tarumi, one of the liquid crystal research/physics team at Merck, is particularly aware of 'Ignorance', without which he might have taken seriously the many arguments against his research project and never achieved the goal.*

In some ways, nothing has changed since the early Otto Lehmann days. Merck continues to supply high quality products as part of its commitment to contributing to safety, health, and protection of the environment. For liquid crystals, this begins with production and encompasses not only environmentally compatible manufacturing processes but also working and living conditions for employees and the region in which the company operates. This applies to all Merck locations worldwide and was one of the basic principles of Heinrich Emanuel Merck, the company's founder.

Progress from these pioneering days has been made possible by the flexibility of our team, the *esprit de corps* between the various disciplines, and a positive attitude toward market trends. In the early days, when a display could be used only at

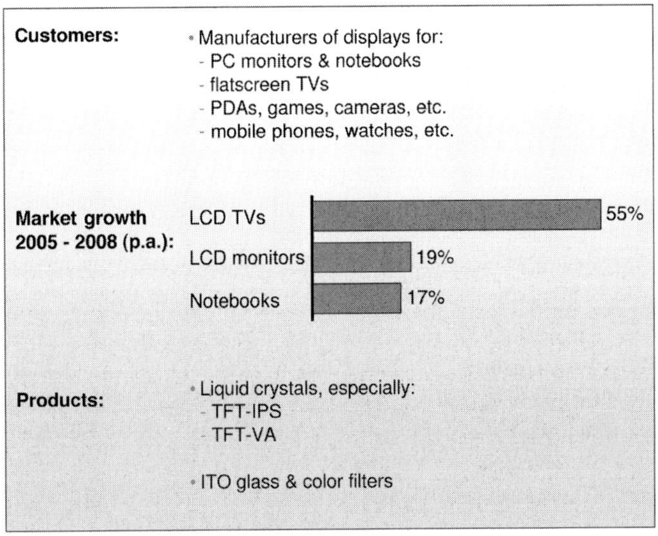

Fig. 4.3 Growth drivers of the global liquid crystal market.

80 °Celsius, one of the greatest achievements was the recognition that this technology had the potential to revolutionize the world of work. Belief in the innovation – and the vision to see beyond obstacles – proved fundamental to the long-term success of liquid crystal technology. This belief came from both years of experience in research, and the willingness and ability to engage in direct dialogue with customers.

In 2004, the LC division had an annual turnover of approximately EUR 580 million, making it the clear market leader. In line with the development of the LCD industry and the display area, growth is predicted to be 30 percent annually on average for the next four years (Fig. 4.3). PC monitors, notebooks and mobile phones all contribute to these figures, but the largest driver going forward is flatscreen televisions. As well as providing specific liquid crystal mixtures for all these applications, Merck also provides special indium tin oxide (ITO) coated glass as well as color filters for color super twisted nematic (STN) displays.

Merck's success in the market is reflected not only in these numbers, but also in its scientific honors and distinctions. Our scientists have received the renowned award of the Society for Information Display several times (1997, 2000, 2002) and in 2003 the coveted German Future Prize from the German Federal President.

4.1.2
Managing an Active Patent Strategy

Ideas and staying power alone will not guarantee success. After all, this is a competitive industry. To deliver sustained value creation, an innovation strategy must factor in effective intellectual property protection. Merck has pursued an aggressive patenting strategy for LCs since the mid 1980s. We developed a closely woven net of patents for both single LC compounds and LC mixtures. Inventions were filed so that cross-license agreements would be possible and we applied for thorough patent coverage, thereby giving our customers long-term protection against imitation products of indefinable quality. Through our commitment to display and application technologies, we also conceived inventions in this area that were disclosed in patents. One prominent example is a basic patent for all active-matrix-driven twisted nematic (TN) displays, which was licensed worldwide in the 1990s.

Merck also regularly realigned its patent portfolio following divestments and acquisitions. For example, in 1985 we acquired a large patent package from Brown Boveri & Cie including basic patents (e.g., the famous 'Maged A. Osman' patent, a very comprehensive liquid crystal patent). This gave us access to some important compound classes and provided substantial licensing income. In 1995 Merck decided to abandon ferroelectric liquid crystals, and we sold the patents to Hoechst. A year later we bought a large patent package from Hoffmann-La Roche, which gave us access to the important class of alkenyl-substituted LCs. These compounds meant that we could address the TN and STN display market segment more effectively.

The story of our patents is ongoing. We have secured our LC business with more than 2,500 patents for the individual substances, their mixtures, and their display applications, but we add about 100 more to this portfolio each year.

4.1.3
A Customer-focused Approach Built on Expertise in Production and Application Technology

Naturally, to turn innovation into business success a company has to understand its craft very thoroughly. Liquid crystal production can require very complicated, multi-level syntheses, and mixture development is far from a trivial process. But any innovative company – no matter how skilled – will struggle to create value from its innovations if it does not listen to and work with its customers.

Merck's LC division has a particular challenge as it has to operate in the context of today's fast moving electronics industry. Displays and related technologies are being developed ever faster, which means that more efficient liquid crystal mixtures are needed ever more rapidly.

Taking the example of LCD TVs, it is easy to see how customer demands shape innovation. PC monitor technology could not simply be transferred to TV screens because the LCDs switch too slowly. Television signals broadcast rapidly changing images, and if LCD monitor technology were used then some images would smear – a fast moving tennis ball would appear to have a comet-like tail. However, with new superfluorinated liquid crystals for multidomain vertical alignment (MVA), we have developed exactly the right LC mixtures for TV sets. Display manufacturers can now mass-produce LCDs from 40 to 165 centimeters measured diagonally. These offer a higher resolution, are lighter, have shorter switching times (compared with the present value of about ten milliseconds), and a broader viewing angle of up to nearly 180 degrees. It was this technological advance that brought us the German Federal President's Future Prize (Fig. 4.4).

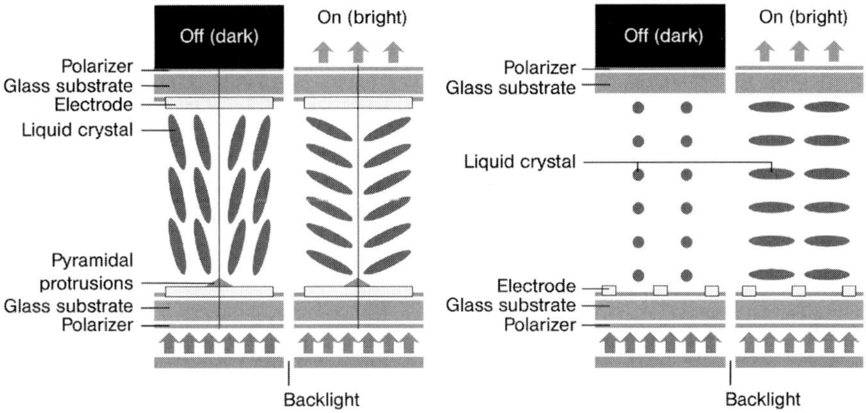

Fig. 4.4 MVA and IPS technology.

MVA and IPS technology

In an LC display for a TV set based upon vertical alignment technology, a thin film of liquid crystalline mixture lies between two glass plates. In the off-state, the liquid crystalline molecules are oriented almost perpendicularly between the display's glass plates (vertically aligned). Because of this perpendicular alignment, the rear illumination is almost completely transmitted and then "absorbed" by a second polarizer turned by 90 degrees, causing the display pixel to be dark. As soon as a voltage is applied to the electrode, the molecules align horizontally. The backlight is then able to traverse the liquid crystalline layer and the pixel is bright.

In addition to MVA technology, in-plane switching technology (IPS) is also of great interest for large format displays. The difference between normal twisted nematic displays and IPS is that the liquid crystals are oriented "in plane", i.e., parallel or horizontal to the glass plates, and by the application of a voltage, they rotate unhindered by 90 degrees in one and the same plane. The light is transmitted or blocked accordingly. A viewing angle of 170 degrees or more is attained using this technique, which competes favorably with cathode ray tube monitors. Unfortunately, this method still has technical difficulties: due to the switching method, the comb-like electrodes are attached to the rear glass plate, reducing light transmission. Thus, a stronger backlight is needed to restore the required brightness. This technology has been applied in televisions and PC monitors, and to a lesser extent in notebook computers and navigation systems.

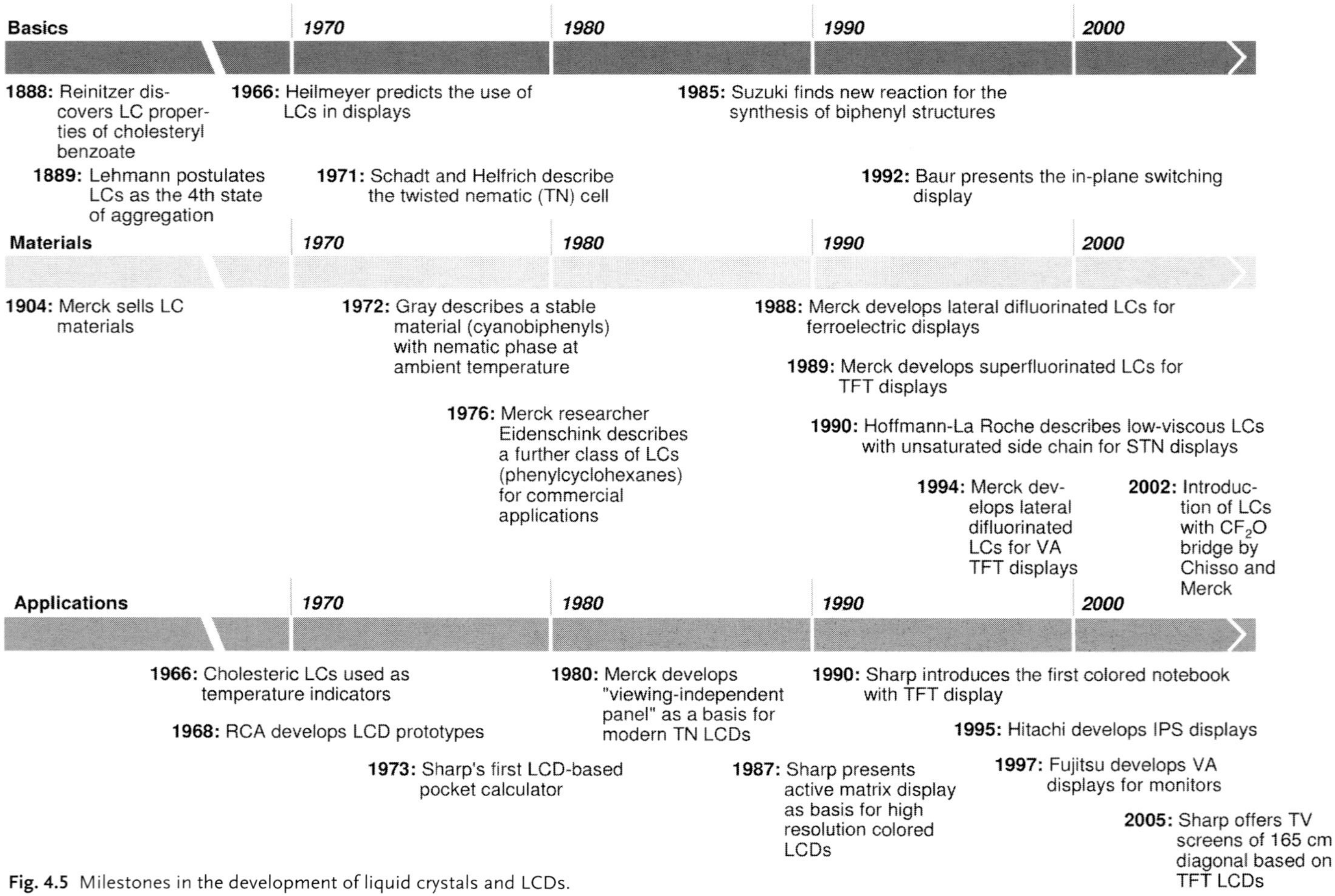

Fig. 4.5 Milestones in the development of liquid crystals and LCDs.

Figure 4.5 shows that each new generation of liquid crystals has triggered developments in the electronics market. As technology leaders, we have enabled LCD manufacturers to adopt new ideas with new high-tech liquid crystals. Other new concepts have often arisen from these for which no suitable LC mixture was available.

At the beginning of the 1990s, new LCs enabled the development of notebook computers, making mobile computing a reality. At the end of the decade, new LCs made flat, thin monitors feasible, which produced a minor office revolution. Flatscreen displays have a smaller footprint than cathode ray tube monitors and produce less heat. This reduces the need for office space, potentially lowering rents in some cases, and reduces the cost of air conditioning. Flatscreen monitors also consume significantly less energy – just a third of that needed for a CRT monitor. It is said that if all the CRT devices in Japan were replaced by LC monitors, it would save energy equivalent to the production of four large power stations. A new generation of superfluorinated LCs has made the development of flat LCD televisions possible since about 2002.

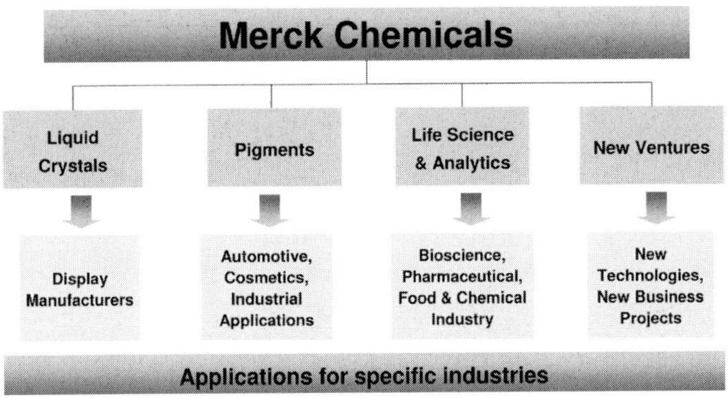

Fig. 4.6 Merck Chemicals' customer focus.

Meeting these customer demands means that both the innovation and the go-to-market cycles shrink (Fig. 4.6). So, to stay a step ahead of customers' needs (or at least keep up with them), Merck and its affiliated companies in Asia always develop LC mixtures in close cooperation with them. Moreover, Merck invested early on in in-house display development; first so that we could evaluate the properties of our LC materials for display applications, but also in order to obtain a better understanding of the final LCD products so that we could remain the technology leader and adapt our various product and applications portfolios to market requirements. In addition, needless to say, it is vital always to have adequate time and production capacity for all orders.

> *"It is the art of mixture development that makes the crucial difference. We sample 'bouquets' with diverse seasonings to ensure that mixtures emerge at the end that have the characteristics desired by the electronics manufacturers." Kazuaki Tarumi*

4.1.4
Timing is Everything: Internationalization and Diversification

Over the history of LC development, Merck's timing has been impeccable. As soon as the first display was introduced in the late 1960s, we were ready with local assistance for our customers. However, global growth needs local partnership and we have already seen how critical customer relationships are to pushing the boundaries of the technology and supporting and developing new applications.

> *"We often have to supply as quickly as possible – even overnight – several kilograms of a liquid crystal mixture with which the customer can, in just three months, bring a technically improved flatscreen to production." Kazuaki Tarumi*

Cooperating with customers is much easier if all parties are close together. Despite the communication revolution we mentioned earlier, some things – such as the exchange of knowledge – are still done better in person. Ninety-five percent of the world's LCD production is in East Asia, and thus Merck shifted its LC activities to the region very early on. In 1980 we moved operations to Japan; and since then we have added mixture development and production operations in South Korea and mixture development as well as, a few years later, mixture production in Taiwan. The individual compounds are still produced in Germany. This is partly because it is hard to establish polyproduction overseas, and it helps to have close proximity to basic R&D functions.

Making these moves early has allowed us to cement our position as the market leader in LC technology because we have ready access to our customers. Trying to do everything from Europe would not work, and delaying the decision to move some operations to Asia might have given competitors a chance to take market share, and might have limited the value we could extract from the technology.

In the 1990s, we decided we no longer wanted to stand on just one leg. We wanted to go beyond just offering customers liquid crystal mixtures, so we diversified into tailor-made system approaches. Our strategy was to strive for horizontal integration and supply other components for manufacturing displays. It was a bold step, so we started ITO coated glass production with a partner and then took it over once the business was on a sound footing. Being prepared to innovate not just within our products, but also in our relationships and our own position in the value chain allowed us to extract even more value from our knowledge-based competitive advantage. We have moved from a pure LC supplier to a manufacturer and supplier of semi-finished materials.

4.1.5
The Best Intellects

Supporting our tradition of innovation means drawing on the intellectual resources of extremely talented, experienced, and enthusiastic people, who cooperate closely in research, development, and production in interdisciplinary teams.

It is through their dedication that we are able to develop innovative hi-tech products such as liquid crystals that are high quality, safe for the user, and environmentally compatible.

These teams are characterized by high flexibility and creativity, and moreover they think and act with a business focus. Unfortunately, such teams do not fall from the sky. The secret lies in recruiting the best intellects, and then supporting them with education and further training. Merck's abilities here mean that its reputation in laboratories secures a stream of highly talented individuals coming into the company.

"Learning is like swimming against the tide. It forever drives you back if you stop!" says an old proverb. We understand this and are committed to lifelong learning.

Due to the international dimension of our business we pay particular attention to cross-cultural management development. This consists of three process steps: identifying potential candidates, developing them, and reviewing the results. Candidates nominated in each region participate in a regional development center, which then provides a pool of employees identified for management tasks. The development centers help us to ensure that everyone is working in accordance with best practice. We have perfected several tools for developing our leadership teams. In international management training, for example, leadership personnel are prepared for their current and/or future tasks. Further development possibilities include Learning by Projects, Learning by Internal/External Speakers, Learning by (Self-) Reflection or studying for a European Partnership MBA. The Merck University, our program for top management, sends candidates to one of three business schools in America, Europe, and Asia. Last but not least, the Liquid Crystal Training School was established in 2005. This training program for new LC division employees gives them a better understanding of our complex business and its context.

4.2 Outlook

The European chemical industry is an engine of innovation. If this is to be sustained, then the European Union's policies relating to the industry need to support R&D. At the moment, regulations on trading emissions, among others, might put off prospective (and current) investors. We believe the rigorous dismantling of legal and administrative obstacles to innovation is urgently required.

The chemical industry is tied to its local innovation network. The ability to recruit the brightest people, as well as to draw on valuable academic research, actually strengthens the university sector, which itself is concerned with excellence in training and research.

Of course, external influences on innovation are only one side of the equation. Chemical companies themselves have a responsibility to place research and innovation at the very heart of their strategies. This might mean looking beyond their usual supply of talent, or going down the value chain to develop innovations that can have a positive impact elsewhere in the business.

It also means preparing the market for change. Chemical innovations and new products manufactured from them will succeed only if they are widely accepted. Merck supports educational programs which seek to develop interesting, well thought-out scientific instruction in schools.

We read and hear a great deal about the drift of the chemical industry to cheaper wage zones such as China. The shift of standard product production into these regions is already a fact but, if the basic conditions are put in place for manufactured, high-tech, research-intensive chemicals such as liquid crystals, then the economic success of the chemical industry in Europe can be secured as the engine of innovation in other industries.

4.3 Summary

- In Germany, the chemical industry invests on average about eight percent of its revenues in research and development. It supplies other industries with highly innovative products and thus performs as an engine of innovation for them.
- This is demonstrated by the success story of Merck's liquid crystals, high-tech chemicals with unique properties. Discovered over a century ago by German scientists, they caused controversy in the scientific community at first, but – with the help of Merck's highest-purity substances – liquid crystals were eventually accepted as a fourth state of aggregation.
- Nonetheless, they became a mere laboratory curiosity for decades until technical applications in thermography and displays were suggested in the USA and Merck resumed its interest in them.
- In this paper we discuss five key factors which have made this highly innovative product class a success story as well as their key application, liquid crystal displays, the dominating communication device technology of today.
- The paper also shows how the persistent development of new advanced product generations has fertilized other industries.

5
Today's Challenges and Strategic Choices

Florian Budde, Utz-Hellmuth Felcht, and Heiner Frankemölle

The chemical industry as a whole has weathered stock market storms pretty successfully over the last 25 years. The Standard & Poor's 500 of 1980 contained 14 chemical companies. The same index in 2005 also boasted 14 chemical players – merged, upsized, downsized, but still there, with a good third of the old names still around.

That is quite an accomplishment. Granted, some industries have made immense fortunes over the same period, which many chemicals players have not: in 1980, Microsoft was just launching MS-DOS, Cisco Systems had not yet been founded. The advance of computerization, electronics, and new technology has seen the core businesses of a number of sectors transformed – apart from the IT world itself, pocket calculators, cash registers, and the world of home entertainment are cases in point. The pharmaceutical industry and financial institutions also created a great deal of wealth in this period. However, some other major players, particularly telecommunications, have seen enormous value destroyed following the post-deregulation boom.

In terms of total returns to shareholders (TRS), too, the chemical industry has proved its staying power in recent years. Although it suffered badly during the Asian crisis, and the ensuing technology and Internet bubble at the end of the twentieth century did not bolster chemical stocks and at the same time lured investors away from more traditional plays, the new millennium has seen the chemical industry continuing to deliver TRS that is almost in line with the market (Fig. 5.1).

This is an achievement to be proud of. In the face of enormous challenges from the oil crises, the flow of capital to more attractive – but maybe also more risky – industries, the emergence of new players, and the growing management dilemma posed by rising new technologies, the industry appears to be more than holding its own. Knowledgeable investors such as private equity companies seem to find it really quite attractive, solid and promising if not stunningly alluring. The reality, in a word, is a far cry from the frequent perception of the industry as consolidating and unattractive, with its participants constantly anticipating the onset of the next downturn.

Value Creation: Strategies for the Chemical Industry. 2nd Edition. F. Budde, U.-H. Felcht, H. Frankemölle (Eds.)

ISBN: 3-527-31266-8

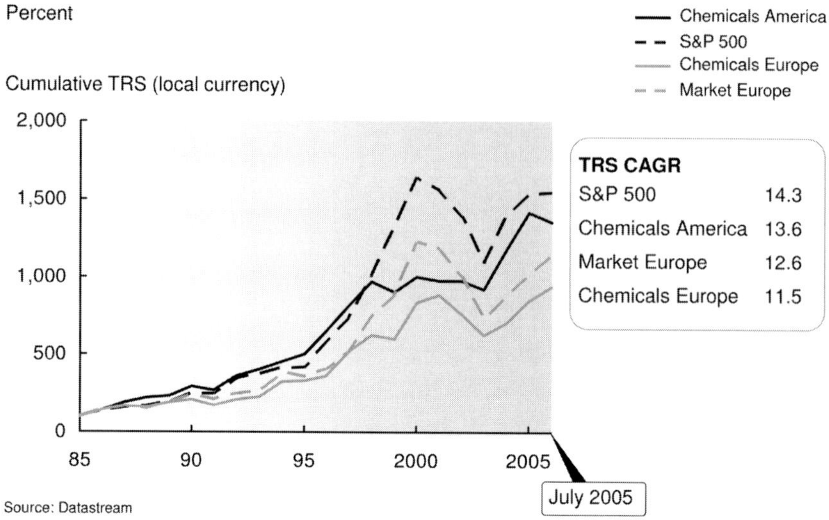

Fig. 5.1 TRS performance 1985–2005.

Can we, however, extrapolate a relatively bright future from past performance, or are we waiting for a singularity to hit, and the whole game to change?

Well, maybe not quite that, but no one can deny that the chemical industry could be facing some major shifts in geographic emphasis and very possibly technology. The landscape is certainly likely to change, but the speed and extent of the transformation is as yet unclear. Excellence will depend on alertness to trends and agility and flexibility in management: executives in the chemical industry will have to keep their ears to the ground, their eyes on the stars, and their hands to the wheel – while probably dipping their toes in new waters as well. In other words, while keeping a watchful eye on the new developments, managers will have to continue to tap into the existing and new opportunities in their current businesses, as well as continuing to play the functional excellence game which will help ensure their cost competitiveness, enable them to cope efficiently with global structures, and provide the cash to explore new pastures.

5.1 Overall Outlook Stable

On a global industry level, there is no reason to expect the industry's prosperity to flag. For one thing, everybody needs chemicals. From consumer products through electronics and mechanical engineering, manufacturers everywhere need chemical products, and there is no sign the situation will change. In many cases where substitution is possible, as in the search for alternative fuels, the substitute for one chemical would be another. As a result, the customer base is very solid indeed. In addition, no discontinuity is foreseeable in the development of demand: it is steady in the traditional markets of North America and Europe, and growing vigorously in the emerging economies, particularly those of Asia.

Second, the industry is not vulnerable to typical landslide shifts such as regulation or deregulation. For example, although the tightening of environmental, health, and safety rules is leading to cost- and expertise-related headaches, this is having nothing like the cataclysmic effect that the deregulation of telephony or postal services had on the industries concerned.

The chemical industry is also highly versatile, owing its success partly to the richness of its mixture. Highly fragmented, with tens of thousands of product lines, it gives its participants immense strategic degrees of freedom to penetrate new product segments and new regional markets. Production resources can be highly versatile, too: very often, chemical companies can come up with different products for the same application that are produced in very different ways, based on different feedstock for the same application – from natural gas and naphtha through sugar.

Furthermore, although it is clear that classical product innovation has slowed down in the last three decades, other forms of innovation are emerging. Chemical companies are seeking opportunities in new products, services, and applications, and are also increasingly having to reach out to cooperate with other industries, as chemical innovation moves out of the chemical companies' R&D labs and into the users' province, and consequently takes place at the interface with other sciences such as biology, physics, engineering, and medicine. In addition to these developments, however, it seems that the blockbuster source of future value from innovation may well lie in new technologies such as biotech, which are not only becoming economically viable, but even make products possible that could not previously be manufactured.

Finally, another and much less positive constant of the chemical industry is the price-cost squeeze, which we expect to remain in the range of approximately two to three percent of sales annually as it has been for the past 20 years, driven by labor and other factor cost increases and by ongoing price pressure, resulting from productivity increases that are passed on to the customers.

5.2 The Value Kaleidoscope

The chemical industry as a whole, then, has staying power. Nevertheless, chemical players are becoming increasingly exposed to fundamental trends that may change the picture drastically at the level of the individual company. Booming demand coupled with a developing chemical industry in Asia, particularly China, is driving a geographic shift in the industry's focus. Soaring feedstock prices in the USA and the increasing use in the Middle East of the natural gas produced in association with crude oil that was previously flared ("stranded gas") are also changing the chemical industry map. New technologies, particularly biotechnology, may have the potential to transform the cost, speed, and environment-friendliness of production processes, and the once-hazy outlines of their future prospects are sharpening up rapidly. A further driver of value and of change is the continuing wave of mergers and acquisitions, which allow companies to regroup and restructure for increased focus, scale, and efficiency – and also, by trading business units, to generate cash. Closely connected with this ongoing M&A activity, private equity players are also increasingly taking stakes in chemical companies.

It is the nature of new moves to involve risks as well as opportunities, and some value in the industry may well be radically redistributed among new winners and losers.

5.2.1
East, West, Is Home Best?

Demand is set to grow approximately in line with GDP, and although this is low in the traditional markets of the USA and Europe, business is booming in Asia, with high single to even double-digit economic growth expected in China in years to come (see Chapter 1). Local and incumbent players both have some hard work to do before they can compete successfully here:

- Opportunities are opening up for local newcomers (particularly in China and India) to grab their piece of the pie. Initially, they will supply the local markets: China, for example, has enormous gaps in local supply and will remain a net importer of many chemicals for years to come. However, as these local players gain in technology expertise, experience, and scale, they will be increasingly capable of attacking foreign markets. They are already doing so, as we are seeing in fine chemicals and other areas. Chinese entrepreneurs have already proven their ability to build business locally and then launch into foreign markets, in areas such as consumer electronics and PCs, as evidenced by the acquisition of IBM's PC division by Lenovo. They are also starting to do so in chemicals, as evidenced for example by China National BlueStar and Rhodia: in May 2005, the two companies signed a Memorandum of Understanding on the formation of a global strategic alliance in the silicones business. The joint venture is to be in place by mid-2006. However, Chinese players have to learn how to deal with Western markets, just as the Western players have to adjust to Eastern ways. Overcoming the language barrier, for example, is a major hurdle for both sides.
- The picture for the incumbents is much more complex. They could find themselves on the back foot, but they have many strengths that the local Chinese and Indian players do not yet have. However, in many cases, their assets are in the wrong locations to serve Asian markets. Also, these assets are older and therefore less efficient and facing renewal much sooner. On the

> other hand, the incumbents have the advantage that some capital investments are already written off, yet still earning cash returns. Their home bases in first world countries are, as we all know, more costly than many of those of Asia. By definition, they are also less familiar with the new markets than the locals are. On the other hand, they still have the technological edge, greater expertise and scale, and local markets that account for the lion's share of chemical consumption and will continue to do so – according to BASF, the Asia-Pacific markets in the year 2015 are likely to account for only around 34 percent of the world total. The old hands' capabilities and their financial strength place them in a good position to occupy space in the Asian markets and produce for both local markets and export.

Different companies are finding different answers here – while initially cooperation with local market players seemed the way for Western players, wholly foreign owned businesses are now mushrooming in China. Investment projects amounted to around EUR 20 billion of contractual foreign direct investment (FDI) between 1999 and 2003, yet the involvement of multi-national companies has a long way to go before it matches the importance of the market. They also need to take swift action if they wish to prevent local competitors from becoming the established suppliers to the Western customer industries which are increasingly setting up in Asia, such as electronics and automotive players. For a detailed discussion of the problems and opportunities particularly for Western players, see Chapter 32.

5.2.2
The Feedstock Rollercoaster

It will be no news to most of our readers that North America has become a high-price region for feedstock since the turn of the millennium. In addition, stranded gas in the Middle East is being more aggressively utilized, making North American feedstock options largely uncompetitive. This will result in a redistribution of wealth between North America (mostly) and players from the rest of the world on the one hand, and gas owners on the other. In addition, both costwise and geographically, the Middle East is better placed than the USA or Europe to supply the booming Chinese market for all products of the C_2 chain. Players from outside the Middle East have a limited chance of mitigating their situation by joining the boat early and exchanging market and other relevant know-how for participation in new plants in the area.

In addition, the rising cost of feedstock will of course drive price increases in other areas and has the potential to change consumer behavior, creating or reducing opportunities for certain chemical products, for example increasing the need for domestic insulation material because of higher energy prices. Even if individual markets are small, flexible players can benefit from grabbing these chances.

5.2.3
Biotechnology – Looking into the Seeds of Time

Biotechnology has, of course, been around for many years. It has faced an uncertain future, unacceptable to many chemical industry players because of the vast amounts of cash and long times required for developments. It seems, though, that biotechnology might just be about to come into its own. Development times are shrinking. According to Laane and Sijbesma (Chapter 29), the development of some industrial biotechnology products is now taking two years, where not long ago it would have taken five. In addition, the rise in feedstock prices will make competing biotechnology pathways economically more competitive, so that those companies which have to date shied away from it may become tempted to get in the biotech game.

For traditional players, who have an intrinsic incentive to defend their assets and the conventional technology, major challenges will lie ahead. At the very least, it would be unwise to ignore recent advances in technology, as the large chemical companies have realized – as Chapter 28 describes, several leading players, including BASF, DuPont, and Ciba Specialty Chemicals, are now investing substantially in the field. Biotech routes can simplify processes enormously – for one BASF-produced vitamin, from eight steps to one – and are also much more environmentally friendly. However, there are still products for which no bio-routes are yet available, or for which the conventional route is still more cost-efficient. In addition, it will take large capital investments to make the switch from conventional to bio production routes.

In contrast, nimble new companies with no legacy assets can also build competitive technology and introduce it. However, such companies would need considerable amounts of cash to start from scratch – although there are already some biotech-only players around in chemicals. Two of the largest of these that are already generating profits include Novozymes, which according to the company website posted sales of around USD 986 million in 2004, and Genencor International with USD 410 million in revenues that year. Most of the rest are very small. Incumbent chemical players with significant industrial biotechnology interests are turning over significantly more than they are in this area, for instance DSM, with currently nearly EUR 1.5 billion per annum in sales of biotech-derived products, according to the company's own information.

There are still a number of imponderables, such as the sustainability of the current high prices of conventional feedstock, and the availability of competitively-priced bio feedstock. However, it would certainly seem advisable for all incumbents to be investing in some biotechnology intelligence, if only to understand the developments. Alliances with companies with biotech expertise might also be wise. If biotechnology has the potential to become an 800-pound gorilla, say, five years down the track, it may be best to treat it with respect and courtesy now.

5.2.4 Mergers and Acquisitions

Finally, immense value can be created – but also destroyed – by mergers and acquisitions. Successfully handled, they can be a huge source of wealth, as the case of UCB, for example, has shown (see Chapter 26). In a series of acquisitions and divestments, the company created three strong business entities and increased its stock value from EUR 17 per share in September 2002 to EUR 38 in March 2004. Although some players are now finding the scope for acquisitions limited in their home markets by anti-trust considerations, there are still many deals going on. But as we all know, fewer than half of mergers prove successful and create value for their shareholders. As Bartels and Koch (Chapter 25) demonstrate and the UCB and Degussa (Chapter 9) stories also show, excellent integration management addressing both operations and strategy and, crucially, people can help to optimize the results of mergers: the best deals can transform the approaches of both companies and take them to a higher level of performance.

In this context, the increasing number of professional private equity companies effecting deals in the chemical industry is also worth noting. It appears that some of these companies manage to generate more value from chemical assets than industry players can (see also Chapters 30, 31). This is due in considerable measure not only to their superior skills in financial engineering, but also to their additional rigor in enforcing operational improvements after the merger. Chemical industry players can learn several lessons from their approach.

All these activities will continue, be they driven by the chemical industry or by financial investors. These developments will provide opportunities for those who are skillful at valuations and post merger management, and who have the funds available whenever the opportunity arises. And over time, they will create a substantial advantage over those who are not successful with their acquisitions and have to write them off, or who do not have the money required in the first place.

5.3 What Happens Next?

Although virtually every chapter of this book discusses solutions to problems in individual segments or functions, the big picture is a little obscure at present, throwing up a number of questions. The viability and sustainability of some of the new developments appear too uncertain, their emergence as serious influencing factors too new to give a really sound foundation for prediction: we do not know whether, and how significantly, the chemical landscape will be redrawn. As the saying goes, strategy means being prepared for when the opportunity arises.

However, a wait and see policy is the last thing we would recommend. The lines for action are already sharply drawn, and some focal areas are outlined above as well as in the remainder of the book. It is plain that, as in the past, a huge variety of choices remains open. In our view, two fundamental guiding principles emerge:

1. *Companies need to create strategic headroom by increasing profitability.* To invest in new areas, they will need money: functional excellence therefore remains a key priority. Players should continue to seek improvements in operations and other functions of the business system (as described, for example, in Chapters 18, 19, 22, and 23). They have plenty of headroom to influence their return on invested capital, and by and large this is not a zero sum game – every company can improve its efficiency without others necessarily losing out, particularly if chemical companies focus more strongly on keeping the benefits in house instead of passing them on to the customer. Functional excellence is a huge source of value creation given the extraordinarily wide performance spread of chemical companies. This is evident in the market-to-book ratio of chemical companies in comparison with that of other asset-heavy industries (see Chapter 3). In addition, excellent performance in every function along the chemical business system has become more and more important as large companies increasingly have to manage global networks to follow their customers and operate at the lowest feasible cost. An absolute precondition underpinning functional excellence and the ability to adapt to changing circumstances will be the ongoing development and refinement of the chemical players' organizational setup, as well as the further refinement of their capabilities in developing skills and therefore people as a crucial resource (see also Chapter 24).
2. *It is time to strategize.* More than ever, chemical industry players need to deepen their insight and foresight in the arenas relevant to their companies.
   - Commodity companies have to take a conscious view of feedstock price developments, and consider a wide range of options from financial hedging and optimizing flexibility to taking stakes in Middle Eastern companies to benefit from their stranded gas assets (see Chapters 7 and 16)
   - Specialty companies in particular have to understand their customers in detail, in order to generate value by product and service innovation that is based on a thorough understanding of customers and markets. In addition, they must keep every class of contender under close and constant scrutiny and evaluate competitive moves, particularly those from upcoming Asian entrants
   - All players need at least to form an opinion on which biotech pathways might be relevant to their activities, at what cost, and within what time frame.

Above all, companies will be forced to make strategic bets, and invest proactively in the development areas they believe in. We hope that the scope and reach of the information in this book will provide a basis of information, insights, and advice that will help chemical industry executives to plan their futures successfully.

5.4 Summary

- Chemical companies have managed to hold their own fairly well in the capital markets over the last twenty-five years.
- On a global level, the industry is likely to remain stable despite a continuing price-cost squeeze.
- At the company level, however, fundamental trends may lead to a radical redistribution of value among new winners and losers, mainly emerging from four key trends:
  - The boom in the Asian markets, specifically China
  - Feedstock volatility
  - The development of biotechnology
  - Mergers and acquisitions
- To come out as winners in the shifting landscape, chemical industry players chiefly need to create strategic headroom by increasing profitability, and develop more insight and foresight than ever in their strategic choices.

6
An Approach to Determining the Long-term Attractiveness of Commodity Chemical Businesses

Scott Andre, Sunil Sanghvi, and Thomas Röthel

Contrary to conventional wisdom that commodity chemicals is an unattractive industry, the sector has earned returns above its weighted average cost of capital (WACC) across a number of cycles, outperforming other commodity industries. Much of the conventional perception derives from the industry's extreme cycles – long stretches of low margin "troughs" with seemingly random, short periods of high profit "fly-ups". These lead to huge swings in sentiment that make investment decisions difficult for managers and deter investors.

This conventional view risks losing sight of the industry's fundamental attractiveness. Although fly-ups and troughs may obscure any short-term predictability in the sector, we contend that fly-ups have resulted from the "invisible hand" of the market, ensuring that producers supply enough capacity to meet demand growth. The higher the cost of new capacity needed to meet demand growth, the higher the margins provided by the "invisible hand"; these higher margins also benefit existing plants. For many commodity chemicals, margins during troughs are below this "required" level, and the resulting under-investment leads to tight supply and margin fly-ups. Thus, the long term potential for repeated fly-ups depends on future demand growth and the cost, including capital return requirements, of new capacity to meet it.

While the futility of trying to foretell chemical industry profitability cycles and time investments to them is well established, we believe it is possible to gain insights into the likely long-term margin evolution of the industry's various sectors, notably by identifying which sectors will continue to experience fly-ups. In this chapter, we show that by taking a detailed look at how the industry will add capacity to meet demand growth (i.e., whether new capacity will be low or high cost), it is possible to roughly estimate average prices and margins across future cycles. The same approach identifies which sectors are likely to see fly-ups and which, on the other hand, risk a long-term margin slide or even degeneration into extended profit troughs. These insights can provide a key input to CEOs' and investors' decision processes, by helping select the operations and locations that are likely to earn satisfactory returns in the future and should be backed, and those for which divestment or other options have to be considered.

Value Creation: Strategies for the Chemical Industry. 2nd Edition. F. Budde, U.-H. Felcht, H. Frankemölle (Eds.)

ISBN: 3-527-31266-8

In the following sections, we discuss the extreme cyclicality of the commodity chemical industry and the challenges this presents to management and investors. We contrast the higher profits of the chemical industry, driven by periodic fly-ups, with those of other commodity industries. We present a framework for how producers and investors can evaluate their businesses' prospects, taking bearings from the returns that producers in each segment will require to consider reinvestment, and the full cost economics of new capacity needed to meet growing world demand for chemicals. We then use the example of the ethylene market to highlight how these key drivers are likely to play out in a key petrochemical sector, as well as the risk of technology- and feedstock-driven disconnects. Finally, we discuss how producers will need to take into account particularities of their submarkets to assess their prospects.

6.1
Looking beyond Cyclicality

Is the commodity petrochemical industry worth staying in? The severity of the downturn from 2000 to 2003 posed that question with a vengeance for players in the industry, who saw their worst results since the aftermath of the second oil crisis in the early 1980s. Profits subsequently recovered for many companies to record levels. But any chemical company top management experiencing such troughs and peaks must ask itself, "Will these cycles continue and perhaps even improve, or will the next trough last indefinitely? Is this a business segment on which I want to build the future of my company?"

CEOs of a number of long-standing players appear to have decided the answers are negative, on the evidence of some major divestments of commodity chemical properties in recent years. These include BASF's and Shell's decision to divest polyolefins maker Basell, BP's spinning off its Innovene petrochemicals unit, DSM's sale of its petrochemicals to Sabic, and DuPont's divestment of its Invista fibers business, which includes leading petrochemical stakes, specifically in nylon and polyester intermediates.

The acute cyclicality of the commodity petrochemical industry's profitability makes it a difficult business for investors to evaluate: the lack of steady earnings means that investors tend to undervalue commodity chemical companies at the trough and overvalue them at the peak (see also Chapter 2). It is also a difficult business for chemical companies to manage, in that timing of investments is both of critical importance to profitability and extremely difficult to get right.

Commodity chemical demand continues to grow steadily: consumption of ethylene, the biggest volume petrochemical, for example, has grown from 90 million m.t. in 2000 to 105 million m.t. in 2004, and is projected to rise to 135 million m.t. in 2010, primarily driven by growing wealth in developing countries translating to higher per capita consumption of chemicals. In timing their investments, however, managers of commodity chemical companies face major challenges.

- They have to contend with overall economic cycles, which drive acceleration and deceleration of demand growth and customer stocking and destocking behavior, as well as upsets such as the 1997–98 Asia crisis.
- They also have to contend with factors internal to the industry – the commodity chemical industry's own cycles of investment. Similarly to players in many other commodity industries, petrochemical makers have for decades fumbled the challenge of how to keep up with steady growth in world demand for chemicals, yet not outstrip that demand and see prices sink and their profits evaporate due to oversupply. Industry players tend to simultaneously invest in new capacity in the wake of product shortages that trigger dramatic price and profit increases – referred to in the industry as "fly-ups" – ultimately resulting in the next price slump. Investment is then starved until demand overtakes supply, and the cycle repeats. Profitability tends to closely track the capacity addition cycle: profits fall after capacity additions.

Our research has shown that attempts to successfully time investment decisions to the cycle – not only the chemical industry's but broader economic cycles – are extremely difficult (Butler, P. et al.), and that other tactics such as maintaining advantages in cost and operational excellence across all functions provide a surer return.

However, if chemicals are irredeemably cyclical, does that make commodity chemicals an unattractive business? In our view, the key is to look beyond individual cycles, at the long-term profitability of the sector. This reveals a perhaps surprising judgment, in view of conventional wisdom about the sector's unattractiveness: commodity chemicals have earned returns above their weighted average cost of capital (WACC) across a number of cycles, making it very different and a much better performer than most other commodity sectors.

That, however, is history, and the CEO needs to look to the future. The key question for many CEOs with assets in high cost regions is, "Will the sector continue to have the periodic margin fly-ups that must occur for me to earn target returns across the cycle?"

6.1.1
Two Types of Costs Needed to Anticipate the Future

To assist CEOs in making this judgment, we have developed an approach that combines the well-understood industry concept of the producer cash cost curve with a detailed look at what types of capacity are needed to meet future demand growth. New capacity can be of many "types", e.g., simple debottlenecking or "creep" that has low capital requirements, greenfield plants with advantaged costs compared with existing plants due to low cost feedstocks or new technology, or new plants that have costs and capital similar to existing ones. We have also occasionally seen new capacity that was unlikely to deliver good returns but was built for other reasons, such as providing self-sufficiency to a nation or profits to the project developers.

We believe that this information can be used to directionally forecast the long term average margins and severity of troughs and peaks in the future, and thus enable incumbents, potential investors, and new entrants to better assess their long term prospects.

The concept of the producer cash cost curve, that aligns and ranks producers in ascending order based on their cash costs of production and the intersection of which with market demand identifies the marginal – and therefore price-setting – producer, is universally known in commodity industries and requires no further explanation here.

Executives are also familiar with a variant of this concept, the "full cost" curve which includes, along with the cash cost, the cash margin required for a new investment to earn its cost of capital. Thus, the "full cost" is equivalent to the target price needed to justify new investment, although other external factors such as the timing of the start-up along the cycle and the industry price/margin pattern across the cycle also significantly affect the required average price and margin to earn cost of capital. As explained below, the "full cost" of new capacity needed to meet demand growth provides guidance on whether and how much future cyclicality can be expected for a chemical segment.

6.1.2
The Role of Reinvestment Economics

In many commodity chemical sectors, as noted above, the boom and bust pattern – long stretches of low margin troughs with no investment followed by short periods of high profit fly-ups during shortages, leading to new investment – is a sure indicator that producers seek a certain level of return on their investments before making new ones. Fly-ups after troughs of poor profitability contribute the returns that a producer needs in order to justify investment in new capacity; the fly-ups effectively top up the returns to the level necessary to justify reinvestment to meet ever-growing chemicals demand.

This behavior is well recognized in the commodity chemical industry, and is referred to as "earning reinvestment economics" or "achieving cost of capital returns", where players in a market sector invest only when they expect to earn enough margin to satisfy their capital return requirements. We contend that fly-ups result from the "invisible hand" of the market to ensure that enough new capacity is built to meet long-term demand growth. Without that new capacity, the market will tighten and a fly-up will occur. If overbuilding results, prices and margins will likely fall below levels required to justify new investments and the industry will enter a trough of underinvestment; the trough will continue until again new investment is stimulated by the expectation of being able to earn cost of capital. If too many players invest ahead of an anticipated fly-up and prevent its occurrence, there is unlikely to be additional new capacity until another fly-up occurs.

This of course points out that timing of investments can have a significant impact on overall returns for new investments. We are not addressing that issue in this chapter. However, the timing of margin patterns should be considered along with our messages on long term average margins.

What determines how high a fly-up can be expected or whether there will be a fly-up at all in the future? Applying the "invisible hand" concept further, we believe that the average margins across a cycle will approximate to the level needed to satisfy the return requirements of the lowest cost capacity needed to meet demand growth – otherwise investors will not build and eventually margins will rise, which will eventually attract investment to meet the demand growth.

The level of average margins will be determined by the types of capacity needed to meet future demand growth.

- If demand growth is low and can be met by low-cost debottlenecks (also known as capacity "creep"), the average margin required to meet demand growth will be low and we would expect to see low margins across a cycle. (Other factors such as a strong unexpected demand spurt like that seen in 2004 to 2005 can create short, serendipitous fly-ups, but these cannot be counted on when making long term portfolio decisions).
- If demand growth is high and cannot be fully met by low cost capacity, such as simple debottlenecking, or by supply from highly cost advantaged regions or new technologies, we expect that through-cycle margins will be higher and at roughly the level needed to justify higher cost capacity to meet demand growth. The actual investment pattern will be driven by investors' reactions to current or anticipated margins, but the end result will be that their returns will be impacted by the level of returns needed to attract enough new capacity to meet demand growth. Not all investments will have equal full cost, and the lower-cost investments will achieve returns above cost of capital.

The types of capacity additions needed to meet future demand growth will provide guidance on average margins across the cycle and the severity of cycle troughs and fly-ups. The average long term margins will equilibrate at roughly the lowest level needed to stimulate enough additional capacity to meet demand growth. However, prices at any given point in time will not necessarily settle at these margin levels. Instead, when there is oversupply, prices will be set by marginal producer economics. If this "trough" price level results in margins below the "average margin" needed to stimulate new capacity to meet demand, then we expect that at some point a fly-up will occur. The lower the trough price level relative to the "full cost"-based price needed to meet demand growth, the higher and/or longer the fly-ups will have to be to stimulate new capacity (Fig. 6.1).

Given the low technological differentiation in commodity chemicals and – since much chemical production is ultimately based on crude oil – worldwide raw material price homogeneity, the cost curves for many chemicals are relatively flat.

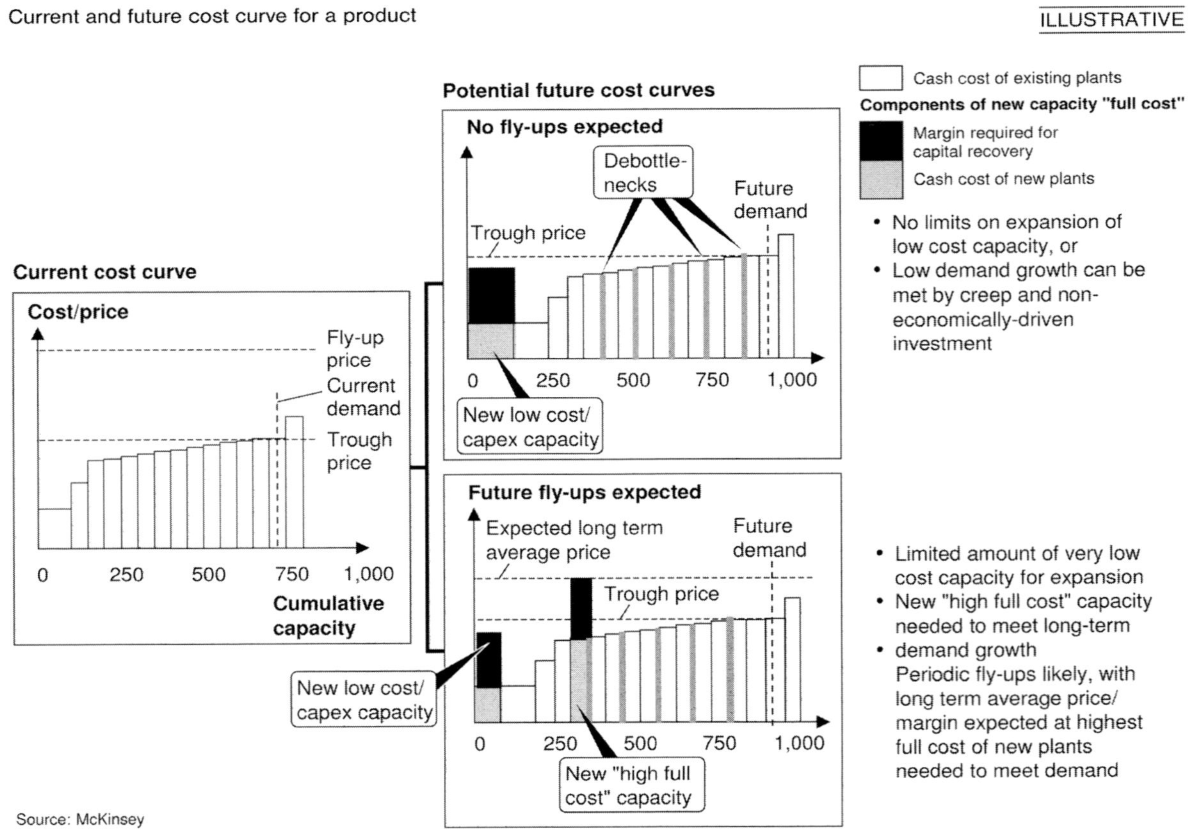

Fig. 6.1 Assessing the potential for future fly-ups.

Moreover, most commodity chemicals are highly capital intensive and require high margins in order to deliver cost of capital returns. Thus, we typically see that the full cost of new plants is well above the marginal producer's cost (i.e., the price expected during troughs) for most commodity chemicals. Combined with robust demand growth requiring significant capacity additions, this implies that periodic fly-ups can be expected and that the long-term average price will approach the relatively high full cost of needed new plants.

An exception to these typical conditions in commodity chemicals, and it is an as yet limited but important one, occurs where new chemical production can be sourced from much lower-cost raw materials, such as natural gas that is "stranded" far from consumer markets. This is the case with most methane- and ethane-based chemicals, where a producer able to secure stranded gas supplies can obtain a major cost advantage, enabling it to enter a market at a low point on the cost curve. As a result of this discontinuity, the "attacker" can harvest market growth potential and earn healthy profits that achieve reinvestment economics, even if no fly-ups occur in the future. It should be noted that an attacker who has access to new technology with very great cost advantages is in a similar position. In fact, in segments where all future demand growth can be met by these low cost capacity sources, the required prices to meet demand growth could be at or below marginal producer cash costs, so that fly-ups might never again be seen. We see this as very possible for methane-based chemicals, but much less likely for most other petrochemicals.

Thus, owners and managers of commodity chemical companies facing new low-cost competitors should carefully examine whether there are limits to how much of that low-cost capacity can be built. If there are limits, long-term margins may not drastically decline if demand growth still needs some higher-cost capacity. The higher margins will be delivered through periodic fly-ups. In this sense, cyclicality is "good" if the alternative is a permanent trough condition.

6.1.3
Learning to Love Fly-ups

The power of these concepts to evaluate future chemical industry scenarios can be demonstrated by using them to help explain long-term performance differences among various industrial commodities (Fig. 6.2). The pulp industry has good demand growth but can meet this with plants based in regions with much lower costs than Western producers', e.g., leveraging low-cost, high-yield forests along with low-cost labor in South America. These new plants can earn cost-of-capital and justify expansions at prices that are at Western producers' cash cost breakeven points. As a result, new low-cost capacity constantly is added and no product shortages – and therefore no fly-ups – are expected.

The aluminum smelting industry has similar conditions to pulp. Demand can be – and has been – met during the last decade by very large-scale, low-cost new capacity in countries endowed with privileged natural resources and/or low conversion costs such as India, the CIS, China, and the Middle East. These low-cost additions to the cost curve have harvested all the market growth potential and kept supply ahead of demand. As a result, until the exceptional take-off of Chinese demand in 2004–05, aluminum smelting has not experienced major fly-ups.

The steel industry presents another model: the low demand growth of the past 20 years has been met by the combination of capacity creep and investments by nations wanting to have steel capacity – without regard to reinvestment economics. The full-cost pricing for such additions is lower than marginal producer cash costs, i.e., trough pricing. This results in capacity growth exceeding demand growth, so marginal cost – and thus trough price – falls as the cost curve shifts to the right and, as a result, we do not see fly-ups except during unexpected demand spurts or supply disruptions.

We noted earlier that commodity chemicals have earned superior returns to other commodity industries. Closer analysis that we have undertaken of the key differentiating factor behind chemicals' higher profitability compared to other commodities shows it to be the recurrence of fly-ups, resulting from robust demand growth, limits on low-cost new capacity, and relatively flat cash cost curves.

Comparative performance of industries during the 1990s
Western producers

Industry	**ROIC minus WACC*** Percent	**Global demand CAGR** Percent	**Highest year's utilization**	Ratios for US players: **Gross margin ratio – peak** vs. trough**	Ratios for US players: **Percent of total cycle cash margin earned during peak**
Commodity chemicals	1.3	4.9	96% USA 93% global	3.4	45
Aluminum smelting	-0.1	2.0	< 90% global	1.7	No major fly-ups
Pulp	-1.1	3.1	86% global	1.9	No major fly-ups
Oil refining	-2.2	1.3	< 90% global	1.8	No major fly-ups
Steel	-4.4	0.5	77% global	1.3	No fly-ups

* Based on 1990 - 2000 reported ROCEs for major companies with most assets in USA and WE.
** Highest 12 month period to lowest 12 month period over 10 year period
Source: Datastream; Compustat; statistical yearbooks; McKinsey analysis

Fig. 6.2 Commodity chemicals have outperformed most basic commodities.

6.2 Putting the Model to Work

We have used the concept of the full cost curve and determination of the margins required to meet demand growth in several situations where investment and divestment decisions are involved, particularly when segments have faced major discontinuities. The following are the main steps in using the framework:

- Assess industry demand outlook scenarios for a base case as well as in the event of potential discontinuities.
- Determine the relative economics (cash cost and full cost) and available future supply of product from various regions and technologies. Consider potential technology developments and potential changes in relative prices of feedstocks.
- Assess what type of new plant or combination of new plants is needed to meet demand growth at the lowest full cost (including cash cost and capital recovery) for various combinations of demand and supply scenarios (Fig. 6.3). If the cash cost and full cost of the "highest full cost" plant among the expected new plants are similar to the cash cost and full cost of prior plants needed to meet past demand growth, and it is in the same region as prior new capacity, players can likely expect margin trends in the future similar to those in the past. If the "highest full cost plant" changes to one with somewhat lower costs or reduced capital intensity (e.g., economies of larger scale), long-term margins will likely trend lower than in the past. If it changes to a plant with much lower costs or lower capital investment, companies may end up facing indefinite trough-like conditions. A shift of the "highest full cost" plant to a different region may impact long-term margins favorably or unfavorably, depending on relative trade flows among regions. Shifts to higher-cost marginal sources are possible but less likely.
- Assess how existing or potential plants' economics will evolve, during trough conditions and across the cycle.
  - Trough prices and margins link to the industry's marginal producer (on cash cost basis). Expected changes in trough prices and margins for an existing plant will derive from any changes in the differential in cash costs between the existing plant and the present versus future marginal producer plant, plus any changes in product price differentials between the plants. Therefore, if the pattern of new capacity or feedstock price changes results in a shift in the location, technology, or scale of the industry marginal producer, you could see a change in trough prices and margins for existing plants.
  - Fly-up margins for the existing plant will relate to the average price/margin needed by the "highest full cost" new plant

(among the new plants/expansions needed to meet demand growth). Specifically, determine: 1) what will be the magnitude of fly-up (i.e., the combination of margin height and duration) needed by the highest full cost new plant so that its average price/margin over the entire cycle is equal to its full cost/reinvestment margin, and 2) how the cash cost and product price at the existing plant compare to the cash cost and price for the "highest full cost" new plant during the fly-up conditions.

– Then the average long-term prospects for the existing plant can be estimated from the weighted average combination of trough and fly-up margins.

The insights from this analysis can be used in asset investment and divestment decisions, as well as assessments of long-term competitiveness to other players, and product strategy decisions such as determining whether to emphasize specialty grades if margins from commodity grades are in danger of permanent weakening over the next few years.

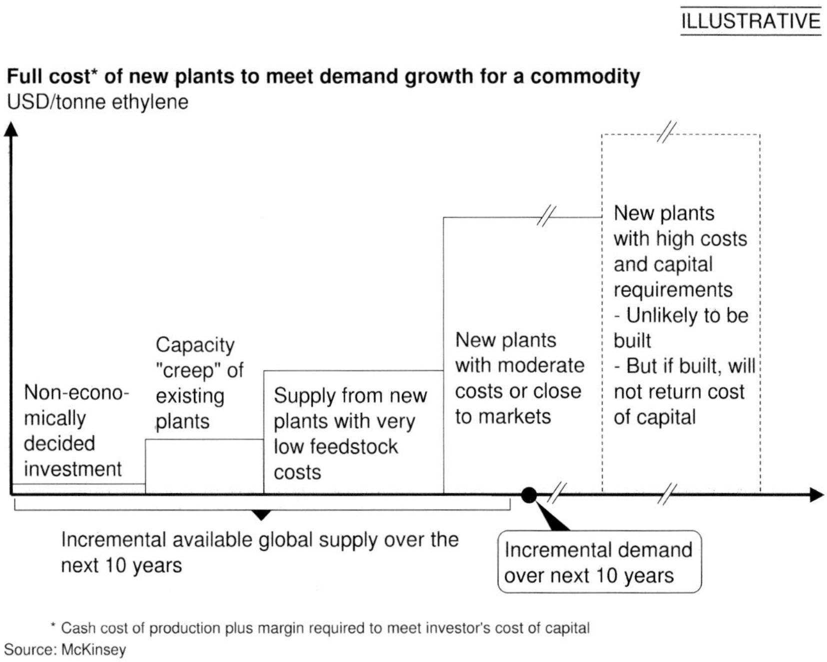

Fig. 6.3 Determining which new plants will earn cost of capital and drive long-term prices/margins for the industry.

6.3
The Example of Ethylene

The ethylene market is worth a closer examination using our approach. Ethylene is the biggest petrochemical market by volume, and its behavior is fundamental to the industry's profitability. The ethylene market is being affected by major changes in terms of new demand growth – primarily in China – and inter-regional shifts in cost positions due to feedstock price changes (in response to crude oil price movements and North America natural gas supply/demand tightening). It is helpful to examine in detail how the dynamics of reinvestment economics and fly-ups, and the impact of new low-cost capacity, are likely to play out in the ethylene sector (see also Chapter 7). The ethylene market also shows why producers need to be alert to technologically- and feedstock-enabled disconnects.

6.3.1
The Rise of the Middle East

As we have noted, world demand for ethylene continues to grow, forecast to rise from 105 million m.t. in 2004 to 135 million m.t. in 2010. If this growth could be fully met by Middle East plants based on very low cost ethane feedstocks, we could predict the end of cyclical "fly-ups" in the ethylene chain. That seems unlikely in the short-to-medium term. The Middle East has huge quantities of ethane in its hydrocarbon reserves, but it is available only as a small fraction within natural gas and oil supplies, so ethane supply growth will depend on natural gas and oil production growth. We believe that the investments underway in the Middle East and other stranded gas locations to launch liquefied natural gas (LNG), gas-to-liquids (GTLs), and other demands for gas and oil will only generate enough low-cost ethane to be able to cover perhaps 40 percent of projected new demand through 2010, and a smaller portion of global ethylene demand growth after 2010.

As a result, it will be necessary to build additional capacity based on traditional, easily transported feedstocks such as naphtha and propane to meet world ethylene demand. These new crackers will most likely be built in China, given that country's leading share of projected new demand as well as construction and operating cost advantages over other Asian and Western nations (Fig. 6.4). However, they will have to buy feedstocks at world prices and thus need much higher product prices than Middle East ethane crackers to generate reinvestment economics. They are therefore likely to emerge as the "highest full cost" plants to provide additional product in the world ethylene business. Previously, the "highest full cost" plants were ethane/propane crackers in North America and Western Europe. These had slightly higher full costs than the China plants will have, and as a result there could be some deterioration in long-term margins of existing Western plants. However, the new China plants will need to see periodic fly-ups to earn cost of capital, and so we would expect to continue to see industry fly-ups in the long run as long as the investors into those crackers will build only if they expect to earn – and eventually do earn – cost of capital returns.

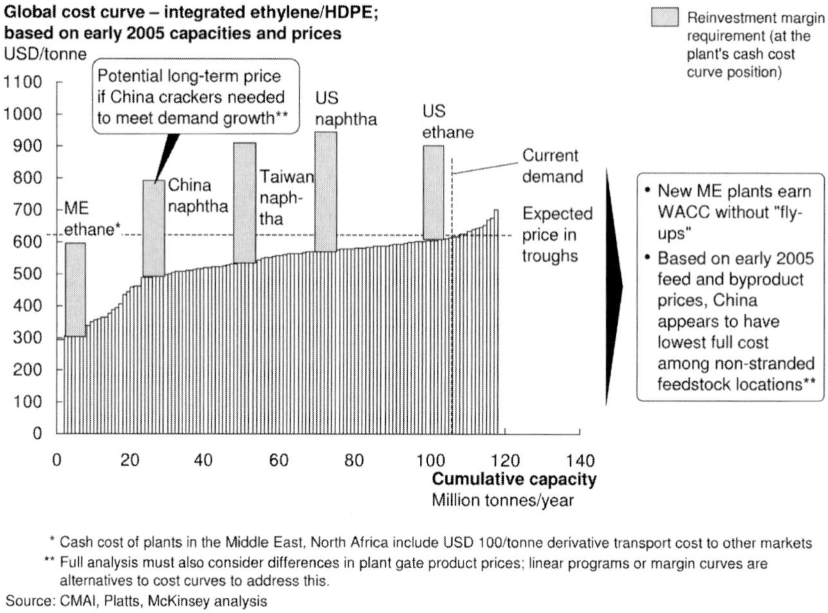

Fig. 6.4 Relative attractiveness of potential new ethylene plants.

6.3.2
Awash with Stranded Gas?

The wild card that ethylene producers must watch for is the emergence of new technologies that could tap other low-cost feeds, particularly if crude oil-linked feedstock prices stay high. Foremost is conversion of methanol into olefins (MTO), using low-cost methanol sourced from the world's abundant supplies of stranded natural gas. While this technology is as yet unproven on a large scale, in theory the world's huge volumes of untapped stranded gas could be used to produce methanol which could be converted to olefins and polymers *in situ*, or shipped to end markets such as China and converted to olefins and polymers there.

Just as new low-cost capacity increments have done in pulp and aluminum, these plants could meet all emerging demand, reducing prospects of fly-ups and leaving high-cost producers beached and vulnerable at the top of the cost curve. The second new technology development to watch is a renaissance of coal as a chemical feedstock, an attractive option in coal-rich centers of new demand such as China. Again, this new low-cost option could siphon off all new demand growth and snuff out the chance of the future fly-ups that established producers depend on for their through-cycle margins.

6.4
Tailoring the Model

Determining the full cost of the type of capacity needed to meet demand growth is the cornerstone enabling managers to assess how attractive the returns are likely to be for their business in a given commodity chemical sector. However, it is also possible to dig deeper into the circumstances of individual plants and product lines, production sites, and geographic regions to build up a more detailed picture of the prospects for a business. Here are a number of key drivers that need to be taken into account.

- Petrochemical producers should consider, for example, that some segments have seen long-term declines in margins during troughs as cash cost curves have flattened. This could likely lead to more severe cycles: flatter cash cost curves imply lower trough margins for low-cost producers, yet since new capacity must be able to expect cost-of-capital returns in the long run, a higher portion of the cycle's margins must then come from fly-ups. Thus, the trade-off between the rate at which cash cost curves flatten and the rate at which cost of capital requirements for new plants flatten will determine whether cycles become greater or less severe in the long run.
- Another trend to watch is the impact of vertical integration, which has tended to pull margins down as it usually results in a lower cash cost – and full cost – of marginal production compared to a non-integrated chain where the overall chain's marginal cost is the sum of the marginal cost of production at each step of the chain. Again, this would tend to contribute to lower trough margins for any given producer, but possibly more extreme fly-ups.
- Shifts in relative prices of feedstock and byproducts among regions will also impact the economics of producers. Take the case of propylene: as ethane takes a greater share of the world's ethylene feedstock mix, less co-product propylene will be produced, which will contribute to propylene gaining in relative price versus ethylene and changes in propylene trade flows and regional price differentials as higher-cost sources of propylene are needed. This may swing relative costs of ethylene among regions because of changes in cross-region values for byproduct propylene.
- Regional supply/demand balances of a product can also impact economics. If a regional market becomes a net importer, this can create a valuable pricing umbrella for domestic producers. Take the hypothetical but not unrealistic case that North America holds back from adding new ethylene capacity and yet demand grows, so that North America shifts from a net exporter of ethylene derivatives to net short, and imports from Asia or other

regions. If that happens, then margins of North American plants will improve. If new Chinese naphtha or propane crackers are required for the industry to meet demand growth, then long-term returns for existing plants in North America could remain attractive over a cycle given that the China plants will need fly-ups to justify new investments and North America prices could be equal to or higher than Asia prices. The risk is that new capacity may then come on stream in North America and destroy the import pricing umbrella.

6.5
Conclusion

The global chemical industry has long struggled with high cyclicality. We do not see that cyclicality going away: if anything, cycles may become more extreme. In addition, investors will likely continue to be nervous about whether or when the next "fly-up" will occur. But we do feel it is possible for chemical companies and investors to better understand long-term cycle-average prospects for their businesses, and that savvy use of concepts like those described in this chapter can help companies and investors to better anticipate – and act on – opportunities and threats in cyclical chemical segments.

6.6
Summary

- The commodity chemical industry is highly cyclical, but earned better returns over the last 20 years than other commodity industries. Underlying the chemical industry's performance have been profit fly-ups during product shortages, which assure producers of the returns they need to reinvest.
- The key question facing companies in the sector is whether this fly-up-driven performance will continue.
- World demand for chemicals continues to grow and new capacity will be needed, but the emergence of new petrochemical-producing regions with significant feedstock cost advantages contributes to uncertainty about the future profitability of established producers.
- Producers can gain insights into the future evolution of their sectors by considering how future capacity additions to meet growing demand will be made. Will all future demand be covered by capacity creep and by new low-cost supply? Or will demand growth be such that it also requires capacity with costs similar to those of existing plants, that will need to earn full cost of capital –

which is likely to perpetuate the fly-ups pattern? Analysis of the full cost of likely future capacity will point to the profitability evolution of each sector.

- Producers can further refine the analysis by taking account of the specifics of markets and geographies in which they operate, which will influence future margins.
- Commodity chemical companies should use this approach to evaluate their prospects. Those with high cost positions that will become vulnerable should consider their exit options. However, due to limited supply of low-cost feedstocks for many chemicals in the short-to-medium term, many high-cost producers in traditional producing regions should continue to see returns – if evaluated across multiple cycles – that are satisfactory.

References

Butler, P., Berendes, R., Elliott, B. *Cyclicality: Trying to Manage the Unmanageable.* In: Budde, F., Farha, G.A., Frankemolle, H., Hoffmeister, D.F., Krämer, K. (Eds.), *Value Creation: Strategies for the Chemical Industry.* Pp 197–210. Weinheim, Germany: Wiley-VCH Verlag 2001.

7
Middle East: Opportunities and Challenges from the Rapid Emergence of a Global Petrochemical Hub

Christophe de Mahieu, Christian Günther, and Jens Riese

The global petrochemical landscape is evolving. New companies are appearing and expanding rapidly, taking advantage of low feedstock cost in the Middle East and low labor cost and fast-growing demand in Asia. Long-established Western companies are exiting, shrinking, or moving east, often in partnerships, to try to defend their stakes. Meanwhile, on the horizon, looms a set of new technologies that may transform the industry still further. This is all combining into a paradigm shift in petrochemicals as the West is abandoned for the East.

Such a shake-up will result in new winners and losers. Expanding upstream and downstream capacity, broadening the product portfolio, and extending geographic reach are all options for players active in the Middle East. At a time of rapid change and uncertainty, companies need to think clearly if they are to make the right strategic choices. The management tools are familiar: good business intelligence, clear microeconomic thinking, sound risk management, distinctive value propositions, and superior partnership skills; and management must also keep a close eye on the organizational challenges such as leadership and capability development, localization of the workforce, and changes in people's mindsets and behaviors for superior performance and execution. All are interlinked and therefore need to be applied rigorously and simultaneously if players are to succeed.

This chapter looks at the key structural changes in polyolefins, how the industry could make best use of attractive feedstock in the Middle East, and the implications for companies across the world (see also Chapter 6).

7.1
Turning to the East

The center of gravity of the polyolefin world is shifting eastwards, and chemical companies that want to stay in this business need to move and transform before their entire economic rationale is swept away (Fig. 7.1).

Value Creation: Strategies for the Chemical Industry. 2nd Edition. F. Budde, U.-H. Felcht, H. Frankemölle (Eds.)

ISBN: 3-527-31266-8

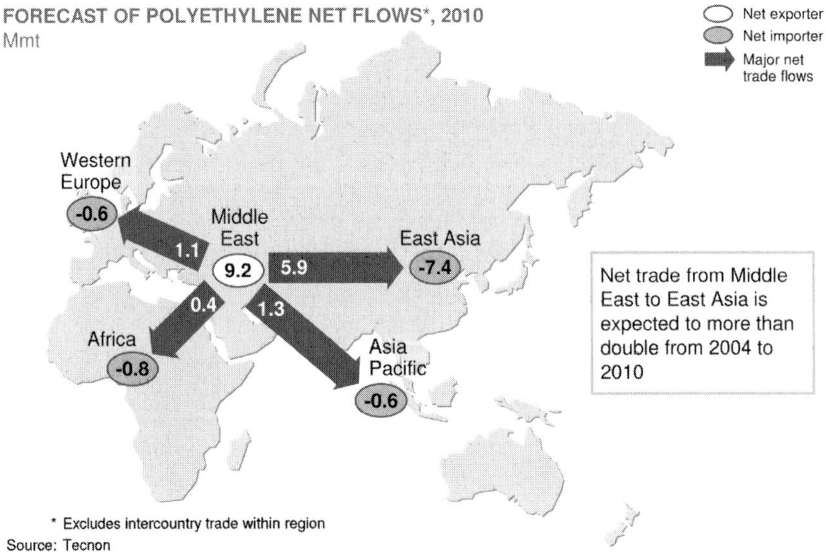

Fig. 7.1 Petrochemical trade flows from the Middle East to East Asia are expected to increase significantly.

There are several well-known causes – in both supply and demand – of this redirection.

7.1.1
Booming Demand in Asia-Pacific

Asia-Pacific petrochemical demand is catching up with that of the USA and Europe, and should draw level with the two combined by 2010. China dominates the growth: polyolefin demand alone is expected to grow at around ten percent per annum (HDPE: 9%; LLDPE: 12%; PP: 10%).

There are three main reasons for this impressive Asian growth:

1. *Demographics*: 56 percent of the world's population lives in the Asia-Pacific region,[1] and the share is growing at 1.1 percent p.a. China alone accounts for a fifth of the global population, and India already has more than 1 billion inhabitants;
2. *Domestic demand*: China's GDP is growing at seven to nine percent per annum, a trend expected to continue for several years, certainly well beyond the 2008 Olympics. Economies in a relatively early phase of development typically focus on industrial developments, creating a disproportionately high growth in demand for basic chemicals. China and India are no exception;

1) China, Taiwan, North and South Korea, South East Asia. Excludes Japan, Australia, Oceania.

3. *Export*: Export growth in China is running at a staggering 20 percent p.a. The country already produces 21 percent of the world's textiles, driving demand for synthetic fibers. Other chemical-intensive export-driven industries include consumer electronics, children's toys, packaging, household appliances, and basic pharmaceuticals.

7.1.2 High Input Costs Render Western Companies Uncompetitive

While demand booms in Asia, petrochemical production in the USA and Europe is flat or declining, especially for C_2 products such as ethylene and polyolefins. Plants are becoming less and less competitive compared with both Middle East and Chinese facilities, which are able to operate from a cheaper cost base (Fig. 7.2).

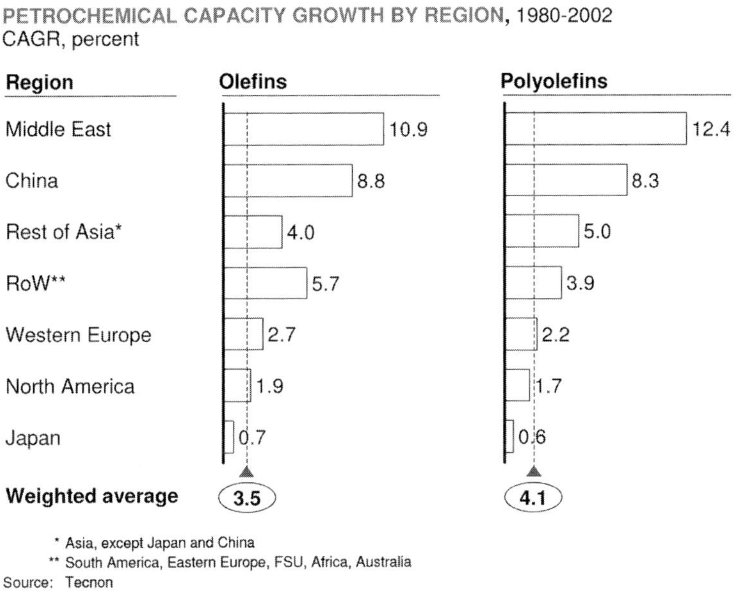

Fig. 7.2 Middle Eastern and Chinese capacity expansion is driving supply growth.

These structural cost disadvantages place Western players under increasing pressure. Beyond feedstock disadvantages, they also suffer from poor productivity improvements, having exhausted most of their cost-cutting options in the 1990s, expensive workforces, and subscale assets. This means they will have a hard time coping as the price-cost squeeze accelerates.

Middle East: the 'stranded gas' advantage: Basic hydrocarbons account for 60 to 80 percent of polymer cost, making them by far the most important competitive element in this business. Both natural gas liquids and naphtha have dramatically increased in price, affecting Western companies, which crack these feedstocks;

Middle East companies use much cheaper gas. With Middle East ethane at USD 0.75 to 1.00/MMBtu, crude oil would need to drop below USD 15/bbl for Western producers to be competitive for LLDPE in Asia. The cost advantage for Chinese players producing LLDPE from naphtha is smaller, but break-even is still at USD 20/bbl.

The cost advantage and volumes in the Middle East are not enough to justify Western companies' shutting down capacity, but new capacity would fail to earn back the cost of capital. Already, the Western petrochemical industry is declining. The US share of global trading in ethylene derivatives has fallen from 30 percent in the early 1990s to less than ten percent, and will probably be less than five percent by 2010. Over the same period, the Middle East is likely to increase its share to more than half the globally traded volume. The industry landscape is responding to this trend: DSM sold its petrochemical business to SABIC in 2002, BP is carving out its Olefins & Derivatives business (Innovene), and Basell is for sale at the time of writing.

Global ethylene capacity is predicted to increase from 114 to 145 million metric tons (mmt) from 2004 to 2010, and the Middle East's share of the market is expected to grow from nine to 17 percent (Fig. 7.3). This implies that about 40 percent of new ethylene capacity will be built in the region. Major capital investment projects are already under way, including Sumitomo's joint venture with Aramco of Saudi Arabia. Aramco-Sumitomo will take over Aramco's oil refinery and crack 1.3 mmt of ethylene and 50,000 tonnes of propylene a year, together with 850,000 tonnes of propylene from a separate facility. Borouge, a 60/40 joint venture between Abu Dhabi National Oil Company (ADNOC) and Borealis, is

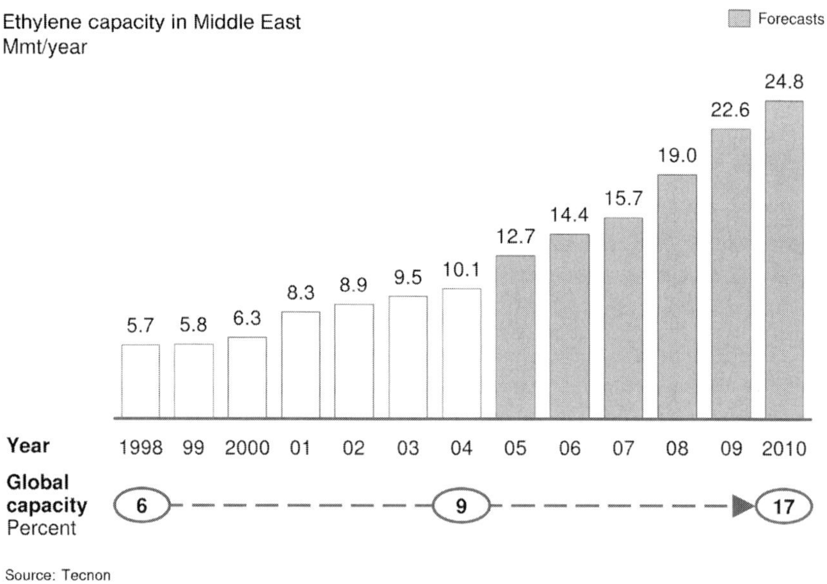

Fig. 7.3 Cracking capacity in the Middle East is expected to boom.

investing in a new petrochemical complex with ethane cracker capacity of 1.4 mmt a year. Borouge is also investigating the possibilities of producing propylene. Iran, Qatar, and Oman all have aggressive plans to expand their capacity and build new crackers.

Conventional thinking assumes that the USA, Europe, and particularly Asia will remain competitive in the longer-chain hydrocarbons business (e.g., propylene and PP, butadiene and synthetic rubber). However, new technology and petroleum strategies might change the game here too, as we shall discuss later.

China: cheap labor and booming demand: Labor costs in China can be just a tenth of those in the USA. Constructing plants is 20 to 50 percent cheaper, and plant operations enjoy a 30 to 40 percent fixed cost advantage over their US counterparts. Western companies find it hard to compete in the Chinese market, with a labor cost shortfall of more than USD 100 to 150/t of LLDPE, and shipping costs from Europe and the USA of USD 90 to 100/t.

Local capacity additions cannot meet the increase in Chinese demand, opening a window of opportunity for foreign companies. Middle East supply will meet about half the demand for chemical imports from China, and Western companies may react by moving capacity into China, often in joint ventures with local producers. But we are also seeing emerging alliances between the Middle East and China, combining the former's cheap feedstock with access to the growing Asian markets. Saudi Aramco, for example, has a 25 percent equity stake in a joint venture in Quanzhou with Fujian Petrochemical Company Limited, which holds 50 percent (the other shareholder is ExxonMobil). The venture will start production in 2007. Other major Middle East players are also contemplating asset-based entries in China.

New technologies could alter the playing field: Regional cost base differences could be exacerbated if some long-awaited technological developments come to fruition. It remains unclear when they will actually break through, but when they do the implications for the industry will be enormous (Fig. 7.4).

The liquefied natural gas (LNG) market is expanding globally. The value chain is on a steep learning curve, and cost reductions are happening fast. This is resulting in more gas development and therefore more ethane supply, which would limit regional price differences. However, while gas prices will vary between USD 0.5 and 1.3/MMbtu in the Gulf, in the USA they will be around USD 6 to 8/MMbtu.

In gas-to-liquid (GTL) technology, stranded gas (mostly methane) is converted into diesel, kerosene, and naphtha, which are easily transportable. Qatar aims to become the world's GTL center, with production from multiple facilities at the Ras Laffan Industrial City reaching 400,000 barrels/day by 2010 to 2012. Qatar is well positioned thanks to its huge reserves of very attractive gas, lack of oil, and the desire to diversify its product portfolio, and is progressing quickly with its partners ExxonMobil, Shell, and Sasol. The GTL case is harder to make for other regions.

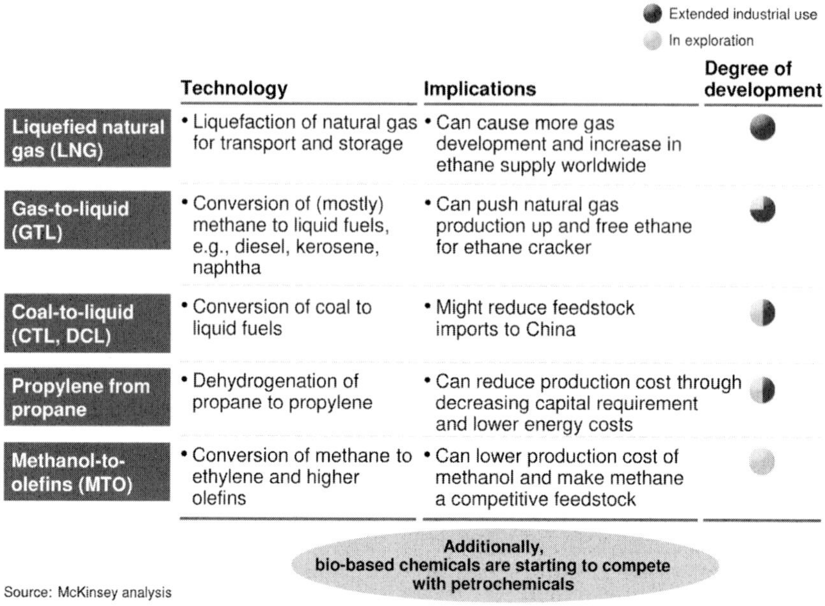

Fig. 7.4 New technologies may lead to further changes – but some are yet to be proven.

Coal-to-liquid technology (CTL or DCL – Direct Coal Liquefaction) might reduce imports to China because it would open the country's huge coal reserves for chemical production. The technology has not been competitive historically, but improvements have been claimed that could change the economics. Towards the end of 2004 the Shenhua Coal Liquefaction Plant went online in China's Inner Mongolia province with a capacity of 845,000 tonnes of oil products. However, capacity increases in Chinese coal exploration are already at their limits to meet the increasing energy demand. Furthermore, most of China's coal is low btu, making it less attractive for CTL.

Looking further ahead, methanol-to-olefins (MTO) technology may provide value-added use for stranded methane and could change the world of petrochemicals. Methane is already used to produce syngas, which is converted into methanol and finally into ethylene and higher olefins. MTO proponents claim that capital investments, process costs, and minimum plant scale can all be reduced significantly. Further breakthroughs might eliminate the methanol step altogether. Middle East methane could become a competitive feedstock for a whole range of olefins and polyolefins beyond ethylene (e.g., propylene and polypropylene, butadiene and synthetic rubber). This might be competitive against naphtha as a feedstock at around USD 30/bbl. However, the technology remains far from proven.

Propylene is also obtained by dehydrogenating propane, which could open up a new product stream for the Middle East. Again, this technology needs to be improved before the process economics are competitive with recovery from steam crackers or refineries.

Last but not least, Saudi Aramco has recently announced an innovative and counter-intuitive refining strategy. Instead of building oil refineries close to the markets, it is proposing to build large, deep-conversion refineries in the Middle East for oil product exports rather than simply exporting crude oil. Structural differences between light and sour crudes would support such a strategy. If the Middle East builds large and complex refineries for export, it will automatically open the whole C_3 and C_4 petrochemical value chain to the region.

7.2 Pathways to Growth for the Middle East

Middle East chemical companies have announced aggressive growth targets, building on their feedstock cost advantage and booming demand. They can pursue growth along three dimensions: capacity, portfolio, and geography (Fig. 7.5).

	Local capacity expansion	Portfolio expansion/ forward integration	Geographic expansion
Rationale	• Fully exploit Middle Eastern "stranded gas" cost advantage	• Combine ethylene with other refinery products or energy-intensive materials	• Gain access to new markets and skills • Growth not limited by local ethane
Options	• Expand alone • Form partnerships	• Expand into – Downstream products – Specialty chemicals – New technologies	• Buy Western assets • Form joint venture • Greenfield expansion to Asia
Constraints	• Ethane availability currently limited to 15% of global demand	• Middle East cost advantage decreases when ethylene is not used as feedstock • New skills need to be acquired/built up	• Clear economic rationale (beyond top line growth) needed • Sophisticated understanding of new markets required

Source: McKinsey analysis

Fig. 7.5 Main pathways to growth for Middle Eastern players.

7.2.1 Local Capacity Expansion

The Middle East should exploit every ton of stranded gas available. The heavy investments in crackers (primarily ethane) and downstream products (mostly polymers) already under way will bring the region close to this level. Iran's NPC is building the two largest crackers in the world in the Pars Special Economic Energy Zone, one of which is a naphtha cracker that will also produce 900,000 t/year of propylene. SABIC, meanwhile, is planning the start-up of its 1.3 mmt/year cracker by 2009 in Yanbu, Saudi Arabia; and a joint venture between Chevron

Plant operator	Site, country	Ethylene capacity Thousand tons/year	Start-up year
Amir Kabir PC	Bandar-Imam, Iran	520	2005
Arya Sasol Polymers	Asaluyeh, Iran	1,000	2005
Jam Petrochemical	Asaluyeh, Iran	1,350	2005
Marun Petrochemical	Bandar-Imam, Iran	1,100	2006
Ilam Petrochemical	Ilam, Iran	320	2007
NPC Iran	Kharg Island, Iran	500	2008
Olefins II	Shuaiba, Kuwait	950	2007
Ras Laffan Ethylene	Ras Laffan, Qatar	1,300	2008
Sharq	Al Jubail, Saudi Arabia	1,500	2008
United II	Yanbu, Saudi Arabia	1,300	2008
Borouge	Ruwais, Abu Dhabi	1,400	2009
Oman PC Industries Co.	Sohar, Oman	1,000	2009
Rabigh Project	Rabigh, Saudi Arabia	1,300	2009

Source: Tecnon

Fig. 7.6 Cracking capacity in the Middle East is expected to boom through several construction projects – examples.

Phillips, Qatar Petroleum, and Total is investing in a 1.3 mmt/year cracker in Ras Laffan, Qatar. However, because ethane production is linked to gas and oil production, the Middle East ethane supply will be able to meet only around 17 percent of global ethylene demand, which is expected to be 135 mmt/year by 2010 (Fig. 7.6).

There may be a constraint on gas development in the Middle East and thus on ethane availability. However, considering the gas reserves, the early stage of exploration and development of gas, the enormous demand from power and desalinization, the unique opportunities for developing other gas-based industries such as aluminum, glass, or steel and their respective downstream activities, and the governments' requirements to generate employment to cope with the region's fast-growing population, one can expect that more ethane, propane, and butane will be made available. Furthermore, rapid expansion of GTL and export-focused deep-conversion oil refineries may offer additional, diverse, and attractive new petrochemical feedstocks.

7.2.2
Portfolio Expansion and Forward Integration

An obvious extension of today's business is to expand into other chemicals. However, the Middle East's cost advantage diminishes when ethylene is not used as a raw material. The greatest value is therefore created in product groups that combine ethylene with other refinery products, methane derivatives, or products that require high energy such as chlorine/caustic soda. For example, polyethylene terephthalate (PET) production could make use of cheap ethylene glycol (EG), combined with PTA based on xylene from nearby naphtha crackers, or acetic acid produced from methane via syngas. Even though it is still cheaper to ship EG and produce PET in China, PET resin for packaging and bottles could be shipped around the world at a competitive cost. Polyester resin for fiber is less attractive (except for Egypt's textile industry). Direct spinning lines in China, which account for most of the polyester fiber growth, use PTA and EG directly as a starting material. Supplies of aromatics are limited in the Middle East, and until recovery capacity is expanded, which is planned in Kuwait and Saudi Arabia, demand might be met by imports. Other product groups to consider are the styrene chain and polycarbonates.

Expanding into specialty products is feasible, but requires a different set of product development and marketing skills, especially as – outside oil field chemicals – the local market is small. There is therefore no strong competitive advantage for companies in the region. Foreign companies active in specialties could be attracted to the Middle East if chemical production clusters were created around refineries, crackers, and a port. This would require local governments to provide infrastructure, legal frameworks, and incentives. The chemical site in Map Ta Phut, Thailand, is a good example of the systematic development of such a cluster.

Investing in new technologies such as MTO is another expansion option. The Middle East has over a third of the world's proven gas reserves, and MTO would open up new opportunities, assuming the technology delivers.

7.2.3
Geographic Expansion

Geographic expansion means access to new markets, more sophisticated skills (e.g., marketing, product development, process technologies), and new talent, as well as broadening the product portfolio.

Middle East companies could expand by buying Western companies' assets, as we saw with SABIC's acquisition of DSM's petrochemical businesses. Another option is to form joint ventures. SIIC, for example, pursued this, entering into an agreement with Chevron Phillips Chemical (CPC) to start operating an integrated styrene facility and expanded aromatics plant in Al Jubail, Saudi Arabia, in 2007. Other examples include Qatar Petrochemicals, a JV owned by Total and Qatar Petroleum, Saudi Polyolefins owned by Basell and NPIC, as well as ventures shared by Dow/Oman Oil/the Government of Oman and CPC/Qatar Petroleum/Total.

Greenfield expansion into Asia needs to be considered carefully as Middle East firms would lose their feedstock advantage and would need a more detailed understanding of the Chinese and other Asian markets. For companies that do want to pursue this, the process would typically have three stages: initially, the focus should be on selling products to China via distributors and direct sales. Then, equipped with a deeper knowledge of the market, Middle East companies can select Chinese partners who will help them boost their local marketing and sales capabilities, and give them better access. Finally, once sufficient distinctive market knowledge has been acquired, companies could consider an independent expansion into China. Joint ventures do not have to mean physically moving to China – Chinese Petroleum Corporation (Taipei), for example, has been invited recently to build capacity in Saudi Arabia.

7.2.4
Hurdles to Overcome

Expanding ethylene and derivatives capacity appears to offer solid returns on investment. Success is predicated on fully exploiting economies of scale, typically reached at capacities of 800,000 to 1 million t/year. However, clever management of capital expenditure, cyclicality, the supply chain, and customers will make a significant difference in the level of returns. To succeed in expanding the product portfolio and geographic reach is even harder, and Middle East petrochemicals companies must tackle several challenges.

Managing risks and uncertainties: Historical swings in demand and prices are likely to be amplified as the unpredictable Chinese market grows in importance (see Chapter 6).

Today, all expansion plans are geared towards meeting Asia's increasing demand. However, Wen Jibao, China's prime minister, struck a note of caution when he said in March 2004 that deep-seated problems and imbalances in the economy over the years had not been fundamentally resolved. This does not mean Middle East companies should stop their capacity expansion, but rather implies that they should consider redirecting it towards the stagnating Western markets, where they could compete against established players for market share. The potential implications for Western companies are more fundamental: should they remain in the business, or exit?

Feedstock, transportation, and labor cost are all key drivers of regional competitiveness. The cost of different types of feedstock is the least predictable variable, so it is crucial to develop well-informed scenarios before making capacity expansion decisions. The evolution of the competitive landscape, technological discontinuities, and industry conduct (pricing discipline) also need careful monitoring.

Managing growth: The first challenge is to decide which route to organic growth is most suitable. Companies need to decide where to compete (e.g., which product groups and regions) and how to compete (e.g., which business model, partnerships, and value chain position). Selecting the right partnerships or acquisition targets alone is a difficult task (more than half of all mergers and acquisitions fail). Companies should consider joint ventures in order to split the risks and costs associated with entering new markets or investing in new technologies or assets. Joint ventures can realize 50 to 75 percent of the synergies compared with M&A and are an alternative if the premium for an acquisition is too high, or the partner of choice is not for sale. However, joint ventures are more complex to manage because of the shared ownership, and overall failure rates are as high as for M&A. Nevertheless, they are suitable in turbulent market conditions. Companies should also consider looser forms of alliances – the benefits of a joint venture or M&A do not always outweigh the added investment and risks.

In building a presence in Asia, personal relationships, trust, and respect are critical. BASF and Shell are good examples of companies that have understood the importance of having the same executives based in China for many years.

Organizational performance: Middle East chemical companies are catching up with their more established Western counterparts in critical skills and capabilities, especially in marketing, supply chain management, product development, and innovation. Although partnerships can help bridge these gaps, Middle East companies ultimately need to build these core skills themselves. Internal academies or universities have been (re-)discovered as an effective and efficient way of building or expanding such capabilities. These are usually set up with the aid of external experts who bring deep functional knowledge and experience in training programs. In most cases, their role is to train the future trainers and design the program. Company employees with the necessary experience can then roll out the training across the organization.

7.3 Way Forward for the West

Western multinationals committed to remaining leaders in the chemical industry have to capture a greater share of the growing Chinese demand and tap the Middle East's cheap feedstock resources. This implies building or participating in local capacity. For example, BP's chemicals business is starting production in a 900,000 t/year cracker with Shanghai Petrochemical Co. in Caojing. However, a joint venture is no longer mandatory and there are examples of major stand-alone investments. Bayer is planning a polyether polyol plant in Caojing for 2008 with a capacity of 280,000 t/year, Mitsubishi Rayon is investing in a 100,000 t/year MMA plant in Huizhou, and Celanese's acetic acid plant (600,000 t/year) in Nanjing will be operational in 2006.

Western companies could also build on and radically enhance their strong customer relationship and R&D skills, and shift the portfolio towards innovative and more specialized products for which the underlying economics are more favorable. This would require a high level of commitment to R&D and innovation. Many Western companies, however, are thinking similar thoughts and there is a risk of overcapacity. Diligent market and competitor intelligence and genuine distinctiveness will help smooth the path.

Asia-Pacific companies are likely to base their investment decisions on similar criteria. Participating in Middle East production is certainly attractive in some areas, especially ethane crackers and ethylene production. China is not self-sufficient in crude oil or natural gas and must import them, or the end products. This has driven recent multibillion dollar deals with Iran. For example, in 2003 China imported 237,600 barrels from Iran per day – 15 percent of China's total imports of crude oil. There is a clear shift in the mindsets of Middle East and Asian players as they become willing to develop close relationships and find ways to collaborate intensively in reciprocal cross-border investments. This is another important shift where the Middle East will look more eastwards than westwards in the future.

Some parts of the production process are better located domestically. Naphtha crackers, for example, may as well be located in China, as naphtha could be imported from the Middle East cheaply. It could also be made using CTL technology, if that becomes competitive or is subsidized by the government.

Downstream polymer production (except for derivatives of ethylene, which is difficult and expensive to transport) will generally be located in the home country where Asia-Pacific companies can capitalize on low labor costs while participating in China's strong GDP growth.

7.4 Capturing the Value

In defining a strategy for value creation in these turbulent times, petrochemical companies, regardless of their origin, need to excel along five dimensions: business intelligence, microeconomic rigor, managing risks and uncertainties, organizational excellence, and execution.

7.4.1 Business Intelligence

Companies must limit uncertainty to that which cannot be known, as opposed to that where the company simply lacks the available facts. One diversified chemical company, for example, progressed from having only a vague understanding of the Chinese industry to being able to model the process economics of all Chinese competitors for all major products in a few months. This was a major effort involving a ten-person project team on the ground in China tapping into a broad set of internal and external sources, and sanity checking the results. Emerging markets experience tremendous churn in their customer base and changes in customer needs, potentially requiring product upgrading. Companies need to understand as much as possible about Chinese downstream customer trends, including where they are, who will be the winners, and what their needs are today and tomorrow.

7.4.2 Microeconomic Rigor

The same effort produced a set of robust industry cost curves that outlined the capacity and cost position of each local and import player in the market today and in the future. Such cost curves still provide the best basis for strategic decision making in chemicals. Care needs to be taken that these cost curves are correct: most cover operational expenditure, but only rarely do they include capital expenditure costs as well. A cost curve has to be destination-specific as prices and cost may vary significantly between locations. Based on the cost curve, a company needs to understand what drives the microeconomic behavior and conduct of different players in the market, particularly with respect to pricing. Only then can it assume that it has sufficient understanding of its market's microeconomics to make well founded strategic decisions.

7.4.3 Managing Risks and Uncertainties

Using scenarios and even option value theory is appropriate for major investment decisions. Several measures also exist to help mitigate risks: companies can introduce flexibility to production, and coinvest and integrate condominium sites to allow for shifts in demand and price. Production can also be made more flexible in terms of feedstock choice and product mix. For example, the ability to choose whether to produce ethylene or polyethylene, and to switch between HDPE and LLDPE in a PE swing plant, can improve the ROS by two percent, depending on circumstances. Finally, developing an integrated marketing and sales and production planning process across the entire global portfolio of assets, and optimizing sales and production across the three major markets, can also yield tremendous ROS improvement from the inefficiencies and gaps in response time between markets.

Managing contracts is also key for success in volatile markets. Flexible contract terms allow firms to capture value from the arbitrage between the product contract and the spot market price. Linking sales contracts to the feedstock price, and shifting from quarterly to monthly pricing, can also reduce the cash margin volatility significantly. Companies can also buy a feedstock hedge, or create a 'dirty' hedge from an actively traded commodity to achieve this.

There is significant room for improvement in supply chain management. Outsourcing options need to be investigated systematically in Asia-Pacific as each country is different and new local and regional operators emerge every day, often with distinctive value propositions. Shipping costs are also very volatile and offer tremendous optimization opportunities for those who can master them.

7.4.4 Organizational Excellence

The organizational implications of pursuing a strategic path are often overlooked. Companies can waste time, money, and careers if the formal organization – the processes, incentives, talent, mindsets, and behaviors – fails to meet requirements. Organizational diagnostics help uncover the gaps between current and desired performance, and form the basis for developing a high-performing organization. Capabilities can be bought in, based on a sound understanding of the important issues, but they can also be built internally. Several chemical companies have introduced internal academies around the key functions, in particular operations and marketing and sales, and leading players are increasingly focusing on enhancing the soft sides of their organizations (e.g., leadership capabilities, mindsets and behaviors, motivation, local talent) rather than the 'hardware' (e.g., structure, processes, systems). However, to succeed in Asia and the Middle East, corporations must adapt development programs to the specific cultures and mindsets of local employees and leaders. This is not easy as these are exceptionally culturally diverse regions, and mindsets are often very strongly embedded. Developing local talent and local capabilities is probably the biggest challenge international chemical corporations face. However, those that succeed will be able to enjoy a durable and truly distinctive competitive advantage.

7.4.5 Execution

Rigorous project management and execution are – of course – indispensable for economic success in the Middle East. Foreign companies planning large investments in the region need an awareness of the potential level of government involvement. Many authorities have relatively little experience with very large, integrated, and complex projects such as combined refinery/cracker/polymer/downstream sites with multiple investors including small and medium-sized enterprises (SMEs). However, they are eager to see these sites developed as they can create many local jobs. They are, therefore, prepared to provide infrastructure but might be unsure of the full range of possibilities for the by-products and the optimization generated by multiple production units. Companies must make it as easy as possible for governments to give them the go-ahead. This implies designing projects in close cooperation with the relevant authorities, other investors, and local SMEs to develop a world-class competitive site with strong local job creation. Companies also need to take a more holistic approach than usual (e.g., build the infrastructure into their plans rather than expecting governments to provide it), and finally, to prepare even more concrete project plans than are usually needed in the West.

7.5 Conclusion

The petrochemical landscape is changing rapidly and radically. New players are emerging, taking advantage of low feedstock costs in the Middle East, low labor costs in China, and booming demand in Asia. Western companies are increasingly being priced out of the market and will disappear completely if they fail to react appropriately – and quickly. Strategic decisions are required from each player and although uncertainties abound, there are rules for success: know your market position, remain flexible under different scenarios, and do not overlook the organizational implications.

7.6 Summary

Companies have to make important strategic choices as the petrochemical industry moves eastward to take advantage of low feedstock cost in the Middle East and low labor cost and growing demand in Asia, while new technologies loom on the horizon.

- The move to the East in petrochemicals is being driven by booming demand in the Asia-Pacific region, structural cost disadvantages for Western players competing against the stranded gas advantage of the Middle East, and cheap labor in China.
- In addition, regional cost base differences could be exacerbated if new technological developments come to fruition, such as LNG, GTL, CTL, MTO, and others.
- Middle East chemical companies are planning aggressive growth, and are in a position to pursue it along the three dimensions of local capacity expansion, portfolio expansion and forward integration, and geographic expansion into new markets.
- However, there are hurdles to be overcome, and companies pursuing such growth will have to manage risks and uncertainties, plan their growth routes carefully, and build critical organizational skills.
- Western multinationals committed to remaining leaders in the petrochemical industry have to build or participate in local capacity in order to capture a greater share of Chinese demand and tap the Middle East's feedstock resources. Players could also consider occupying positions where higher-level skills are needed. Building win-win partnerships with local players is key.
- Finally, in order to capture the value, petrochemical companies need to excel along the dimensions of business intelligence, microeconomic rigor, management of risks and uncertainties, organizational excellence, and execution.

8
Survival when It's Hard to be Special: A Perspective on Specialty Chemicals

Eric Bartels, Joël Claret, Sabine Deppe, and Ralph Marquardt

The specialty chemicals business needs to face facts: the drop in profitability that characterized the 1990s has not abated in the new millennium. Competition has intensified, capital productivity has fallen, and the specialties niches are shrinking with ongoing commoditization. This does not have to be the end of the road, but specialty companies have to take a long hard look at themselves and decide whether they are prepared to fight back, and how they can best equip themselves to do so.

Successful specialty chemicals companies can still reap rich rewards and create value in the future, by concentrating on two key factors for success: they have to develop strategies which allow them to shape leading positions in markets or arenas that fit their competitive advantages, and achieve superior operational performance at the business unit level. In this chapter we map out the key strategic levers that specialty players can pull to achieve this, and assess what is needed to fulfill them.

8.1
Performance Has Been Disappointing

Specialty chemicals is a very fragmented market, with more than 30 major segments ranging from small niches such as lubricating oil additives to huge global markets such as active pharmaceutical ingredients (Fig. 8.1). Each segment is represented by a variety of subsegments, with often very specific product/market and competitive profiles.

The top ten specialties companies only have a cumulative market share of about 31 percent (Fig. 8.2). Most players serve multiple segments, with the customer base ranging from large industrial and consumer goods companies to small businesses such as textile manufacturers or construction companies. Specialty chemicals are often tailored to specific applications, sometimes even to the needs of specific customers. Producers typically struggle to maintain the right balance between focusing on small business segments and at the same time attaining sufficient size to use economies of scale and gain access to capital, ideas, and talents.

Value Creation: Strategies for the Chemical Industry. 2nd Edition. F. Budde, U.-H. Felcht, H. Frankemölle (Eds.)

ISBN: 3-527-31266-8

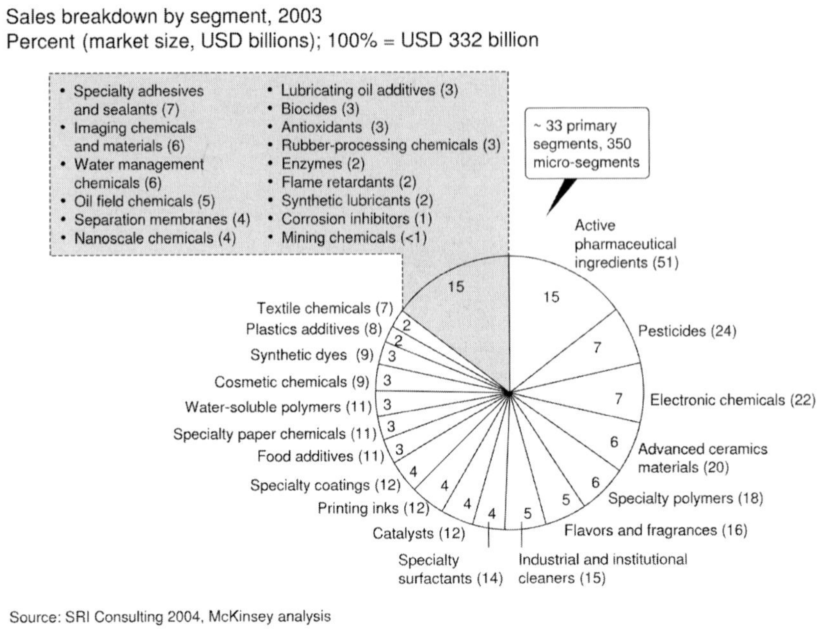

Fig. 8.1 Specialty chemicals market segments.

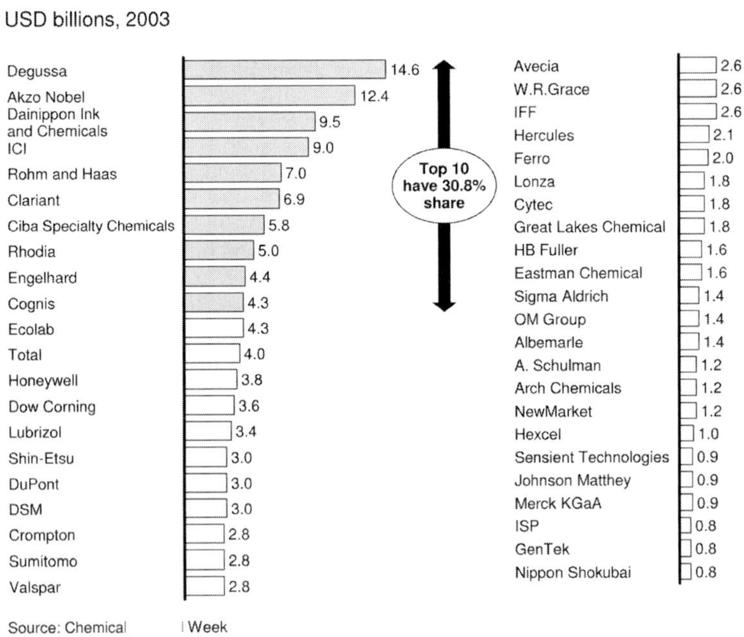

Fig. 8.2 Specialty chemicals sales.

Until the mid-1990s, specialty chemicals were among the most attractive segments in the global chemical industry. Market growth in many specialties was high compared with that in commodities, often two to three times GDP growth. Gross margins and operating margins were also excellent, and most segments were shielded from the vicious supply/demand cycles that plagued commodities markets.

As a result, many chemical companies embarked on major corporate transformations, shifting away from commodities and into specialties. ICI is a classic example. During the 1990s, it sold most of its commodity businesses and made large purchases of specialty-focused companies, including National Starch & Chemical and Quest.

Then, however, the market started to change. With the arrival of the new millennium, the specialty chemical business did not look so special anymore. Volume growth in many segments slowed as markets matured. Prices and margins came under more pressure due to increased commoditization and lower innovation rates. While prices declined, costs went up (Fig. 8.3). In Germany, for example, factor costs for chemical production increased by 2.9 percent p.a. over the last 15 years, while at the same time product prices declined by 0.4 percent p.a. Though these numbers include commodities, the trend holds true for specialty products as well.

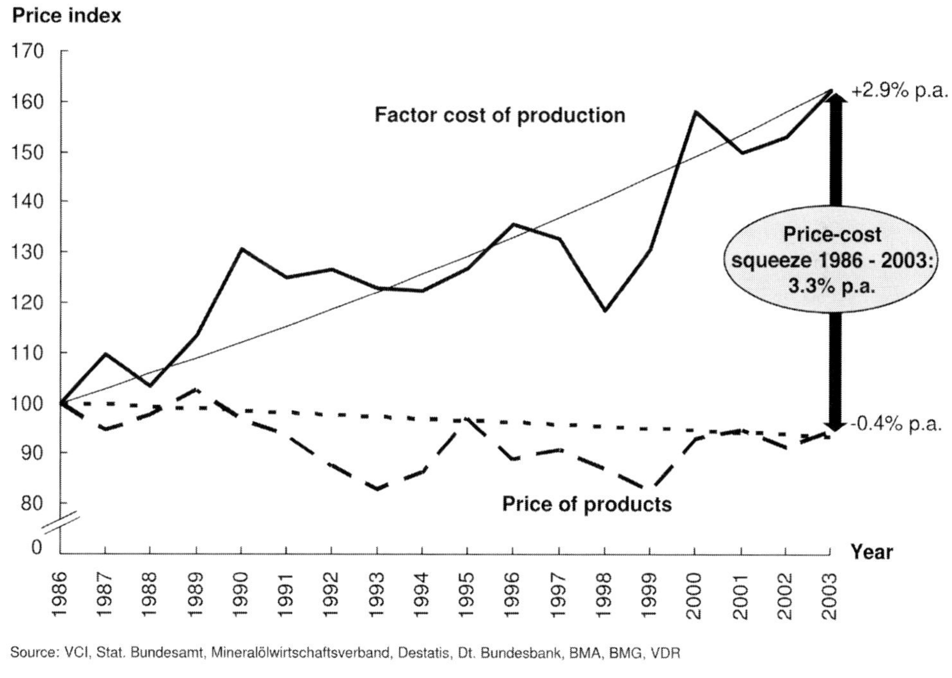

Fig. 8.3 Development of the price-cost squeeze in the German chemical industry.

Some specialty companies started to respond to these threats by mergers and acquisitions. However, many deals were ill conceived or too expensive, pulling down ROICs and market valuations. In fact, on average specialty chemical companies have not fared well over the last decade (Fig. 8.4). The median ROIC of European and North American specialty companies declined from approximately 20 percent in 1995 to 15 percent in 2004. Though the industry at the same time managed to maintain – and temporarily even improve – median EBITA margins, it was not enough to compensate for a significant drop in capital productivity, mainly attributable to ill-fated, goodwill-heavy acquisitions, strong price erosion, and lax capital discipline leading to overcapacities.

Despite this general decline, some companies are still doing excellently: in 2004, the top three players boasted an average return on net assets (RONA) of more than 25 percent, while the bottom three players had to settle for less than six percent on average (Fig. 8.5). This spread, a factor of around four, is much bigger than in other industries. The top companies in specialty chemicals seem to know the formula for success, and we will share below our perspective on the specific challenges that need to be addressed in order to remain or become a top player in this increasingly problematic competitive arena.

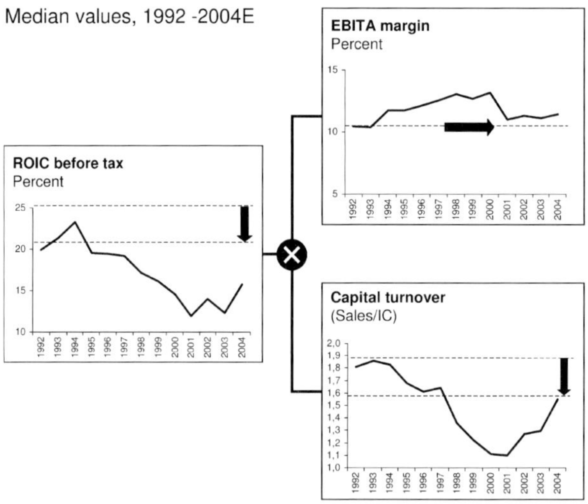

Source: McKinsey proprietary chemicals long-term performance database (CLTPD)

Fig. 8.4 ROIC decomposition of specialty chemicals.

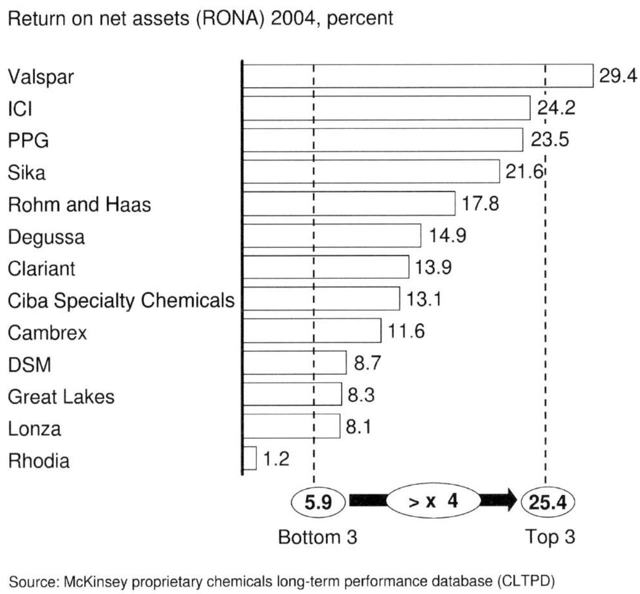

Fig. 8.5 Profitability of specialty chemical players.

8.2 The Market Is Not Getting Any Easier

The golden days of high margins and consistent earnings growth for all are not likely to return. We expect specialty chemicals companies to remain under continued performance pressure. Three issues in particular will plague the industry in the coming years: (1) maturing demand in established markets, (2) a shift in geographic demand, and (3) ongoing changes in the industrial landscape.

8.2.1 The Price-cost Squeeze Is Getting Tighter and Demand Is Maturing

For years, the specialty chemicals industry could rely on innovation and growing customer demand for top line profit growth. However, opportunities for large-scale innovation are less obvious today. Often enough, customer needs can be satisfied with existing products. R&D investments have decreased to near commodity levels and many companies have diverted their R&D dollars to incremental innovations. Decreased differentiation and increased competition have become the name of the game in many specialty segments, such as vitamins, zeolithes, or superabsorbers. An analysis for one specialty chemicals major revealed that 40 percent of the product portfolio can no longer be differentiated by product properties. As customers are not willing to pay premium prices for standard products, price erosion is likely to continue. In the future, more and more specialty products will be sold mostly on price rather than on performance.

Furthermore, consolidation and globalization will continue to increase the purchasing power of chemical buyers, adding to the existing price pressure on chemical producers. In segments such as automotive, personal care, and food, the customer base is rapidly concentrating. Advances in supply chain management, strategic sourcing, and e-commerce also contribute to higher price transparency and customer awareness. Specialty producers will have a much harder time passing on rising raw material and labor costs to maintain their margins.

The continuing price-cost squeeze will put increasingly strong pressure on industry profits. At the same time, customer demand will be further maturing in most segments and established geographies, leaving specialties producers with significantly less room to maneuver: simply outgrowing performance problems by selling additional volumes is no longer an option. We expect the average annual growth rate of specialties to decline from 3.8 to 3.1 percent between 2003 and 2008, reaching roughly GDP level.

Though the perspective of the industry in general looks rather disheartening, some niches will remain comparatively attractive – generally the more complex and fragmented ones, such as active ingredients for personal care, nanoscale chemicals, or specialty polymers.

8.2.2
Customer Industries Are Migrating

Sales of specialty chemicals are concentrated in developed markets, but the fastest growth can be found in emerging geographies such as Eastern Europe, South America, and Asia, where the GDP is growing rapidly and the per capita consumption of chemicals is still quite low. For a representative chemical product portfolio, the expected annual growth between 2001 and 2008 is between five and eight percent for Eastern European and Asian countries – compared to a meager two percent in Western Europe and the USA.

This shift in regional demand will accelerate as customer industries such as textiles, electronics, and automotive vehicles keep moving to emerging markets to take advantage of the low labor costs and the proximity to their customer base.

BASF reportedly expects demand for chemicals in China to double from 2002 to 2015 – and the share of specialty chemicals needed for that region to increase sharply. The company estimates that specialty chemicals will grow to 35 percent of the total Chinese chemical market by 2015.

Specialty chemicals producers must find an answer to the exodus of customers to emerging markets, which not only threatens to break up established supplier-customer relationships, but has the potential to destabilize entire segments. This is particularly true for formulation-based segments such as admixture chemicals for the concrete industry, which depend on close interaction with customers. What happened to textile dyes in Europe is just one particularly striking example: the migration of the European textile industry to Asia and subsequent emergence of Asian textile dyes producers serving the industry locally left established European players with stranded production assets.

8.2.3
The Competitive Landscape Continues to Change

Reshaping the competitive landscape has long been a mantra of the specialty chemicals sector. However, the influx of market entrants from low-cost countries as well as the level of concentration reached in some segments is bringing a new competitive dimension to the industry.

New competitors in low-cost countries are on the rise, wanting to take their piece of the pie. Profiting from significant labor and investment cost advantages (e.g., the hourly labor wages in the chemical industry differ between Western Germany and China by a factor of almost 90), they may turn out to be a serious threat to established players (and already are in many segments). Chinese producers of specialty chemicals, for example, started expanding abroad almost a decade ago, using the domestic market as a springboard to international success. This new competition is likely to have a noticeable effect on price and market share in many segments, adding to the profitability pressure on specialty companies in both their home and export markets (see Chapter 32). Market segments such as dyes, pigments, fine chemicals, and vitamins have already been affected. It is safe to assume that they will not be the only ones.

A striking example of new competitors gaining global clout is Nicholas Piramal, the second-largest player in the Indian pharmaceutical industry. In 2004, the company made its first overseas acquisition, purchasing the global inhalation anesthetic business of Rhodia Organique Fine, UK. The company will gain complete access to Rhodia's sales and marketing network of exclusive pharma distributors in over 90 countries – including the USA, Europe, Japan, and Australia. Nicholas Piramal is planning to use the acquisition as a platform to build a global critical care business.

In addition to the emergence of new competitors, consolidation within segments will continue. Deal activities have shown no signs of a slow-down in the chemical specialties industry, and the last five years have seen a number of megadeals such as Bayer's acquisition of Aventis CropScience and the purchase of Nalco by Suez Lyonnais (and its subsequent sale to a private equity consortium), as well as some ill-fated acquisitions within the fine chemicals segments. While segments such as printing inks and pesticides have been in the limelight so far, flavor and fragrances, paper chemicals, and others are likely to follow suit.

In the future, medium-sized players in particular could be consolidation targets. They often suffer from cost disadvantages compared to mega-players, and lack the dedicated niche players' strong position. Private equity firms in particular have developed a taste for the role of industry shapers. They accounted for more than 20 percent of all deals between 2000 and 2004, and over time they have built quite sizeable specialty players such as Rockwood Specialties and Hexion Specialty Chemicals, which combines Borden Chemicals, Resolution Performance Products, and other businesses.

8.3 Meeting the Challenge – Five Key Levers to Pull

In order to survive in the specialties environment, companies need to act quickly. However, given the high diversity of the specialty chemicals business there is no one right answer. Each player has to find its own way to create value at the business unit level against the backdrop of its business portfolio and the dynamics of the segments served, as well as its own strengths and competences.

However, despite the strong specificity of the solution for the individual company, our experience indicates that, now more than ever, specialty companies should focus in particular on five critical strategic levers: they should (1) revisit and focus their strategies, (2) find an answer to the opportunities and threats arising from emerging countries, (3) strive for top-class operations, (4) rediscover innovation beyond the development of new molecules, and (5) upgrade the performance management of their various business units.

8.3.1 Revisit Strategies

Since all segments of the specialties market have changed considerably over the last few years and this process is set to continue, companies wanting to beat their competition need to develop superior strategic insight and foresight. This is true in every area of the chemical industry, but particularly so in specialties, as the underlying industry dynamic of each segment can be very complex and difficult to assess against the background of customer migration, value chain disruption, regulatory changes, and new technologies such as biotechnology and nanotechnology. Developing such insight and foresight includes gaining a unique understanding of the opportunities and threats a company's markets will offer in the future: what is the life cycle status of markets? How will their growth evolve over the next five to eight years? Will they become, for instance, global and technologically-driven niches or cost-driven, mainly locally served businesses? How are competitors likely to react? How will the customer base change? What are the risks from potential new entrants?

The answers then have to be mirrored against the company's strengths and competitive advantages: in which industry segments or arenas can the company gain critical mass and take the position of a leader and shaper – and in which segments can it not? Danisco, for example, is building a leading position in food ingredients, Degussa in construction chemicals. Private equity firms are reinforcing this trend with their buy-and-build strategy, such as EQT in flavors and fragrances.

Depending on the answers to these questions, companies must shape a leading edge portfolio in which the corporation is the best owner across all of its businesses. Our studies show that focused specialty players can boast a higher return on invested capital. Companies with too diverse a portfolio and a lack of critical mass vis-à-vis customers and suppliers will increasingly suffer from competitive disadvantage. Such a focused portfolio can be built in a number of ways: by making carefully selected acquisitions and divestments, by consolidating fragmented segments of the industry, or by seeking new kinds of synergies between businesses. Examples such as UCB's transformation into a pure bio-pharma company (see Chapter 26) or the reshaping of Degussa into a clear-cut specialty company (see Chapter 9) demonstrate that portfolio management continues to be a key strategic element. Proficiency at acquisitions and divestments will remain indispensable.

Once a specialty chemical company has shaped its portfolio, it needs to begin the hard work of maximizing the performance of each business. Successful players tend to choose one single business model for any given segment rather than attempting to implement a mix. We distinguish six archetypal strategies or business models in the specialty chemicals industry:

- *New product developer:* Such businesses develop (and patent) novel chemicals and formulations that provide additional product performance benefits or that have lower cost-in-use than the current alternatives.
- *Applications developer:* In this model, businesses develop improved products for new applications hand-in-hand with customers.
- *Systems/solutions provider:* These businesses provide an integrated bundle of products and services tailored to single customers' specific requirements, maximizing synergistic effects.
- *Process technologist:* Here, businesses develop (and patent) novel process technologies.
- *Value-chain integrator:* In this model, businesses physically and/or virtually integrate with their customers' value chains.
- *Low-cost producer:* Businesses employing this model produce at the lowest cost, providing their chemicals to customers at attractive prices.

The business model chosen must be continuously re-examined and adapted, especially as a segment matures. Once again, the key factor for success will be the ability to anticipate future market developments.

8.3.2
Find a Winning Answer to Emerging and Developing Markets

Among emerging markets, China takes on a special position. If present growth rates continue, and they are expected to do so, China will be the number one market in the chemical world by 2015, if not earlier. While many Western executives now see China in terms of a once-in-a-lifetime growth opportunity, the Chinese chemical industry is also a potential rival. Specialties are particularly at risk here because, in the past at least, the higher product prices have meant that logistics costs are not such an important issue, thus facilitating long-distance exports.

In any case, the market is developing so rapidly that the wait-and-see approach some companies have followed is no longer appropriate: companies will have to decide right now whether to build up plants in China, develop existing local players, or fend off attackers and seize opportunities in other ways. As mentioned above, the right response will vary by segment or product. Two basic questions help to determine where the opportunities and threats lie: how attractive is the market (size, growth, profitability)? And: is the product's scope of supply global (with low transport costs and customs duties relative to production costs) or local? Options based on the answers are indicated in Figure 8.6.

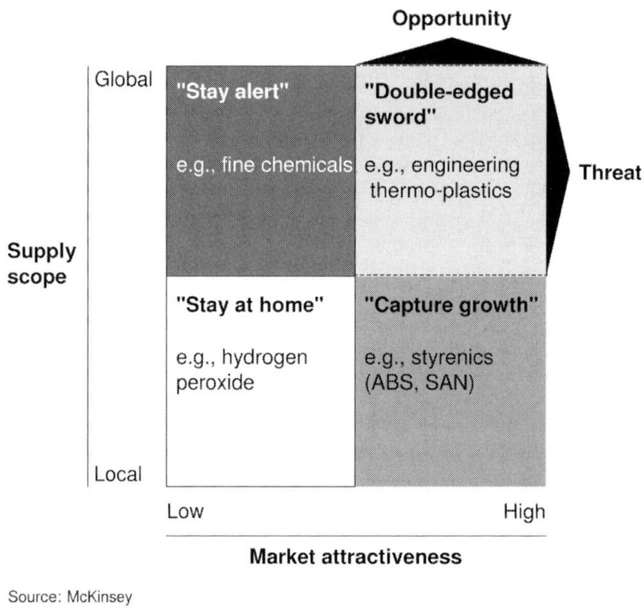

Fig. 8.6 Supply characteristics vs. market attractiveness.

This framework does not only apply to China, but can also be used when evaluating opportunities and threats in other emerging markets such as Eastern European countries. This model also highlights some key differences between different emerging economies: while logistics might be less of a challenge compared with China, the economy in Eastern Europe is already much more developed, offering less spectacular growth rates to specialties producers.

In our view, it is critical to really understand and quantify the magnitude of threats and the potential of new low-cost demand and supply markets for a five- to ten-year horizon very early, ahead of competitors, for instance through dynamic performance gap analysis and cost curve comparisons. This will either help in taking up first mover advantages (e.g., building new capacity, forming local alliances, making local acquisitions) or lead to earlier business restructuring, hence enabling companies to realize profit whilst in a downward price trend caused by low-cost competitors (as in the case, for example, of vitamins and textile dyes).

8.3.3
Strive for Top-class Operational Performance

We have explained why it is virtually impossible to grow out of the sector's performance problem. This, however, means it is mandatory to pursue excellence in operations, an area easily neglected in the days when growth was still the answer. In fact, operational efficiency is already high on the agenda of all major specialty chemical companies. Clariant, Degussa, and Schulman are only three among many companies that have embarked on efficiency improvement initiatives. Key for success will be to match the level of sophistication in each step of the business system with the business model(s) selected.

Operational excellence is particularly difficult to achieve in specialty chemicals, where the high complexity and diversity of the business makes operations particularly demanding, with more choices of levers to pull and in general higher capabilities required than in assets-driven businesses. Overall, we see mainly two crucial areas to address:

- *Lean operations in production and related processes:* companies need to be flexible and cost-efficient – this "ever-true" statement is especially important in the resource-intensive production processes for specialties (batch production and related manual handling). Experience shows that adopting the principles of lean manufacturing from assembly industries can be a powerful operational improvement lever. Lean management, first introduced by Toyota, appeals because it holds out a promise no other approach can make: not only does it cut costs and improve quality, but it also stabilizes operations and matches supply with demand. A companywide lean transformation program structured around the improvement of Overall Equipment Effectiveness (OEE) may unearth a hidden capacity that is equivalent to adding another plant to the manufacturing network. Lean, however, should not only be applied in production, but across the entire value chain. Supply chain management processes as well as application technology and service operations can benefit from the introduction of this concept as well (see Chapters 18 and 22).

  Another ongoing task in operations management is structural cost optimization: producers of chemical specialties typically have numerous production locations with many sites in high-cost countries. Optimizing the global production network – using the opportunities of low-cost production in emerging markets and declining logistics costs (see also Chapters 19 and 32) – can significantly improve a company's competitive position. The typically fragmented asset base is in fact a very significant opportunity to optimize product-to-asset allocation, and hence to leverage lower-cost locations and reduce fixed cost structures.

- *Excellence in marketing and sales:* based on pristine market insight, companies must develop a thorough understanding of customer needs (see also Chapter 21): what is the value of a product to the customer? How much is the customer willing to pay (what is the cost to serve, margin contribution by customer)? What are the key requirements at the customer interface? Though the exact focus and amount of resources needed in marketing and sales will depend on the business model selected for a particular segment, it is safe to assume that most companies will have to significantly improve their performance: the chemical industry has never been in the vanguard of developing state-of-the-art marketing and sales expertise.

  What makes improving marketing and sales performance a particular challenge in specialty chemicals is the fact that there are so many levers to pull – from value marketing and selling to transactional pricing, key account management, channel management, product positioning and branding, sales force effectiveness – that companies have to pick a few for which they want to become truly world class. The crucial issue here is the development of distinctive institutional capabilities (including processes and methodologies, tools, and individual skills and people) on a par with the best comparable marketers in other, more "advanced" industries. Many specialty chemicals companies still have to recognize that top-notch marketing and sales skills can be the deciding factor between outperformance and failure in an ever tougher market. Dow Chemical, BASF, and DuPont are recent examples of companies that have initiated programs for improving marketing and sales capabilities.

8.3.4 Rediscover Innovation

Innovation is needed to steer the market back into specialty territory: customers will only grant a company specialty margins if they perceive differentiation. Innovation is also the best strategy to raise the barrier against low-cost market entrants. Or, as one CEO put it bluntly: "Commoditization only happens when you let it".

However, sources of innovation in the specialties industry have started to change – from developing new molecules and discovering new applications to less obvious forms of creative customer solutions. BASF Coatings, for example, no longer only sells paints to automotive customers, but manages their entire paint shops. Ecolab has become a provider of a broad range of cleaning products and comprehensive services to institutional and industrial end users. Degussa has developed a new automotive tire generation in a joint effort with customers in return for further deepened relationships.

Depending on the business model(s) chosen, companies need to define their specific innovation "headroom". This includes answering questions such as: how important are innovations in the segment? What do strategic customers need? Should companies go for innovation in technology, service offerings, or in the value chain? What does that imply for R&D capacities and skills? How can R&D best collaborate with marketing and sales, (potential) customers, and external research institutes? What level of external versus internal R&D is required?

All research initiatives have to be prioritized according to their inherent risk-weighted revenue and sustainable growth potential. Resources have to be allocated in tune with a rigorous stage-gate process which receives steady input from (internal and external) market intelligence to check projects' alignment with market targets. DuPont, for example, has cut the number of ongoing R&D projects from more than 4,000 to about 2,000 since 2002. The company also launched a program called Top 75, designed to shift financial and management resources to those projects most likely to generate revenue and sustainable growth (see also Chapter 14). About one-third of R&D resources now goes to the Top 75. DuPont is well on its way to achieving its 2005 objective of one-third of its revenue coming from new products introduced within the last five years.

8.3.5
Upgrade Performance Management

Specialty chemicals "conglomerates" typically have a large spectrum of different business units that need to be operated on a global level. In the face of such complexity, it seems easier to define the same financial criteria for all business units and use a standardized performance dialogue. All units are thus measured against the same EBITDA/top line growth targets and sales/headcount ratios.

However, the easiest way is not always the best. Depending on the business model, key factors for success can differ considerably – especially along the dimensions of performance metrics. These differences need to be taken into account: project attrition/win rate, for example, is an appropriate performance metric for a people-intensive applications developer with several small local subsidiaries. A global new product developer with low overall headcount but high external R&D spend should, instead, be measured by the value of its product pipeline and the intellectual property owned. Given the different risk profiles of those two businesses, their ROCE targets, too, should be individually set.

Besides a financial performance dialogue, specialty chemicals conglomerates also need to establish an operational performance dialogue. The corporate center can help drive operational excellence by challenging and coaching the individual business units, setting benchmarks, and fostering know-how transfer between the units.

Specialty companies are facing an ever tougher environment. Eventually, the extent to which a company will manage to outperform other players, be they existing players, financial investors or new low-cost entrants, will depend on its ability to develop a strategy which will enable it to shape a leading competitive position in a market or arena that can allow it to tap into its advantages, and to achieve superior operational performance at the business unit level.

8.4 Summary

- Specialty chemicals have lost some of their appeal: the days of high ROIC margins and consistent earnings growth are over.
- Despite a continuing price-cost squeeze as well as maturing and saturated end markets keeping specialty producers under performance pressure, the best performing specialty companies achieve very impressive results across segments.
- As numerous customer industries are migrating to emerging markets in Asia and Eastern Europe, new low-cost competitors are arising from these areas.
- For many specialty chemical companies to continue to create value in the future, they will need to accelerate their business performance improvement trajectory.
- In particular, they will have to develop strategies which will enable them to shape a leading competitive position in markets or arenas where they can tap into their advantages, and to achieve superior operational performance at the business unit level.
- They need to be clear about what it will take to compete in tomorrow's world with low-cost chemical players around, and act accordingly ahead of the change.
- They need to strive for operational excellence in particular in production and marketing and sales to improve their market position.
- They need to refocus innovation to fight commoditization and low-cost market entrants.
- They need to upgrade their performance management by acknowledging the intrinsic differences between their various business models.

9
Creating the World's Leading Specialty Chemicals Company

Utz-Hellmuth Felcht

When VEBA and VIAG announced the merger of their chemical affiliates Degussa-Hüls AG and SKW Trostberg AG on May 15, 2000, the markets were less than optimistic about the chances of success of this venture. Most analysts agreed that the new company – later to be named Degussa AG – was an unfocused conglomerate of isolated activities, with diverging cultures and no inherent strategic logic. The two companies' complementary portfolios ruled out significant operating synergies, and the lack of a strategic rationale would make it hard for the new company and its management to define a common identity that could be credibly communicated to customers, capital markets, and its own employees.

Thus, when the management of the new company was appointed a week later, it faced a threefold challenge:

- It had to transform the conglomerate into a focused portfolio that gave the new company a credible "story" of value creation for customers, shareholders, employees, and global communities;
- It had to transform a medley of contradictory organizational concepts into a structure that allowed both fast and flexible operation in a changing global market environment and strategic alignment towards common value creation goals;
- It had to transform a variety of strong individual corporate cultures into a common spirit of value creation and entrepreneurship which would incorporate the positive values of the "old" cultures rather than streamlining and destroying them.

9.1
The Portfolio: "Creating the World's Leading Specialty Chemicals Company"

9.1.1
Focus on Specialty Chemicals through Divestments

Many years before the announcement of the merger, Degussa-Hüls, SKW Trostberg, and their predecessors had already been following strategies directed toward

Value Creation: Strategies for the Chemical Industry. 2nd Edition. F. Budde, U.-H. Felcht, H. Frankemölle (Eds.)

ISBN: 3-527-31266-8

"specialties". At the time of the merger, focused specialty chemicals companies generally achieved a better long-term ROIC than hybrid or commodity chemical players. On average, specialty markets showed faster growth rates, lower cyclicality, and better returns to shareholders than most other chemical businesses. It was therefore no coincidence that specialty chemicals represented almost 60 percent of the combined sales of Degussa-Hüls and SKW Trostberg – around EUR 9.8 billion. The merger, however, allowed a rigorous focus on this area for the first time: although more than 40 percent of the combined sales volume of both companies had to go in order to achieve this, the remaining business would have a worldwide number one position in specialty chemicals and be guaranteed a listing on the DAX 30 index.

In February and March 2000 a strategy team from Degussa-Hüls and SKW Trostberg had examined various options for the future development of "VEBA/VIAG Chemie". It was decided early on that – despite a low potential for direct synergies – a merger of the two companies would be able to generate more potential than other options (e.g., trade sale of one or both companies), but only if a credible merger concept with a focused specialty portfolio could be defined. The following method was used to identify the future core businesses:

- To give the new company a distinct image as a chemical company, "non-chemical" businesses were sorted out regardless of their attractiveness. These included Degussa-Hüls's units Asta Medica (pharmaceuticals), dmc² (precious metals & industrial catalysts), Dental (teeth fillings), and Degussa Bank, as well as SKW's salt and metallurgy interests.
- In a second step, the remaining chemical businesses were classified according to five factors that, according to the consensus of the team, together defined a successful specialty chemicals business:
  - High added value for the customer
  - High technological and problem-solving competence
  - High rate of innovation
  - High entry barriers
  - Small number of competitors
- Additionally, three supporting criteria were used which are obviously desirable for any business:
  - High market attractiveness (growth & profitability)
  - Leading global position
  - Cost leadership

Several popular standard criteria for "specialty" chemicals – such as high price per kilo, low volume, importance of application technology – were not used, because no correlation could be detected between these criteria and value creation.

This method produced a "non-core" portfolio of 14 businesses, most of which were sold successfully by 2004, generating gross proceeds of more than EUR 3.8 billion.

9.1.2
Development of the Core Portfolio through Organic Growth and Acquisitions

The elimination of businesses that were a poor strategic fit, unattractive, or both resulted in a core portfolio for Degussa with a high percentage of specialties and of leading global positions. However, like many other decisions that formed the "New Degussa", these decisions were not dogmatic, but based on an overall perception of each business's potential for value generation: businesses with questionable or limited specialty character, such as C_4 chemicals, hydrogen peroxide, and feed additives, stayed because they represented a high sustainable value on the basis of leading global positions, outstanding technological competence, and cost leadership. On the other hand, specialty businesses with weak market positions and (in some cases) low profitability such as flavors, bioactives, exclusive synthesis, and water treatment were also retained, because they operated in fundamentally attractive markets and offered opportunities for development into attractive value generators through external growth, i.e., by acquisitions. Finally, some specialty businesses, such as superabsorbers, building blocks, or Plexiglas®, held leading market positions but were operating in mature markets and lacked profitability due to sub-optimal business systems and/or restructuring needs. These businesses were to be developed as indispensable or potentially valuable parts of the overall business system, but needed to be turned around.

As an overarching tool for the strategic development of its diversified portfolio, Degussa introduced an improved Strategic Management Process (SMP), based on methods previously developed in the merged entities: the entire core portfolio was segmented into around 80 units (usually business lines) that are analyzed annually according to market attractiveness and competitive position. Based on its position in the portfolio matrix (Fig. 9.1), each business is assigned a category ("Emerging", "Growth", "Mature" or "Restructuring") as well as growth and profitability targets depending on that category. Capital resources are also allocated by strategic category.

This process enabled the investment focus to be shifted significantly from mature to growing businesses within two years of the merger (Fig. 9.2). In parallel, total capital expenditure was reduced, resulting in an average cut for mature businesses of 20 percent (in some cases up to 50 percent), while growing businesses enjoyed an increase in available resources.

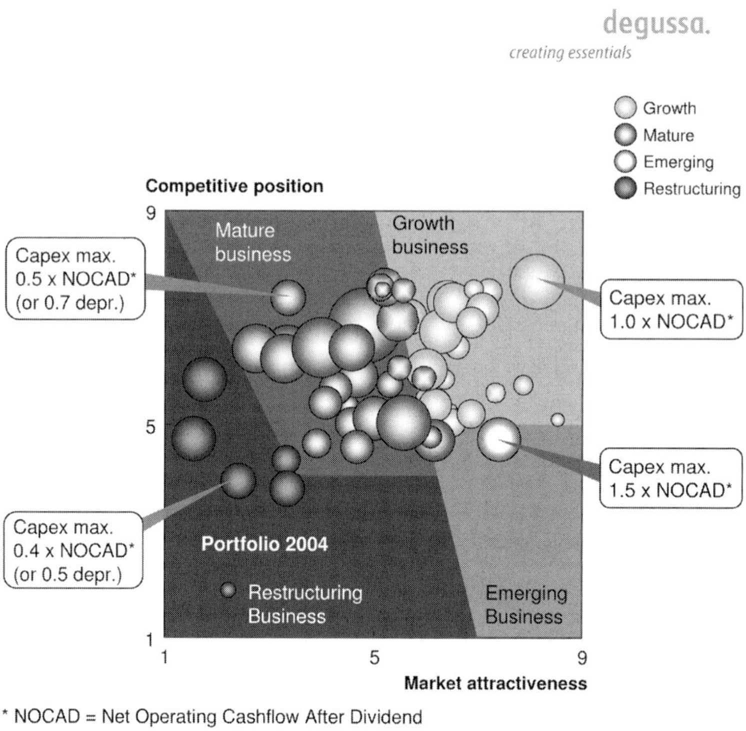

Fig. 9.1 Target assignment and resource allocation based on strategic categories.

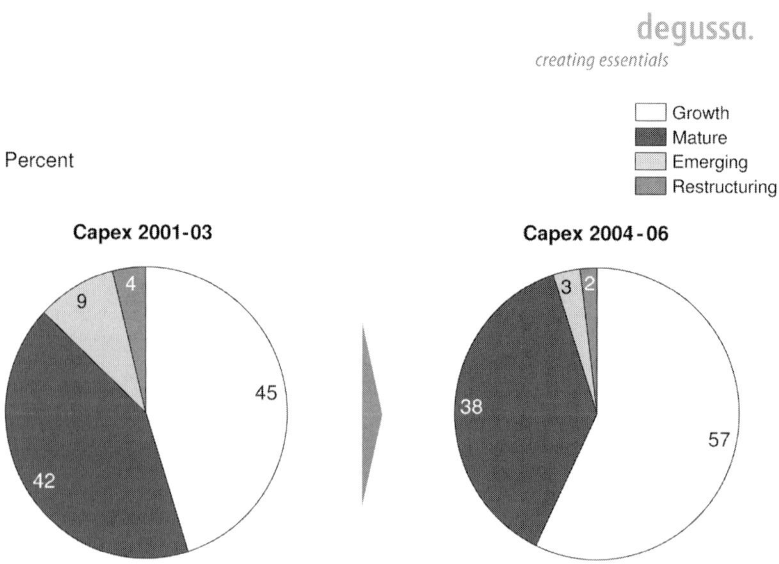

Fig. 9.2 Shift of capital expenditure from mature to growing businesses.

In order to develop businesses that operated in especially attractive markets but had subcritical competitive positions, three "corporate growth areas" were identified and prioritized for large-scale external growth: exclusive synthesis (pharmaceuticals), food ingredients/flavors, and high performance polymers. Pursuing first the strategy on exclusive synthesis, in parallel with the merger Degussa acquired and integrated the British fine chemical company Laporte in 2001. Although the integration itself worked as planned and established Degussa as one of the top three players in this field, a subsequent massive deterioration in market conditions meant that the combined activities were unable to generate the expected value. In 2002, Degussa's attempt to close the gap on food ingredients/flavors through the acquisition of Haarmann & Reimer was abandoned, because the purchase price expected by the owner (and offered by some competitive bidders) would not have allowed Degussa sufficient potential for value generation.

Although the economic downturn of 2001–2003 was accompanied by massive losses on the stock markets, this development did not reflect to the same degree in the prices paid for acquisitions in the specialty chemicals sector. Since at the same time the economic outlook for most specialty chemicals markets offered considerably lower growth expectations than in the 1990s, major acquisitions at high multiples seemed an increasingly unattractive option for value generation. As a result, since 2003 Degussa has concentrated mainly on organic growth. Within this strategy, acquisitions are mainly considered in order to gain access to additional key technologies or attractive market segments or regions, thus offering additional levers for internal growth. On the other hand, this paradigm shift made it necessary to re-evaluate development strategies for those businesses whose weak competitive position demanded major acquisitions. As a result, Degussa's water treatment and food ingredients businesses are currently being prepared for sale.

On top of the portfolio restructuring, it was decided to use the energy released by the merger to initiate a sweeping business process optimization program throughout the new company – "best@chem". The resulting savings have been even greater than expected: although the final evaluation has not yet been completed, we already know that EBIT improvements of around EUR 500 million per annum have been achieved.

9.2 Organization: "as Decentralized as Possible, as Centralized as Necessary"

9.2.1 Market Oriented Business Units: "Business First"

The organizational concept for the new company was defined before the merger, based on a strategic ground rule later summarized in the slogan "Business First". The strategic rationale for this concept was driven by the specific environment which any specialty chemicals company has to face:

- The large number of diverse market segments with distinctive rules which needed to be addressed by a set of distinctive business systems
- Fast changing market conditions which made innovation and market intelligence extremely important factors.

The business therefore had to be segmented into flexible, efficient business units that were large enough to achieve leading positions in their respective markets, but also small enough to adapt their business systems easily to these markets' specific requirements. Clearly, such units had to be managed entrepreneurially, with a distinct focus on their markets and customers. They therefore had to be given responsibility for all key elements of their business systems, such as R&D, production, sales, and marketing.

The consequence of this autonomy, however, is a natural conflict with central strategic alignment. If, on top of that, the business units were managed in line with their legal form as separate legal entities headed by supervisory boards, the resulting organization would be so cumbersome and inflexible as to severely hamper any organizational changes or strategic leadership. It was therefore decided to separate the management structure of the new company completely from the legal structure, although this meant additional interfaces. In order to ensure a clear chain of command, all top managers of legal entities are employed by Degussa AG and delegated to their companies. No Degussa board members are members of any supervisory board of any affiliated company.

The strategic alignment of related business units and the requirements of operational leadership made it necessary to combine the 23 (later 21) business units in six (later five) divisions. However, previous experience had shown that direct responsibility of management board members for operational branches resulted in two problems:

- Conflicts arose between "business" and "corporate" motives for strategic decisions
- Mechanisms for decisions on strategic priorities and resource allocation were often based more on negotiation than on merit.

Therefore, a management board was formed of four members with purely central functions (Chief Executive Officer, Chief Financial Officer, Chief Development Officer, and Chief Human Resources Officer). The divisions were headed by Chief Operational Officers (COO) with supreme responsibility for the operational business of their divisions (including the setting of goals for the BU heads), who were not members of the Degussa management board. Instead, the COOs report directly to the CEO. Together, the management board and divisional heads form the Executive Committee, which meets regularly before board meetings and ensures that board decisions take full account of operational issues (Fig. 9.3).

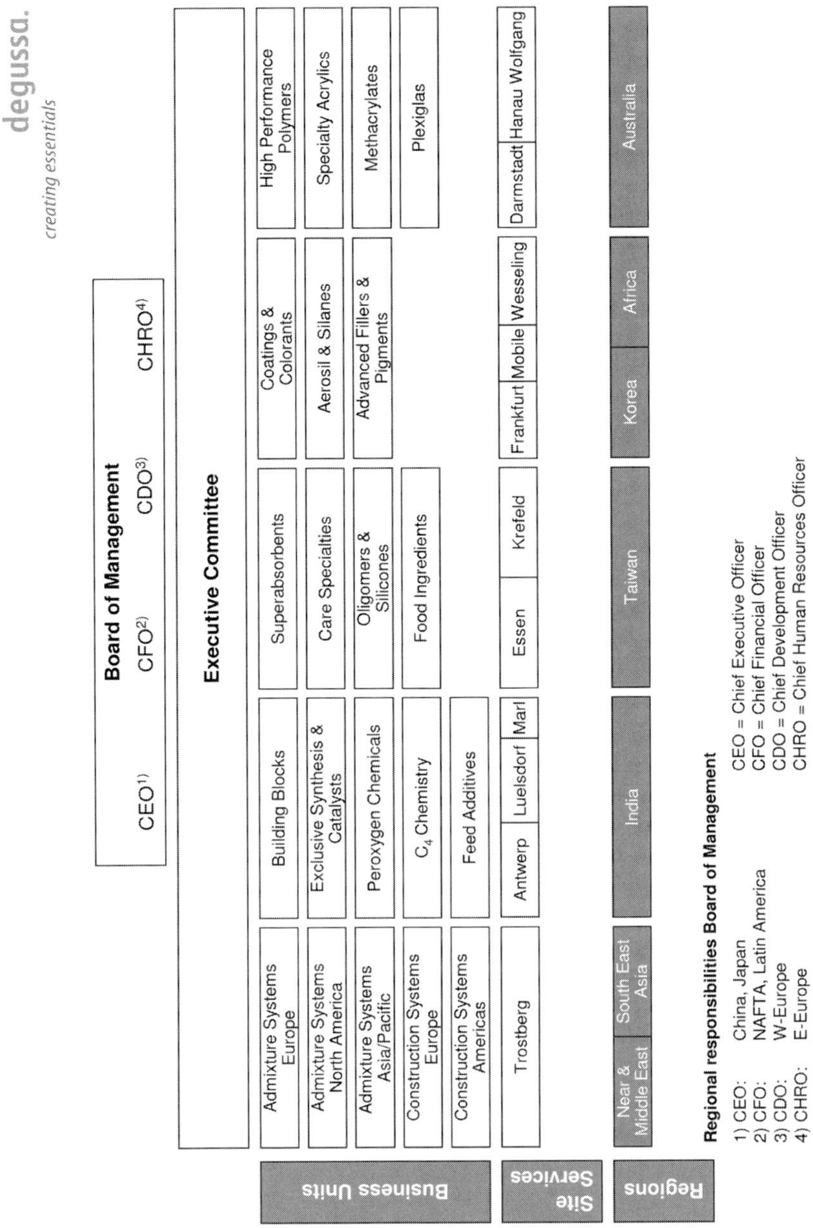

Fig. 9.3 Degussa's operating organization (January 2005).

9.2.2
Small Corporate Center, Independent Shared Services

The corporate center in Dusseldorf was created as a relatively small unit with strictly strategic and administrative tasks, including portfolio management, financial consolidation, the allocation of resources, maintenance of the dialogue with shareholders and the public, management development and the appointment of executives, rules and guidelines (Legal, Taxes, EHSQ, Corporate Governance, etc.), and management of synergies (value-increasing corporate initiatives).

Within this organization, corporate center departments make suggestions and prepare decisions, but they have no management responsibility in relation to the operating units. Corporate center costs are carried separately in the P&L statement, and not allocated to the businesses. Thus, businesses control their total cost without corporate charges.

A major task during the merger was to keep the corporate center organization (which is operated by necessity as a cost center) as small as possible. Administrative cost could only be saved effectively if as many activities as possible were subject to "market" forces, such as competition or performance control by the recipient. This led to the creation of "Shared Services" (Purchasing, IT, Human Resources, Financial, Process Technology/Engineering, Legal/Insurance, etc.). Whereas the separation of genuine management (= corporate) functions from service functions was difficult and took some time and iterative optimization, for some of the new services the transformation of mindsets (from central administrator to service provider) proved even harder and more time-consuming. Conversely, some operating units used their new-found freedom to negotiate service level agreements too extensively. It took some time and a number of cultural changes (see section 3 below) to reduce this antagonism to a sound level.

9.2.3
Site Services

Degussa has more than 300 sites worldwide, including around 160 with production facilities. Only eight sites include significant activities of more than one business unit, but almost 50 percent of all Degussa employees worldwide work on these "Multi-User Sites". Each of them started as a single company's integrated production facility, such as the Hüls site in Marl. As a result of the restructuring of the entire chemical industry, most of these Degussa sites now also play host to companies owned by third parties: in Marl and Wolfgang the combined third-party businesses have become as big as or even bigger than the Degussa activities. Thus, whereas all minor sites (and some major ones, such as Rheinfelden) are run as part of their business unit organizations, a special solution in the form of Site Services had to be found for the multi-user sites. Internally, the most advanced model is Infracor, which – as a separate legal entity – operates the site in Marl as an industrial park with approximately 20 external customers besides Degussa. Typically, services such as energy and utilities, logistics, maintenance, analytics, human resources, etc. are supplied to all internal and external customers on a specific site.

Degussa's site services face strategic challenges from both internal and external sources:

- A number of strategic restrictions make "entrepreneurial" site service development rather challenging. Internally, Degussa site services are organized similarly to business units with their own assets and resources, and are managed entrepreneurially. However, their strategic task is not to optimize profit (return on capital), but to support the Degussa units operating on their sites to the best possible effect. As a rule, they may acquire external business for the optimization of their cost/return structure, but not if this requires major investments or the build-up of additional resources. Even internally, any increase in "sales" that a service may realize with Degussa businesses means an increase in costs for the latter, and is therefore watched carefully - after all, Degussa's core business is manufacturing specialty chemicals, and not running industrial parks.
- Externally, Degussa site services face competition from third-party suppliers of single service functions such as maintenance, but also from third-party sites which try to take away businesses through superior prices, infrastructure, or tax provisions. Especially in a generally mature business environment such as Germany (where six out of the eight multi-user sites and 80 percent of Degussa's site service employees are located), this creates constant pressure to optimize or cut down resources and to manage remnant cost.

9.3 A United Corporate Culture: "Blue Spirit"

The history of mergers in the chemical industry shows that corporate cultures and identities, once firmly established, are much more resistant to change than any "hard" industrial or organizational structure. Different cultures and identities within the same company represent a major (sometimes insurmountable) obstacle to any cooperation between units (i.e., people) that requires old borders to be crossed. For a diversified company such as Degussa, collaboration among units and the linking of knowledge among employees is a major source of value creation. To preserve this potential, a way had to be found for the new company to create a common identity with cultural values that appealed to its managers and staff (all of whom had "grown up" in the preceding cultures), and at the same time resulted in a positive image to its customers, shareholders, and communities.

9.3.1
A New Brand for the Company with a New Name, Vision, Mission, Guidelines

The first decision was made quickly and boldly: the new company would be named *Degussa*. This had obvious advantages: the name Degussa was widely known, respected, and certainly easier to identify with than any fashionable, artificial creation. However, there were pitfalls as well. Whereas the most important cultural goal of the merger was to create something entirely new (and better than the predecessor companies), the chosen name Degussa might signal a fall-back to former Degussa times and structures or – even worse – the intention of one old identity consuming all the others. This was solved on the one hand by the creation of a new logo **degussa.** incorporating elements of all the predecessors (the name from Degussa, the blue from Hüls, the letters and the dot from SKW) and thus symbolizing the creation of something innovative. On the other hand, the speed with which the merger project was executed, the open communication, and the content of the project itself (new management, new structures, new strategy) made it easy to accept that a real change was intended. The relocation of the corporate center away from any of the predecessor companies' headquarters also signaled the creation of an entirely new organization.

Relatively early on, a group of top managers from all the companies involved convened to design vision and mission statements (Fig. 9.4) and guiding principles for the new Degussa. Naturally, these statements contained nothing really new. However, they stood for goals that most Degussa stakeholders could identify with. The guiding principles (Fig. 9.5) again contained behavioral rules that are undoubtedly applied in many other companies as well – the most innovative one is probably "Act with passion for our business". Here, too, it was important not so

degussa.
creating essentials

The vision

Everybody benefits from a Degussa product – every day and everywhere

The mission

- To be the world's premier company in specialty chemistry and more
- To be the benchmark for innovation and profitable growth
- To continuously create value for our customers, shareholders, and employees
- To become a truly multinational company
- To act in a socially responsible and ethical manner
- To act in accordance with sustainability and Responsible Care® principles

Fig. 9.4 Degussa vision and mission statements.

degussa.
creating essentials

Guiding Principles

- Focus on our customers
- Operate according to Responsible Care® principles that protect employees, environment, and the community
- Respect diversity in culture, gender, nationality, and race
- Treat people fairly and with consistency and respect different opinions
- Be open and honest and share information
- Encourage learning, develop people, foster teamwork
- Set clear goals, empower employees, and encourage open feedback
- Take responsibility and lead by example
- Support innovation and initiative, learn from mistakes, strive for excellence
- Act with passion for our business

Fig. 9.5 Degussa guiding principles.

much to create something new, but to show people that the new company stood for something valuable that it was worthwhile to identify with and was already contained in statements they knew from the past, even if couched in different terms.

In 2001, after most of the physical merger had been completed, work started on the new corporate culture. The first step was a corporate conference in September of that year which involved (besides board members and divisional heads) Degussa's top 70 managers. The values of the new corporate culture were communicated under the name *Blue Spirit,* using the vehicle of "corporate theatricals" which made Degussa's internal conflicts visible in a role play – including a radiant finale starring the four members of the management board. In the following months, *Blue Spirit* was rolled out to all units of the company, accompanied by various events such as visits by the board members to all major Degussa sites. The final step at the corporate level was the next corporate conference in February 2003, this time involving the top 250 executives. By the end of this conference, *Blue Spirit* had welded most of Degussa's executives into a team.

9.3.2
Umbrella Brand Degussa: "Creating Essentials"

Blue Spirit provided a set of rules for internal relations and behavior, but another important task remained: to define the values of the new company in relation to its customers and markets. The intention was for the new brand identity "Degussa" to define what was new (because the public identified Degussa mainly with gold & silver), but it could not deviate too much from the values of the predecessor companies, in order to retain credibility.

In a detailed bottom-up process which involved surveys among customers, employees, management, and external opinion leaders, the core contents of the old company brands Degussa, Hüls, SKW, Röhm, and Goldschmidt were analyzed. Internal and external sources turned out to connect four attributes with all five companies: reliability, ability to innovate, uniqueness of products and services, and intelligent linking.

These attributes were therefore selected as the core of the new umbrella brand "Degussa", and summarized in the claim "Creating Essentials".

Beginning in 2003, a concept for the content and background of the umbrella brand was formulated and rolled out throughout the company. A major task was the coordination between the role of the umbrella brand and strong product brands such as Plexiglas®. Externally, the roll-out was supported by a press campaign that presented interesting and innovative Degussa solutions as examples of "Creating Essentials".

9.4
New Challenges and the Next Transformation Goals

Today, the "New Degussa" has become just "Degussa", the world's leading specialty chemicals company and Germany's third largest chemical company. By end of 2004, the best@chem program was completed. The separate listing of "non-core businesses" in the financial statements ended at the same time, ending the startup phase of Degussa. However, this did not end the need for transformation.

Most of Degussa's specialties are used in products and systems for an "enhanced (Western) lifestyle", limiting the company's chances of growth in Western Europe and NAFTA. Like any other specialty chemicals company, Degussa is currently facing the challenges of globalization with increasing competitive pressure in established markets, but also increasing opportunities in new markets in regions with prosperous, fast-growing economies. These challenges demand new points of corporate emphasis, which can be summarized under four headings:

- Solutions to Customers
- Emerging Markets
- Site Excellence
- Human & Corporate Excellence.

9.4.1
Solutions to Customers

Today, producers of specialty chemicals like Degussa are increasingly challenged on home markets as well as abroad by competitors who offer the same chemicals of comparable quality at prices that are – in some cases – even lower than their own production cost. This issue cannot be solved by cost reduction alone. The only valid answer to price-based competition is to move the discussion toward something that is more valuable to the customer than a low price. This is the core of the *Solutions to Customers* (STC) concept. Its major components are as follows:

- A thorough understanding of customers, their value chains, markets and challenges (including those of the customers' customers and so on, all the way up to the end consumers and the trends that influence their preferences and behavior)
- Based on this understanding, a unique selling proposition that is not based just on "products" (i.e., chemical molecules) but on a whole concept of products, qualities, innovations, and services that offers added value to the customer and also allows Degussa to share in this value.

STC demands that each individual business see itself and its customers as an integrated part of the same value chain. This integrated concept also means that every individual function (such as R&D, production, or logistics) is treated as part of an integrated and multifunctional marketing system: "Everyone is in marketing". For example, STC makes frequent contacts between R&D and various customer functions a matter of course, and may even lead to the organizational integration of marketing and R&D functions in STC-based businesses.

Degussa is fostering the concept by the creation of "linking knowledge" elements that overcome the limitations of the business unit organization. For example, a network of industrial and customer teams was founded to identify and exploit the needs of specific industries and customers in various projects that are "owned" by the contributing business units (Fig. 9.6).

degussa.
creating essentials

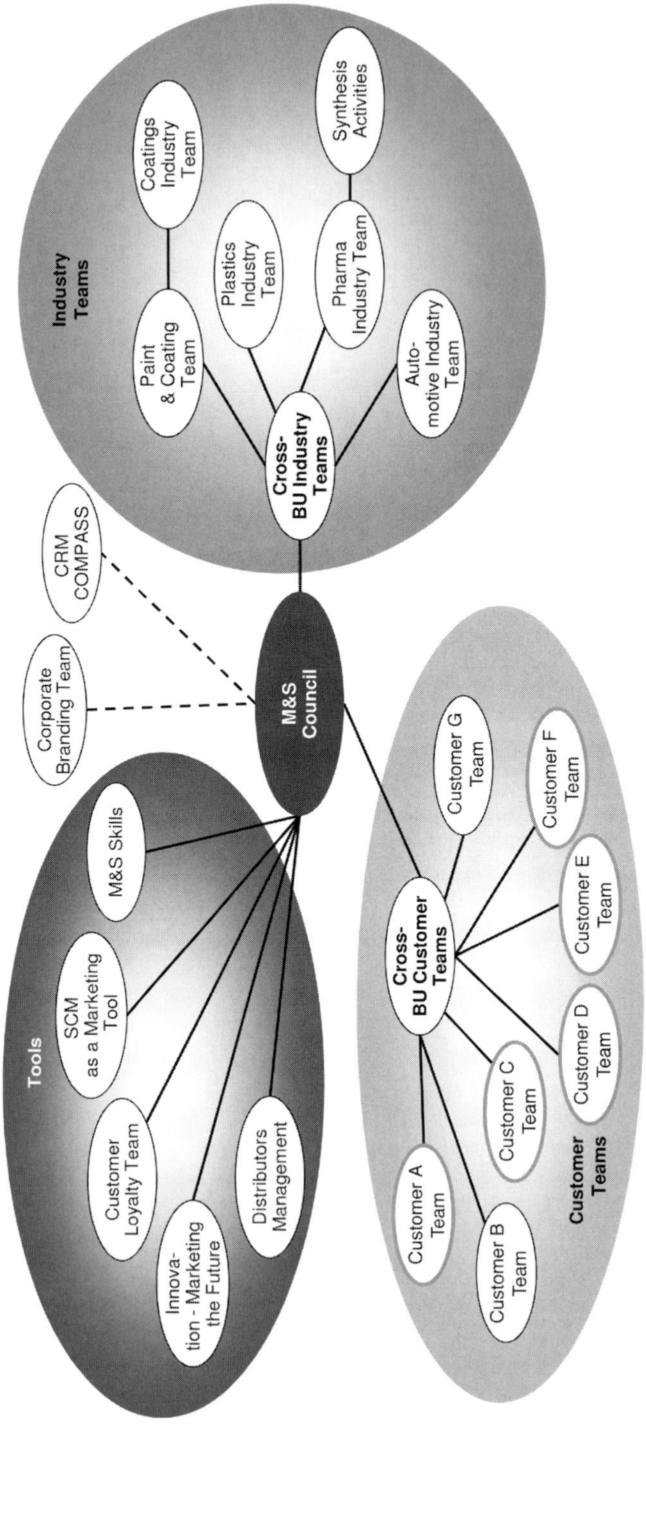

Fig. 9.6 Market-related team organization.

9.4.2 Emerging Markets

The rapid development of regions such as China and Eastern Europe creates chances as well as risks for the specialty chemicals industry, as we have also seen in other chapters. On the one hand, an increasing number of people are reaching a level of prosperity which allows them to purchase valuable consumer products (whose production requires specialty chemicals). On the other hand, an increasing number of local competitors are becoming capable of supplying "specialty" chemicals to regional and global markets at much lower cost.

Degussa's response to this two-pronged challenge is a two-pronged strategy. In the coming years, we will channel a significant share of our resources into establishing leading positions for our attractive businesses in China and Eastern Europe. This will also mean a growing globalization of our workforce, but even more of our management. At the same time we will secure our Western position as far as possible by restructuring our product portfolio (Fig. 9.7): ultimately, Degussa will only have products that can be marketed under a differentiated "Solutions to Customers" system that delivers added value to Degussa as well as to its customers, or products that have a leading cost position, or both. Exit strategies may therefore have to be developed for at least ten percent of our portfolio within the next few years.

degussa.
creating essentials

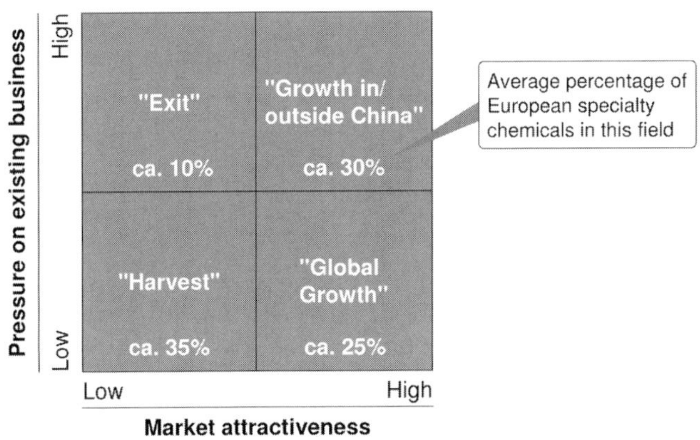

Source: McKinsey (2004)

Fig. 9.7 Portfolio matrix for China strategies.

9.4.3
Site Excellence

The developments in the emerging markets will cause many Western production activities to stagnate or shrink in the coming years. Together with the strategic and competitive challenges for site services described in section 9.2.3, this creates a situation which can only be solved by an innovative strategic approach. Therefore, in the next few years Degussa will implement the concept of *Site Excellence,* an integrated strategic approach which views site services and operating businesses at each site as a community that has to combine to optimize value. As far as feasible, this will also involve third party producers at each site. While continuing to observe the "Business First" rule and the separation of operating and service organizations, the *Site Excellence* concept aims to align individual strategies in order to find the one combination of strategic options that delivers optimum value to all parties. This could, for example, involve a redistribution of tasks between service and operating organizations, or the clustering of services or whole sites (including, where feasible, partnerships with third parties) in new organizational concepts. Innovative work organization concepts such as self-directing teams will also be evaluated.

9.4.4
Human and Corporate Excellence

The strategic challenges described above will also create increasing demands on the quality of our people and our processes. Under the heading of *Human & Corporate Excellence,* we will therefore analyze how further developments at Degussa can be managed, and what processes have to be created or improved in order to foster entrepreneurship and develop the global management resources that will be needed to execute Degussa's globalization strategy within the next years.

9.5
Conclusion

After successfully mastering the integration and startup phase, Degussa is now entering the second stage of its development, focusing on generating value through innovation and profitable growth. Each of the past stages was characterized by specific organizational and cultural transformation needs, and the same applies to the next stage. In fact, transformation within a specialty chemicals company these days is not a project with a fixed beginning and a scheduled end (although any transformation will be project-based), but rather a way of life. Under these circumstances, only those companies will survive and prosper whose corporate culture enables and motivates people at all levels of the corporation to accept this permanent transformation and to shape it proactively to create value. We will make sure that Degussa is one of these companies.

9.6 Summary

When Degussa-Hüls AG and SKW Trostberg AG merged in May 2000, the new Degussa faced the challenge of transforming a widely diverse conglomerate into a specialty chemicals company with a focused portfolio, a flexible organization which was nevertheless aligned towards common value creation goals, and a single corporate culture that incorporated the positive values of the old ones.

- Focus was achieved by restructuring the portfolio by an approach based on eliminating non-chemical businesses, classifying the remaining businesses according to a definition of a successful specialty business, and divesting the non-core businesses over a four-year period. The portfolio was then further developed, mainly by organic means due to market conditions. In parallel, a major operational improvement project was launched, best@chem.
- A market-oriented organization was created, "as decentralized as possible, as centralized as necessary". The strategic rationale was driven by the diversity of market segments and fast-changing market conditions, an environment in which innovation and market intelligence were crucial. The new structure separated the lean corporate center, operating businesses, and services to maximize autonomy and entrepreneurship.
- The corporate culture aimed to create a common identity with cultural values that appealed to the management and employees by drawing on the positive values of the pre-merger entities, while also communicating a positive image to the organization's customers, shareholders, and communities. The choice of name and logo reflected the company's origins, while the processes and content of the merger project drove home the message that a new entity was being created. Further key elements were the *Blue Spirit* project, which laid down a set of rules for internal relations and behavior, and the creation of a single umbrella brand whose attributes were summarized in the claim "Creating Essentials".
- With the startup phase very successfully completed, Degussa is making transformation a way of life, currently focusing on profitable top-line growth through new points of corporate excellence relating to customers, emerging markets, site optimization, as well as people and processes.

10
Prospects for Agribusiness: an Essential Contribution to Global Food Demand

Michael Pragnell and Robert Berendes

Food security is the most basic of all human needs. Since its emergence around 70 years ago, agribusiness, combining crop protection chemicals, seeds, and recently biotechnology, has allowed growers to substantially increase crop yield and thus played a fundamental role in helping to fulfill this need. However, in some areas of the world even the minimum requirements of food supply are not yet secured, while in the industrialized world, demand for improved food quality constantly increases.

Agribusiness, as defined above, has grown strongly into a USD 49 billion global industry and consistently broadened its ability to realize the full potential of plants, helping the world's growers to protect, increase, and enhance plant yield (Fig. 10.1). It has also expanded into related industry segments such as the application of crop protection chemical compounds in consumer-driven markets such as home and garden applications.

Sustained food and therefore agricultural demand will secure the industry's profitable growth in the future. Further expansion towards consumer markets is adding additional momentum for broader growth opportunities.

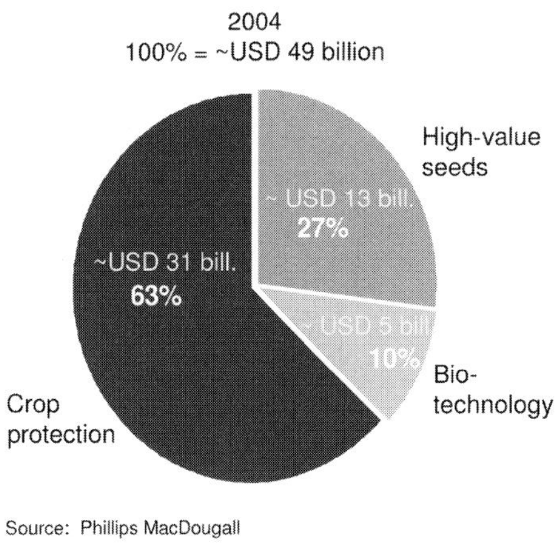

Fig. 10.1 Agribusiness size 2004, chemicals, seeds, biotech.

Value Creation: Strategies for the Chemical Industry. 2nd Edition. F. Budde, U.-H. Felcht, H. Frankemölle (Eds.)

ISBN: 3-527-31266-8

10.1
The Rationale of Agribusiness

The world's population has doubled to 6.5 billion in the last 40 years (US Census Bureau), and is likely to increase by 80 million every year for the next 25 years (according to FAO), mainly in urban areas and in the developing world. Calorie consumption per capita is expected to increase by ten percent over the next 20 years, driven partly by the change towards meat-based diets, for example in China, which will of course lead to a sharp increase in demand for animal feed crops such as soya and corn (Figs. 10.2, 10.3). Due to a combination of these two trends, the total world demand for calories will grow by about 50 percent by 2025. This is a major challenge for agriculture, as the area of arable land worldwide has remained constant in recent decades and is not expected to increase in the future. Therefore, the only way to increase output is by improving plant yield.

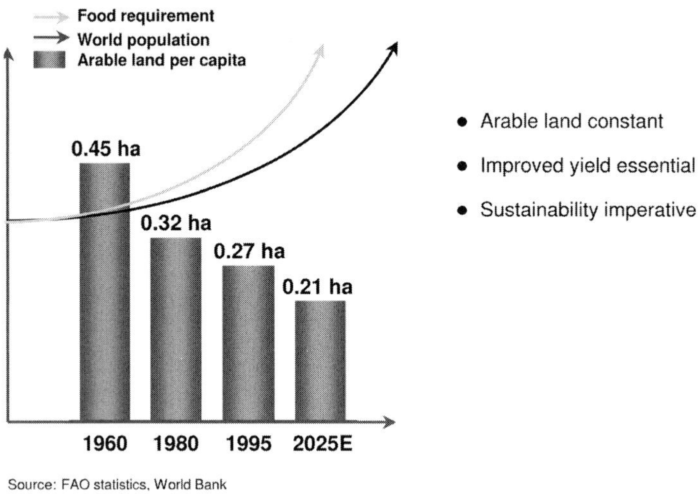

Fig. 10.2 Global calorie demand to grow about 50 percent by 2005.

Continuous yield improvements based on agricultural innovation have secured a higher quality and ever decreasing cost of global food supply; however, nearly one billion people in developing countries still suffer from food scarcity. Additional growth in agricultural production of around 1.5 percent per year is needed to eradicate repeated food shortages, especially in developing countries. This growth rate seems viable given the largely similar rates of improvement in crop yield achieved by agribusiness over the past 35 years.

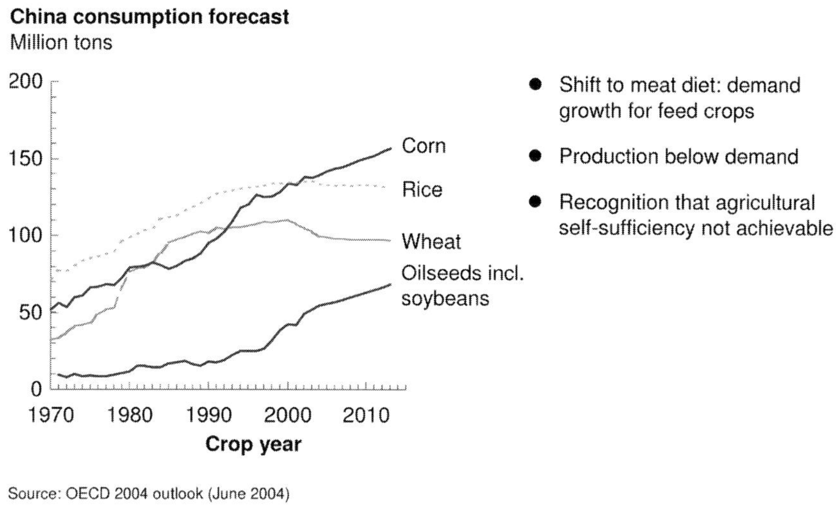

Fig. 10.3 Commodity demand: example China.

Nature imposes still more challenges on the grower and on agribusiness. Pests and plant diseases have devastating effects on yield, and have destroyed entire harvests from time to time throughout history. Over its comparatively short existence, agribusiness has responded well to these challenges with crop protection chemicals to deliver yield increases. In addition, improved seeds have led to better crops with greater resistance to the more extreme elements, with optimal use of water and other nutrients. These produce not only higher yields, but also enhanced food quality.

Agribusiness also makes a considerable contribution to the sustainability of food supply, to the preservation of biodiversity (by reducing the arable land mass required, intensive agriculture makes more land available for protected areas), and to the protection of economically viable farming, not least for small farmers in developing countries. More and greater challenges lie ahead in respect of climate change and water scarcity.

The scale and quality of this contribution underpin the rationale for an economically viable agribusiness. However, there are various and persistent political and economic elements that interact to limit free trade and restrict markets. These have the capacity to interfere substantially with the drive for sustainable agriculture in those areas best able to support it. For a successful agricultural economy, incorporating a strong export potential and stable domestic prices, politics and public perception have to be broadly supportive of the agricultural sector and the technologies that support its viability, particularly in developing countries. Without this broad support, the scope and justification for the protection of agricultural production is much reduced.

Agribusiness has developed over more than 60 years into a powerful engine and guardian of world food supply. Understanding the main developments during this period will help us to assess its strategic requirements in the future.

10.2
Three Eras of Agribusiness

What is now termed agribusiness started in the 1930s with the introduction of the first synthetic crop protection chemicals. Systematic breeding efforts also enabled substantially higher-yield field crops such as wheat, corn, and soya. This "green revolution" largely succeeded in feeding a rapidly growing world population, particularly since 1950. This first era of agribusiness lasted until the mid 1980s, and was mainly driven by innovation. New blockbuster chemicals and improved crop varieties were introduced every year. Overall crop yield grew between 1.5 and two percent per annum. Major industrialized countries in North America and Europe had begun to subsidize agricultural production, first to secure food supply, and then in order to capture export markets. As a result, agribusiness developed into a powerful global industry with revenues of around USD 30 billion (1980), high profit margins, and growth rates that were consistently around ten percent per annum (Fig. 10.4).

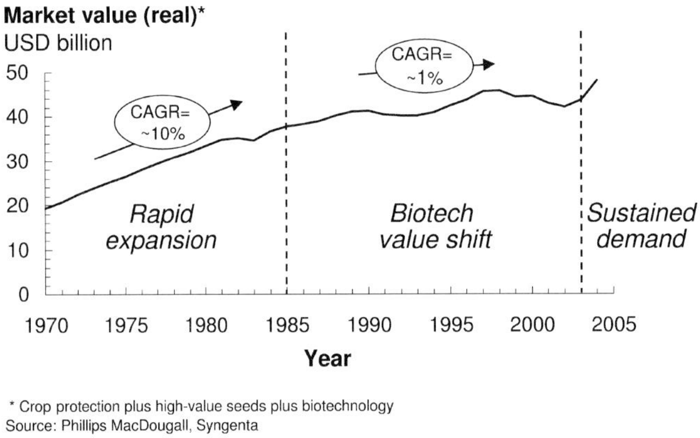

Fig. 10.4 Three eras of agribusiness.

The second era of agribusiness lasted from the mid 1980s to around 2002. Throughout this period, global growth of agricultural demand continued at a slower rate. The regulatory environment became tougher and subsidies were reduced as a result of agricultural reforms in the USA and the EU in the late 1980s and early 1990s respectively. This also drove an ever-growing demand for new chemistry and technologies.

Agribusiness grew by around two percent per year throughout the 1980s to around USD 42 billion in the mid 1990s, when biotechnology solutions to plant protection problems were first introduced. These included field crops resistant to specific non-selective herbicides or harmful insects. They offered substantial benefits to growers in cost and convenience of application. The rapid acceptance of so-called "biotech input traits" by growers, particularly in the Americas, resulted in a significant overall value transfer from chemicals to biotechnology (Fig. 10.5).

During this period, further biotechnology penetration and value competition led to an overall decline in the market value for crop protection chemicals of around USD six billion, only partially restored by biotechnology.

Towards the end of the second era, major industry restructuring was therefore initiated. The "life science concept" of integrated biotechnology platforms supporting pharmaceuticals and agribusiness led to a separation of the two industries and to a concentration of six major companies representing around 80 percent of global crop protection chemical revenues in 2004 (see Fig. 10.6). Three of these companies, Syngenta, Monsanto, and DuPont, are also the three leading seeds suppliers and, in order to drive innovation across chemicals, seeds, and biotechnology, they spend around ten percent of their revenues on R&D. Through major cost restructuring following this concentration, profit margins have improved with EBITDA margins in crop protection chemicals typically at or higher than 20 percent.

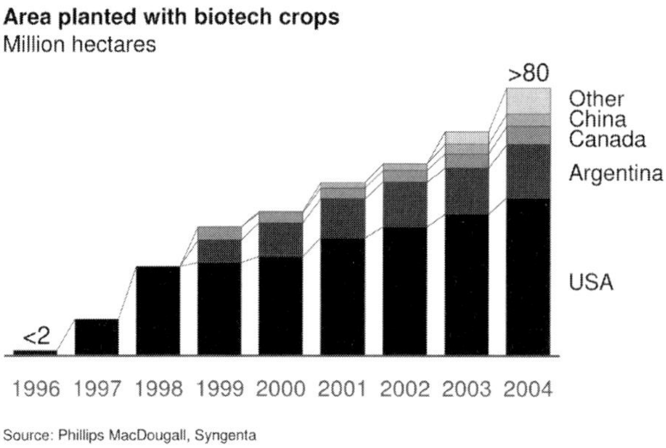

Fig. 10.5 Development of biotech crop acreage.

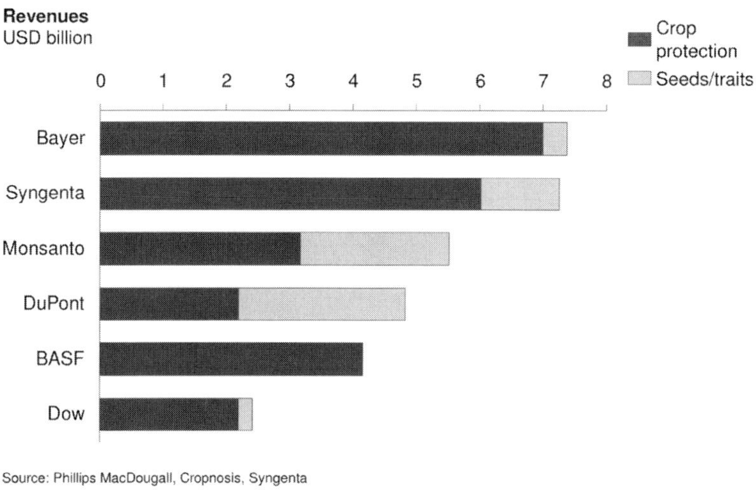

Fig. 10.6 Major industry players in agribusiness 2004.

Agribusiness entered a third era from around 2003 onwards. Demand remained strong, due to crop acreage increases and the need to combat the emergence of new pests and diseases. Although the industry is not expected to return to the very high growth rates of its first era, the market is nevertheless forecast to grow by about two percent per year in the future. In this more constrained industry environment, major protagonists are likely to have to start concentrating their efforts and resources on one or more of three basic strategies:

- They can target the ongoing demand for modern crop protection compounds by introducing new products with new modes of action and higher efficacy.
- They can innovate in crop seeds with new biotech traits.
- They can opt for commoditization and further develop or adopt the generic model by reducing investment in R&D and marketing.

Only the leading companies with a complete agribusiness portfolio, global reach, and local know-how, such as Syngenta, can combine the advantages of more than one option into integrated agribusiness offers to deliver improving benefits to the world's growers. A good example of this is in the US field crops business, where only those companies with a complete agribusiness portfolio can offer maximum choice to growers in both the biotech and conventional markets.

10.3 Driving Plant Yield

In the future, and whichever strategic model they adopt, successful agribusiness companies will continue to deploy their strengths across a broad portfolio to increase plant yield, which can only be done in three ways: plant protection; increasing yields through improved seed quality; and enhancing yield and quality in ways targeted to consumer value.

The application of crop protection chemicals continues to be vital to protect plant yields. This means constant innovation to create and commercialize effective new compounds. Looking at the industry's overall pipelines in crop protection and the strongly innovation-driven areas of seeds or biotech, it is clear that companies are challenged to produce a constant stream of new products due to the limited duration of patents, with long registration periods leaving only ten years or so of protection on the market. In addition, not only do the existing plant diseases and pests affect crop growth in every planting season and require systematic treatment, but Mother Nature is also adept at developing new resistances to chemical compounds in weeds and at spreading new diseases and pests that destroy entire harvests, requiring new and effective treatments. For example, Asian rust in soybeans infested most of Brazil's acreage in only four years, becoming endemic in the 2003/04 season. Spending on fungicides for the disease now exceeds USD 600 million per annum with further substantial potential demand, saving growers a much larger sum in soybean yield and representing an increase in the fungicide market there of more than ten percent (Fig. 10.7). Asian rust has now appeared in other large soybean markets, notably the USA. Finally, increasing public and legislative demand for lower tolerances of crop protection chemicals in feed and food products is leading to the deregistration of old compounds, making the continuous development of modern crop protection chemicals mandatory.

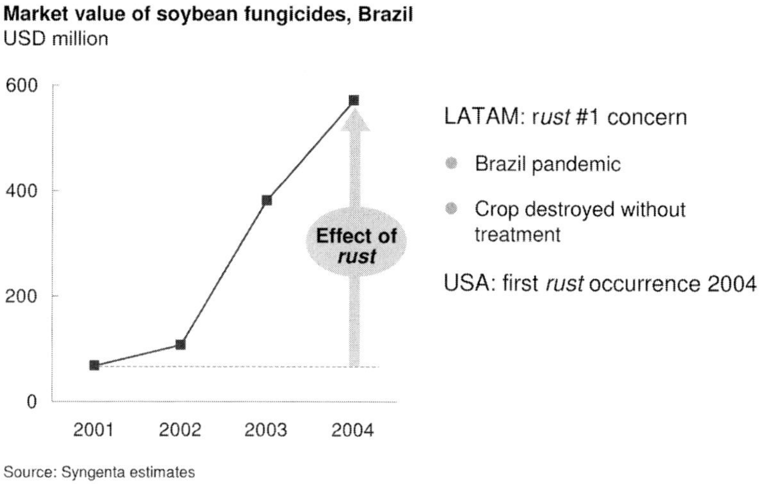

Fig. 10.7 Challenges from new diseases: Asian rust in soybeans.

In vegetables and fruit, modern crop protection chemicals support safe and high quality produce superior to that of the fashionable "organic produce" market because it is free of fungal toxins or insect damage. This has made possible year-round availability of fresh vegetables and fruit at ever lower cost, particularly benefiting low income families. "Organic farming" achieves substantially lower yields because of its inefficient use of arable land and the inherent stress on the unprotected plant.

Substantial yield increases can also be achieved by exploiting the full genetic potential of seeds, whether by modern breeding technologies or genetic modification. Hundreds of new seed varieties with improved characteristics are introduced every year. Over the past 30 years, the yield of major field crops such as corn, soya, wheat, and rice has increased by 60 to 80 percent and further increases are forecasted. Appropriate treatment of those seeds with crop protection chemicals before planting has proven to increase plant vigor overall.

Improved crops with additional consumer appeal also provide new market opportunities, largely in developed countries, although market size is difficult to estimate. Companies engaged in biotechnology have significant pipelines of field crops with increased protein or improved fatty acid content. Unique vegetables and fruit with characteristics such as better taste, texture, and color and also longer shelf-life are being developed and marketed. For example, retail sales of

Syngenta's personal size watermelon quintupled in 2004 alone to USD 35 million.

10.4 Opportunities beyond Yield

Innovation in agribusiness is not restricted to fulfilling agricultural demand. In a global market of around USD five billion in sales, the larger companies, including Syngenta, are using their existing broad portfolios of active ingredients as a springboard to develop markets in various professional segments such as turf, e.g., for golf courses or public parks; the protection of diverse materials such as wood, plastic, paints; professional pest management, e.g., against termites; or multiple applications in home and garden. These consumer-driven markets, typically developing at rates faster than GDP growth, provide additional growth opportunities beyond the traditional areas, in which a growth rate of around two percent would be insufficient in the longer term. Depending on the sub-segment, the agribusiness companies together have a market share in these areas of between 15 percent (in materials protection) and 70 percent (in turf).

Eradication of food shortages to alleviate starvation in developing countries and the consistent supply of high quality, safe food, produced sustainably, remain among the most fundamental challenges for mankind. Now that 70 percent of the world's total research in agriculture is in the private sector (as opposed to 30 percent in 1970), agribusiness provides the means to address these challenges with an impressive track record in innovation, global reach, and local know-how and presence to enable rapid progress. Assuming the appropriate political framework allows agribusiness companies to exploit their strengths, this industry will not only make a crucial contribution to resolving these challenges but will also continue to grow profitably in the future.

10.5 Summary

Sustained agricultural demand will secure the profitable future growth of the agribusiness, which plays a fundamental role in allowing growers to substantially increase crop yield and thus in securing the world's food supply.

- The growing world population and a constant area of arable land call for additional growth in agricultural production. Agribusiness has an important contribution to make here both now and in the future, through crop protection chemicals and also improved seeds which improve yields and enhance food quality.
- The authors define three eras of development in agribusiness: the era of rapid expansion from the 1930s to the mid 1980s, the peri-

od from the mid 1980s to 2002 in which slower growth and the biotech value shift drove industry consolidation, and a new era of steady growth from around 2003 onwards.
- In this new era, three basic strategies can be pursued: the introduction of new and more efficient products, innovation in crop seeds with new biotech traits, or commoditization. Only the leading companies will be able to pursue more than one option.
- Whichever strategic model they adopt, successful agribusiness players will continue to deploy their strengths to increase plant yield through plant protection applications, improved seed quality, and approaches to targeting consumer value.
- In addition to this traditional focus, the larger agribusiness players are now using their portfolios of active ingredients to enter other, faster-growing professional and consumer-driven segments.

11
Industrial Gases – Growth by Continued Self-Renewal

Aldo Belloni and Lennart Selander

Space satellites? Beautifully polished glass? Golden-yellow, ripe bananas? Clean lakes? Few people would connect these with industrial gases. Yet all over the world, every hour of the day, industrial gases are in use in a huge array of applications: welding, freezing, powering, heating, industrial cleaning, providing respiratory relief, testing.

The industrial gases industry dates back more than 100 years. Over this period it has grown – by continuous innovation and, at times, reinvention of itself – to become a USD 40 billion/year business. Its future offers attractive growth opportunities, too, in the form of regional expansion into emerging markets and into new applications in markets like healthcare and electronics. Realizing these opportunities will require new skills and new business models that will eventually change the face of the industry once again.

11.1
'The Invisible Industry': Stability and Profitability

The market for industrial gases includes air gases, synthetic gases, and specialty gases, but excludes hydrocarbon gases such as natural gas. The main products by volume are nitrogen (44%), oxygen (39%), hydrogen (11%), carbon dioxide (4%), plus a variety of smaller-volume and specialty gases (2%) such as xenon, neon, krypton, and helium (Benson, A. et al.).

Worldwide, industrial gases generates annual revenues of approximately USD 40 billion[1], with a geographic spread roughly reflecting the distribution of industrial activity: Europe and North America each account for about one third of the revenues, Asia for another quarter.

The products are used in a wide range of applications. Oxygen can be used in healthcare, or as a combustion agent in, for example, the primary metals industry; nitrogen is used in the production of fertilizers and to create inert atmospheres in chemicals manufacturing and food packaging; hydrogen is used in the refining of crude oil and as a 'green' transportation fuel; argon is used in welding applications.

1) Company reports, CSFB, BCC

Value Creation: Strategies for the Chemical Industry. 2nd Edition. F. Budde, U.-H. Felcht, H. Frankemölle (Eds.)

ISBN: 3-527-31266-8

11.1.1 Industry Characteristics

The industry has a very diverse customer base. Industrial gases are sold to almost all industrial sectors: they are used in metal processing and metallurgy, the chemical industry, in the food-processing industry and in environmental protection, in glass production and electronics, in construction, in rubber and plastics processing, in medicine, and in research and scientific discovery. About one third of the annual revenues is generated in manufacturing industries, up to 15 percent each in healthcare, chemicals, electronics, and primary metal production (Benson, A. et al.). This wide span ensures that the industry does not become dependent on the evolution of any one single sector.

Another characteristic of the industrial gases industry is its high capital intensity: an investment of two and a half dollars is needed to generate one dollar of revenue, with one half of the investment going into production, the other half into distribution assets. Distribution is very complex: products need to be pressurized and/or liquefied for storage and transport, making their handling difficult, dangerous, and expensive. Traditional distribution channels for industrial gases are cylinders (also known as packaged gas), tank trucks (also known as bulk or merchant gas), and pipelines (also known as tonnage business, where several production plants may be connected to a pipeline network serving a number of end users located within the same industrial cluster). Pipelines are the most capital intensive distribution channel, but their scale-up factors are the lowest. Pipeline networks are strategic assets because of the low marginal cost of connecting new customers, the resulting economies of scale, and better flexibility in managing capacities. Contracts for pipeline distribution are usually for ten to 15 years. For customers requiring large quantities of industrial gases, the industry offers customized on-site production solutions.

The industry has high barriers to entry due to its high capital intensity, major technical know-how requirements, strong customer relationships, and long customer contracts. Furthermore, having a strong regional foothold tends to reinforce an incumbent's position because it maximizes distribution and production cost benefits, increasing its ability to defend market share.

11.1.2 Key Players

The current industry structure is that of an oligopoly. The five leading players – Air Liquide, BOC, Praxair, Air Products, and Linde – have a combined market share of 81 percent. All five players have significant activities in more than one region, with Air Liquide and BOC closest to be being truly global[2] (Fig. 11.1).

2) Company reports, CSFB, BCC

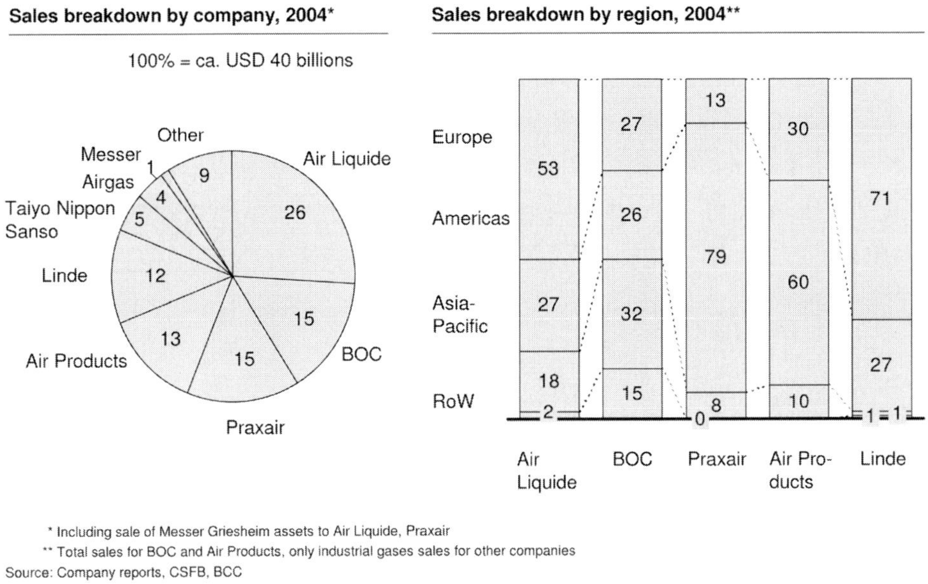

Fig. 11.1 Revenues of major industry players (in percent).

- Air Liquide is the world's largest industrial gases company, with a market share of 26 percent. It derives more than half of its sales from Europe. Air Liquide has been leading the industry in successfully expanding its services offering. Its services now include, amongst others, hospital hygiene, homecare, process control, and energy solutions.
- BOC is the most diverse industrial gases company. Its key markets are the UK, USA, Australia, and South Africa. It is also the leading company in Asia with about one-quarter of the market. Other company activities include semiconductor equipment and distribution services.
- Praxair is the largest industrial gases producer in North and South America. It focuses on the high-growth markets of electronics and health care as well as on service opportunities in metals and food. Praxair Surface Technologies specializes in coatings.
- Air Products and Chemicals is the second largest player in North America. Its main end markets include semiconductors, hydrogen to the oil refining industry, and chemicals. The company has been a pioneer in on-site gas production.
- Linde is the largest gases player in Europe. It has a strong position in the cylinder market, with on-site production – enhanced by strong engineering capabilities – and healthcare gases gaining importance. Linde also has a leading position in material handling.

Taiyo Nippon Sanso, with a strong focus on Japan, holds a five percent market share. The remaining 14 percent of the market is covered by smaller specialized players such as Airgas, a North American cylinder company, and regional companies such as Messer Group.

Each of the five top players has a unique profile, depending on their regional strongholds, their choice of customer industries and applications, and their innovation focus. The large players also hold strong positions in specific areas to reap the benefits of critical mass and focus.

11.1.3 Financial Performance

The industrial gases industry routinely manages to earn attractive return rates. From 1992 to 2004, the average annual return on invested capital (ROIC) before tax was 13.6 percent. The industry's profitability was thus between that of specialty chemicals (17.8%) and commodities (12%), but more stable. In 2004, industrial gases almost came out on top of specialties, due to the latter's declining profitability in the second half of the '90s (Fig. 11.2).[3] Air, its main raw material, is "free", and its products are probably the purest commodities from a chemical point of view. In any case, the industrial gases industry has managed to deliver better and less volatile returns than the commodity chemicals sector.

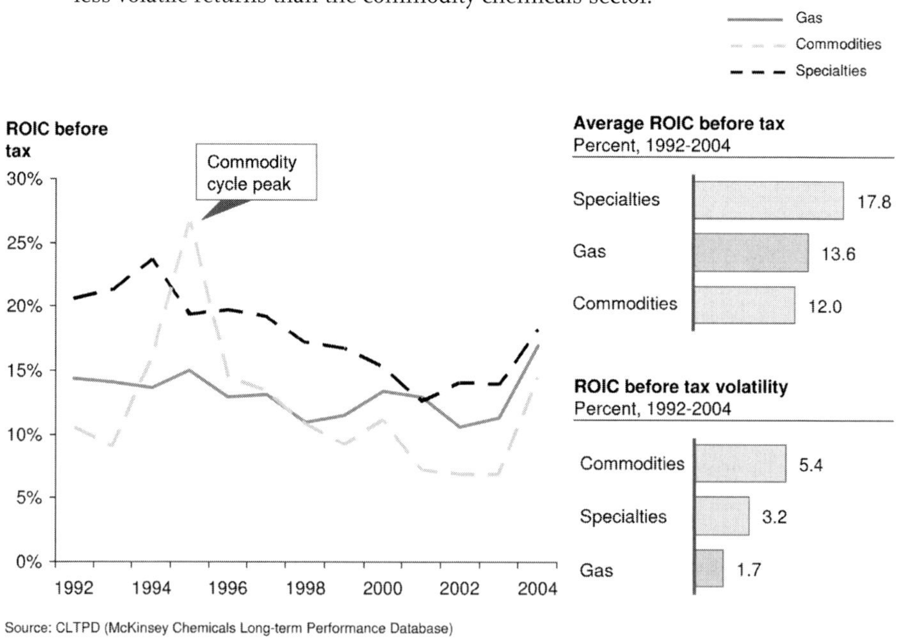

Fig. 11.2 Profitability by sector, 1992–2004.

3) McKinsey Chemicals Long-term Performance Database

From 1992 to 2004, sales in the industrial gases industry increased on average by seven percent p.a. The level of growth has been consistently higher than in specialties (4.9%) and commodities (4.1%).[4)]

Since gas players have invested cautiously despite continuous growth, they have managed to increase their capital productivity and prevent value leakage during the last five years. Compared with specialty chemicals companies, industrial gases players have shown a relatively stable market-to-book ratio over the last ten years. Current valuation levels are driven by a relatively stable operating performance and substantial growth expectations.

11.2 One Hundred Years of Success: the Winning Ingredients

Industrial gases has been a growth industry ever since its beginnings. Keys to success were (and still are) technical innovations as well as an extraordinary ability to adapt to ever-changing market requirements – even reinventing the business in times of market discontinuities.

The industry got its start in 1895 when Dr. Carl von Linde, a pioneer in refrigeration technology, received the first patent for the liquefaction and rectification of air into its major components: oxygen and nitrogen (Almqvist, E., IOMA). Georges Claude soon developed this work further in Paris, France, improving the efficiency of the liquefaction process.

In 1901, Linde AG decided to construct its first plant in Munich, Germany to sell pure oxygen locally. Air Liquide built its first production plant for industrial gases near Paris in June 1903 and a production plant for oxy-acetylene torches in 1905. In Sweden, AGA, which in the 1990s became part of Linde AG, started as a small acetylene filling station in Stockholm, expanding into acetylene lighting for railroads and then for lighthouses. Griesheim got into the production of oxygen when the development of welding and cutting led to increased demand for this air gas.

In its early decades, the industry saw the formation of many companies, often family owned, serving the growing welding market. Restrictions on plant capacity as well as distribution and storage capabilities led to a fragmented industry structure, with players mainly operating on a local or regional level. Nevertheless, Linde and Air Liquide very soon established wholly or partially owned subsidiaries in other countries. The first North American company to commercialize cryogenically separated oxygen was the Linde Air Products Company, founded in 1907. That company later became Praxair.

From the outset, major improvements in technology and distribution advanced the industry. The invention of the double column in 1907, for instance, made the simultaneous production of pure oxygen and pure nitrogen possible. Underwater welding and cutting was introduced in 1924. With time, the availability of argon and helium led to further advances in cutting and welding technology. Liquid

4) McKinsey Chemicals Long-term Performance Database

transport was invented, although it took until the 1950s before the interplay between further advances in this field and the development of applications and customers led to its strong growth. Other inventions reduced the costs of producing oxygen in large quantities, making large-scale application possible. As air separation plants increased in size and the associated costs of oxygen, nitrogen, and argon decreased, the flow of process improvements and inventions continued, allowing further diversification into other customer industries.

After Word War II, the industrial gases industry kept showing impressive growth rates, profiting from the surge in industrial production worldwide. The industrial gases market was a producer's market, with production capacities practically the only constraint on growth. Smaller, local companies were increasingly squeezed out of the market by bigger players that enjoyed the benefits of economies of scale and the abundant investment capital needed for the construction of ever larger plants and pipeline networks.

In the '70s, however, the market underwent a significant change. As industrial growth slumped due to the global economic crisis, the gas industry was forced to change gears – and did so by reconsidering and ultimately reinventing its business: since cutting and welding activities alone would not be enough to secure sustainable growth, the big players started to systematically explore ways in which other industries might possibly benefit from gaseous products and applications. Customer value became the guiding principle – the industry moved from selling established products to developing new business by better understanding customers' specific needs.

Growth after the 1970s was thus driven by the development of new applications not only in metallurgy, but also in new markets, such as food or healthcare. The primary drivers of innovative applications were productivity, quality, and environmental improvements. The big players benefited the most – besides economies of scale in production and distribution, they had the backing of scale in applications know-how.

This new focus on the customer was reflected by new organizational structures: Air Liquide was the first to organize by customer segments, other companies followed. Many companies also started hiring people with experience in the targeted industries, complementing their knowledge of how to produce, handle, and transport gases with a thorough understanding of their new customers.

Volume growth characterized the 1980s. New markets and customer segments developed, e.g., gas-shielded welding became established in automobile manufacturing by the mid 1980s, significantly increasing the demand for argon. The increasing demand for high-purity gases in the semiconductor and fiber optic cable industries also proved economically important.

The 1990s saw yet another change in the market: with commoditization increasing and the purchasing power of customers growing, price pressure became more extreme. At the same time, growth rates from innovation on the applications side slowed. The days of relatively easy wins were gone. Above all, customer industries globalized, encouraging industrial gases players to expand into new regions themselves, and resulting in overinvestment into new capacity. The industry responded with consolidation – in the late 1990s, mergers and acquisitions opened up new growth opportunities: in 1999, Linde acquired AGA, becoming the world's fourth-largest gas supplier. In the same year, an offer was made by Air Liquide and Air Products for BOC, though the bid finally failed. In 2004, Air Liquide acquired Messer Griesheim's industrial gases business in Germany, the UK, and the USA, following up with some divestments to comply with competition laws. BOC and Air Products sold parts of their cylinder business to Airgas in the USA.

With the growth and synergy potential offered by major M&As now largely exhausted, the industry once again needs to look for new ways to generate profitable growth.

11.3 Fit for the Future: Ready to Meet the Challenges

As history has shown, the industrial gases industry has managed to thrive on market discontinuities. Of today's five major players, three belonged to the founding fathers of the industry. Inventive spirit and flexibility will remain the key ingredients of success – helping the industry master future challenges as well. Stock markets are keenly anticipating developments: on average, 20 percent of the industry's valuation is based on expectations of continued earnings growth.[5] With some players that figure is even higher. To maintain its position, the industry has to deliver. The indications are that it can.

11.3.1 Continued Operational Efficiency Improvements

During the last economic downturn, all industrial gases companies carried out large cost reduction projects. The pressure to improve operational efficiency has not lessened. Like the rest of the chemical industry, the industrial gases segment continues to suffer from a price-cost squeeze: while customer industries, especially in the more transactional bulk markets, demand lower prices, factor costs keep increasing. The biggest cost reduction levers lie in production and logistics and distribution (Fig. 11.3).

In production, rising energy costs are one of the major challenges. As energy represents a significant proportion of product costs, energy efficiency has always been paramount for industrial gases companies.[6] Over the last 30 years, the energy required to liquefy a ton of product has decreased by 40 percent.[7] Energy efficiency improvements have to be continued. In addition, players are trying to leverage scale by cooperating to build even larger plants to drive down the unit cost especially of liquid products. As companies approach theoretical limits in production and clear tradeoffs between plant size and distribution economics, further

5) McKinsey Chemicals Long-term Performance Database, broker's reports
6) Air separation is one of the most energy intensive industrial processes. A medium-sized air separation unit can use as much electrical energy as a large airport complex. Source: EIGA 2002
7) EIGA 2002

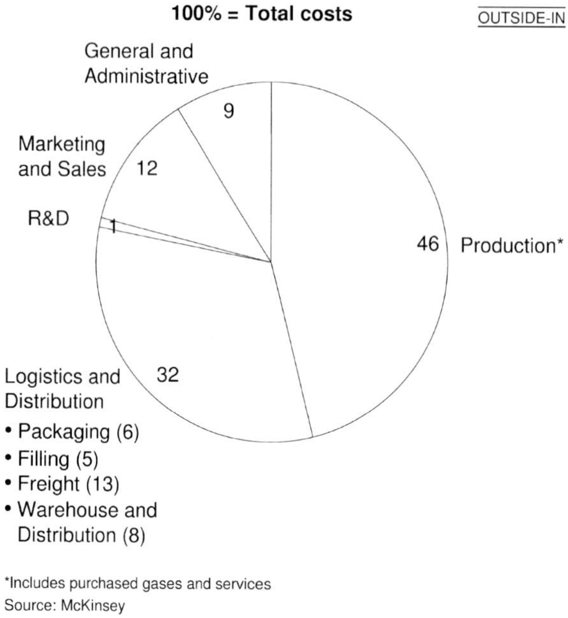

Fig. 11.3 Typical cost structure of industrial gases (in percent).

cost improvements will depend on breakthroughs in plant design, engineering, and operations. Whether new production technologies will again shake up the industry, as non-cryogenic separation did in the 1980s, remains to be seen.

In logistics and distribution, industrial gases players, similarly to logistics companies, continue to utilize technological advances in scheduling and planning to optimize their distribution logistics. In addition, they are increasingly trying out new sales models. In Europe and South America, for example, franchise systems are being tested in the retail segment, with the franchise outlet typically selling gas cylinders as well as other hardware supplies that small customers such as plumbers might need. The industrial gases company maintains control over the point of sale, while at the same time benefiting from the service orientation and higher entrepreneurial agility of local franchisees. Developing such new sales models will remain a major task in the future.

11.3.2
Managing Capital Investments

Another major challenge lies in carefully managing capital productivity, and especially rightsizing investments. The industry as a whole must maintain its discipline and avoid establishing overcapacities that would in the long run hurt all players involved.

On a company level, the evaluation of where and how to invest is becoming more and more challenging: while the construction of new plants is very capital-intensive and investments require long-term planning (assets are depreciated over 15 years but usually operated for 30 or more), markets and industries keep changing with increasing speed. Who can tell exactly how a customer industry and its demand for gases will develop over the next 20 years? Will a specific customer still be around in 2025? Where will global customers invest in the next couple of decades? Are investments into an isolated new plant or into a cluster more feasible? And where can joint ventures be an attractive option?

More than ever before, an industrial gases company needs to learn to develop a competitive investment strategy under conditions of uncertainty. It must identify these uncertainties, assess future market scenarios and their implications, and explicitly consider the roles it might play in influencing these scenarios. The company then must create a strategic portfolio that involves a combination of actions that, collectively, are robust enough to respond to the uncertainty in the marketplace and actively manage this portfolio as uncertainties resolve themselves and the marketplace evolves.

11.3.3 Maintaining Growth

While internal improvement efforts such as cost reductions need to be continued, the main challenge of the industry will be to maintain the stable growth in volumes it has exhibited in the past. With the importance of the old welding and cutting customers slowly eroding and many of the traditional applications maturing, the industry must be ready to look for alternative growth options.

11.3.3.1 Expanding into Emerging Markets

Emerging markets are an attractive opportunity for growth. The main regions and countries of interest to the large industrial gases players are Asia, in particular China, the other BRIC (Brazil, Russia, India, China) countries, and Eastern Europe. China alone, for example, has a market volume of USD 1,200 million, India of 430 million (2003) (Economist Intelligence Unit). In both countries, consumption will show double-digit growth rates over the next couple of years.

The industrial base in these emerging countries has reached a critical threshold. Gases players are following their customers to these areas, particularly customers in the chemical, steel, electronics, and other manufacturing industries. Western players have some competitive advantages over the local ones. They can leverage critical technology know-how and the capital for the necessary investments. Since regional monopolies are likely to emerge in these countries, too, first movers among the major players will secure a sustainable competitive advantage.

Entry strategies for the individual geographies, however, need to be differentiated, since logistics, distribution, and market development largely depend on local conditions. In China, for example, the on-site business model is the key to seizing growth opportunities, as the cylinder business will be difficult to develop profitably due to the absence of a rental fee model and the dominance of local players. Two strategic thrusts arise from this: one is to engage in developing the on-site business by building new plants. The other is to convince captive users to outsource their existing plants to the industrial gases players for them to operate and upgrade. This market entry strategy is quite different from the path followed in Eastern Europe, where gases players started with cylinder gas and moved up through liquid bulk deliveries and small on-site plants to large on-site business.

A thorough understanding of the individual local markets is a prerequisite for minimizing investment risks: due to the high capital intensity of the industry, any decision to invest in an emerging country will automatically have long-term consequences.

11.3.3.2 **Developing New Business Models**

Moving into new countries and regions, however, is not the only strategy promising attractive returns. Industrial gases companies ready to explore new business models in existing as well as in new businesses will find ample opportunities for additional growth.

On-site production is one such opportunity. It has been a strong driver of growth for several years and will continue to be so in the near future. On-site production is a clear departure from the classic business model of a gases player. Instead of merely supplying gas, the gases company supplies a "solution": it not only develops an optimum supply concept for an industrial customer, but also builds, installs, operates, and maintains the on-site gas supply systems.

Though advanced applications development in established industries will remain the most important growth driver in the gases industry, even here a fundamental shift is on the horizon: the more sophisticated application solutions become, the more cost and know-how-intensive they are. Needing to recover their investments, industrial gases companies will push to develop more patent-protected applications, making customers pay, for example, for obtaining licenses instead of recovering their investment purely through volume growth. The future will see an increased unbundling of physical products and technology.

Ultimately, however, success in established industries and markets will not be enough. In the face of high growth expectations but maturing markets, the industry once again needs to be bold and expand its reach beyond its traditional core competences: attractive markets such as healthcare or electronics offer interesting opportunities for those companies ready to develop new skills and new business models. Take healthcare, for example: medical applications for gases – in hospitals and in homecare – have seen double-digit growth in recent years and will continue to grow in the next couple of years. All large players already do business in this area, developing the specific skills needed, such as successfully dealing with governments and insurance companies. Linde's INOmax™ (inhaled nitric oxide) product, for example – indicated for the treatment of newborns with hypoxic respiratory failure – is the first gas mixture that the respective American and European authorities have approved for use as a drug. Several of the big medical gases players have become healthcare providers. Linde, already serving normal hospitals within its business portfolio, recently acquired a hospital/care center in the field of high complexity care for ventilation patients. Praxair is selling kitchen, disinfection, and food services to hospitals in South America.

In the growing homecare applications market, gases companies are also taking market share. The way into this fragmented market is often to acquire family-owned homecare companies that have the core competences of running what is essentially a consumer business. While most of these companies already have a focus on respiratory care, they also sell other healthcare products such as wheelchairs, beds, and infusion equipment. Consolidating these companies offers attractive scale advantages and helps achieve operational excellence. Some of the gases players supply not only oxygen and specialized equipment, but also emergency services, administration management, patient and family training – all in close collaboration with medical teams and reimbursement institutions. Not surprisingly, gases companies typically set up homecare as a separate corporate unit, reflecting the specific entrepreneurial attitude needed in this kind of business.

Electronic gases can serve as another example of new business models. Producers of semiconductors and liquid crystal displays are increasingly outsourcing gases and the supply of chemical auxiliaries. Industrial gases companies take on the role of facility managers, becoming responsible, for example, for supplying wet end chemicals, steam, and energy, and for maintaining the auxiliary infrastructure (e.g., piping and wiring). Gases companies are often the natural owners of these functions: they have transportation, shipping and handling know-how and can use economies of scope by offering one-stop shopping solutions. Value-adding services already common in electronics are increasingly diffusing into other customer industries, e.g., full cylinder management at manufacturing and chemical sites. Other additional services that are beyond the piloting phase are web/IT applications that make remote inventory management possible and allow for vendor managed inventories.

A recent example of what it can take to explore new business opportunities is dry cleaning. Based on its core competence CO_2, Linde developed a technology for dry cleaning in liquid-based carbon dioxide. CO_2 cleans much faster than the traditional method, and eliminates the strong solvents that are currently used. The global climate is not affected because the carbon dioxide is recycled. Since established dry cleaners were reluctant to sign on, Linde opted for an alternative business model: to penetrate the dry cleaning market, the company developed a franchising concept and opened its first European shops in 2004.

Last but not least, industrial gases companies are keeping a close eye on the potential of hydrogen-based fuel. When hydrogen is burnt, the only by-product is water, making it the cleanest fuel available. If hydrogen is adopted as a fuel, the potential growth opportunities are enormous. To give just one figure: if road transportation in Germany were converted to hydrogen-based fuel systems, the demand for hydrogen would be roughly ten times the current entire annual production in Germany. Hydrogen is already being used as fuel in space rockets and has been piloted extensively in municipal buses. For it to become a viable everyday fuel in the future, safe and convenient storage formats need to be developed, and significant expenditure will need to be invested into the infrastructure and fuel stations. Though the timing of a conversion to hydrogen-based fuel is still uncertain, the gases industry is ready to take a piece of the pie.

No matter what industry or technology delivers the growth opportunities – the key factor for success will be innovation and flexibility in identifying these opportunities and in turning them into profitable businesses. The next decade is likely to see yet another reinvention of the industry: the different gas players will experiment with new markets and market approaches, testing and crossing the boundaries of their core business. Each company will develop specific spikes, depending on its know-how, its technology profile, and the people who work for it. After more than 100 years, the success story of the industry looks set to continue.

11.4 Summary

- Industrial gases has been a growth industry ever since its beginnings, successfully managing market discontinuities.
- The industry routinely earns attractive returns, with profitability rates between those of specialty chemicals and commodities.
- On average, 20 percent of the industry's valuation is based on expectations of continued earnings growth.
- Companies must continue cost reduction efforts in production and distribution and – given the high capital intensity of the industry – develop competitive investment strategies.
- More importantly, companies must look for new ways to continue volume growth, since the traditional customer base is slowly eroding and many applications are maturing.
- Emerging markets such as China and Eastern Europe offer an attractive growth perspective.
- Additional growth opportunities can be seized by reaching beyond core competences and developing new business models in attractive markets such as healthcare and electronics.
- The key factors for success will be innovation and flexibility in identifying growth opportunities and turning them into profitable business models.
- In the next decade, the major gases players will each experiment with different new markets and market approaches – eventually changing the make-up of the industry once again.

References

Almqvist, E. History of Industrial Gases, 2003
IOMA, The Invisible Industry, 1997
Benson, A., Blair, P., Burns, I., Heslop, M. and Jacobs, A., Schroder Salomon SmithBarney, Report 25 January 2002
Economist Intelligence Unit Forecast Statistics 2003

12
Perspectives of Chemical Distributors as Partners of Industry

Klaus Engel and Gabriele Roolfs-Broihan

Focusing on the complex area of the market where chemicals are required at short notice and in fairly small quantities, chemical distributors have developed over time into important players in the value chain from producers to consumers in a broad role, serving as supply chain managers offering a wide range of logistics and value-added services.

This mid-sized industry, currently addressing a total world market of roughly EUR 70 billion, is estimated to show a healthy two to four percent growth on an annual average for the foreseeable future.

Three interrelated trends are currently shaping the industry. Chemical and chemical-consuming manufacturing companies are going global to tap into new markets and low-cost production and, at the same time, are facing the added complexity of meeting ever-higher QHSE (quality, health, safety, and environment) standards. Furthermore, the need for efficiency is driving them towards a focus on their core business, resulting in increased outsourcing of warehouse logistics, distribution, and related value-added services. These trends offer major opportunities for value creation and competitive differentiation, especially to the multinational and leading players in the still-fragmented distribution industry.

Distributors with sufficient financial resources, geographic scope and/or know-how could benefit from these three trends and offer both chemical producers and customers significant value by handling outsourced business and services in a cost-efficient and market-effective way.

To do this, they have to meet all the key success factors for traditional efficient service in local markets while at the same time leaping to the next industry S-curve and becoming strategic partners for the chemical industry and for chemical-consuming customers.

Value Creation: Strategies for the Chemical Industry. 2nd Edition. F. Budde, U.-H. Felcht, H. Frankemölle (Eds.)

ISBN: 3-527-31266-8

12.1 From Wholesaling to Supply Chain Management: the Evolution of Chemical Distribution

Historically, the main task of the chemical distributor was wholesaling, buying and selling chemicals in that part of the market that requires quick responses to delivery requests and smaller volumes, where it was uneconomical for the large producers to handle the business. Along with this came transportation, storage, and the preparation of chemicals for further distribution by filling, packaging, and labeling inventory units.

Over the course of time, however, leading distributors like Brenntag developed more and more towards the concept of supply chain management. Today, chemical distributors play a number of essential roles in the value chain between chemical producers and consumers (Fig. 12.1).

The annual global consumption of chemicals, including consumer products and pharmaceuticals, exceeds EUR 1.7 trillion (American Chemistry Council), the vast majority of which is sold directly by chemical producers to the chemical-consuming manufacturers. Roughly ten percent, or EUR 180 billion, of this total is consumed at short-notice request and in smaller quantities, representing the market generally available to chemical distribution. Depending on the region and the specific industry structure, an estimated 30 to 60 percent – accounting for an annual market size of approximately EUR 70 billion at present – is actually outsourced to third-party chemical distributors, leaving a significant opportunity for them to benefit from further outsourcing developments.

Today's chemical distribution market is still highly fragmented, despite consolidation over the past years. A multitude of regional and local distributors with limited geographic coverage and product offerings command more than 50 percent of the overall market, whereas the top five international players account for less than 20 percent of total sales worldwide. With sales of EUR 4.6 billion in 2004, Brenntag, headquartered in Mülheim/Ruhr, Germany, is the world leader: well ahead of Ashland, Chemcentral, and IMCD, and close to Univar.

Although the chemical products purchased from distributors are not usually one of the largest cost items for the customers, the importance of those chemicals to these customers' overall operations is often very high and purchasing a multiplicity of chemicals is often very complex. As a result, important elements of customer satisfaction are timely and reliable delivery as well as appropriate products and services.

The portfolio of services performed by chemical distributors is an important opportunity for competitive differentiation. Specifically, these services include:

- *Value-added services*: chemical distributors provide a number of value-added services beyond transport and storage to both suppliers and customers, such as single sourcing for customers, customized and exclusive mixing and blending, inventory management, worldwide sourcing, sales and marketing, extended technical support – particularly in the area of specialty chemicals and environmental services – as well as after-sales services like drum return handling.

- *Compliance with environmental and safety standards:* distributors are specifically trained and skilled in the handling, transport, and storage of a large variety of chemicals, many of which are hazardous and virtually all of which require strict compliance with regulatory provisions in multiple jurisdictions.
- *Economic delivery:* distributors can usually deliver chemicals (particularly small quantities) to geographically widespread and remote customer locations more cost-efficiently than the producers themselves. Furthermore, distributors hold significant stock locally and are able to supply small customers efficiently and according to their specific needs, mostly at short notice.

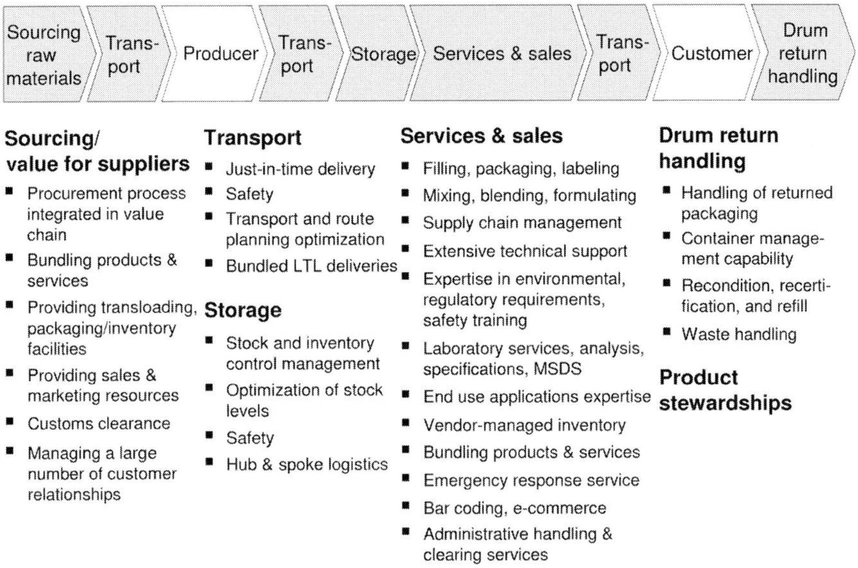

Fig. 12.1 Value chain and performance spectrum of chemical distributors.

Leading distributors offer a broad spectrum of these services. Brenntag, for example, as a full-line distributor of industrial and specialty chemicals, provides a "one-stop shop" of over 25,000 chemical products for more than 150,000 customers, combined with a broad range of related value-added services such as mixing, blending, repackaging, and technical services and support.

12.2 An Industry on the Move – the Major Trends

In the last decades, the chemical industry and its customers have had to cope with a massive expansion of their reach and scale as globalization took effect, coupled with a need to comply with ever more stringent and varied QSHE regulations. To improve operational efficiency in this vast and complex environment, these companies have turned more and more to outsourcing so that they can concentrate on their core competences. The changing market environment offers major opportunities to the distribution companies:

- *Consolidation and globalization:* in the past decades, the chemical industry has experienced simultaneous consolidation and globalization trends (Fig. 12.2). The latest focus of this internationalization is the emerging Asian market, which already accounts for roughly one third of the global market volume (American Chemistry Council). For distributors in particular, the Asian market has gained significant importance from the point of view not only of consumption, but also of sourcing. The globalization of markets is being accompanied by an ongoing rationalization of production sites for cost efficiency. Similar trends and the need to reduce the complexity of procurement also apply to the distributors' customers, the chemical-consuming industries.

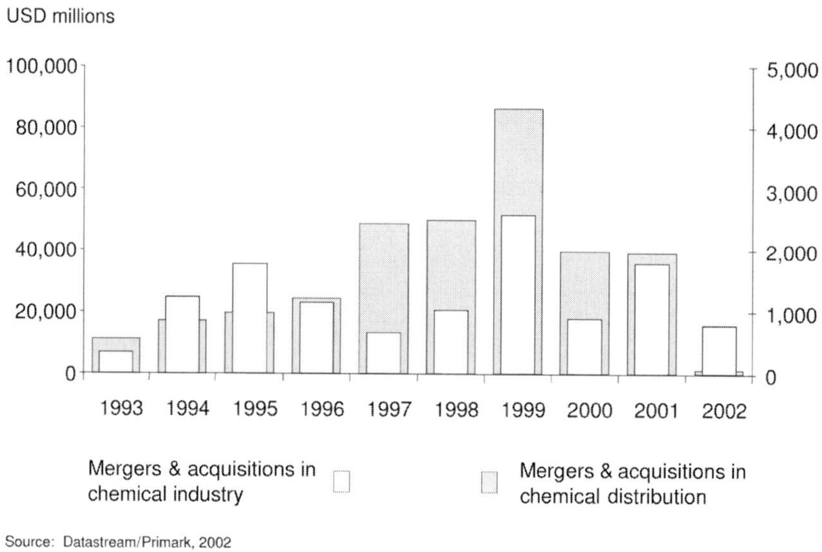

Fig. 12.2 Development of the annual value of merger & acquisition deals, 1993–2002.

- *Rising QSHE requirements:* chemical manufacturers operate within a sensitive environment which is subject to extensive environmental, health, and safety laws and regulations. These vary

slightly from country to country but are, in principle, gaining more and more endorsement from local authorities world-wide. Today the requirements are basically comparable for distributors, with the expectation that laws and regulations imposed on environmental protection and safety in general will become even more stringent.
- *Outsourcing by chemical producers:* as chemical producers faced up to increasing competition over the past decades, they attempted not only to expand their market positions geographically, but also to improve operational efficiency by focusing on their core activities. As a result, suppliers have continuously outsourced sales and logistics functions as well as related value-added services to third-party distributors.

These three trends offer significant opportunities, especially for the large distributors which have sufficient resources to cover the geographical scope and skill requirements, and to create new win-win business relationships with suppliers and customers.

12.3 Perspectives Within Changing Environments

The industry trends discussed above have led to healthy average worldwide growth rates in chemical distribution of up to four percent per annum. Not every player, however, is equally ready to take advantage of the changing market conditions and capable of doing so. To ensure future value creation, chemical distributors have to follow specific strategies, striving for excellence as strategic business partners by delivering on the traditional local success factors and, building on these, the broader-based "partnership-specific" ones as well. This is the key to sustainable competitive advantage in the new, relationship-based environment.

Excellence in the local factors for success forms the foundation of the entire business, and is almost a prerequisite for being in it. Because it is often difficult, costly, and potentially hazardous to transport chemicals over long distances, chemical distribution remains a fundamentally local business strongly influenced by regional and sub-regional dynamics. Furthermore, distributors need to operate warehouses fairly close to customer sites (generally within a 150–200 kilometer radius) to enable them to make frequent, timely deliveries at short notice, as customers often do not store large supplies of the chemicals in question. Chemical distributors therefore have to strive for local excellence by providing a balanced full-line product and service portfolio with adequate cost structures, an efficient logistics system, and state-of-the-art infrastructure. The most profitable distributors are often those with the best skills in leveraging truck and warehouse assets and optimizing route density. In addition, local success is heavily dependent on decentralized but skilled and specialized sales and purchasing personnel, ensuring strong and long-term relationships with customers and suppliers. This applies especially to high-value specialty chemicals, as they require substantial technical expertise on the applications side.

While local dynamics will always remain vital to the profitability of chemical distributors, strategic partnerships with suppliers and customers will play an increasing role in their performance. To ensure substantial value growth, specific key success factors must be fulfilled to enable distributors to leap to the next industry S-curve as partners of the chemical industry and chemical-consuming manufacturers.

12.3.1
Partners on a Global Scale

Globalization has not only triggered changes in traditional trade patterns for chemicals, but also induced significant changes in the cooperation between chemical producers/customers and distributors. In both Europe and the United States, suppliers and customers have begun to rationalize the number of distributors to which they sell and from which they buy, favoring strategic partnerships rather than opportunistic deals. Large-scale distributors are best positioned to make the investments required to meet customers' and suppliers' ever-increasing expectations. As a result, a tremendous consolidation has taken place in the chemical distribution industry during the past years, and is still continuing steadily (Fig. 12.2).

Besides meeting leading global suppliers' desire to deal with fewer but internationally represented distributors, the main reasons for consolidation are additional growth through enhanced geographical coverage and the required financial strength to make investments in IT, provide state-of-the-art logistics, and meet increasing environmental standards. Because of these requirements, smaller and mid-sized – mostly family-owned – local players are giving up their independence, selling their businesses to proactively and successfully consolidating industry leaders.

Brenntag, for example, has completed more than 20 acquisitions since 1998. Today, the company manages an extensive network with nearly 300 distribution sites covering all major markets in Europe, North America, and Latin America. In addition, Brenntag has enjoyed a presence in China for almost ten years through an exclusive sourcing agent, ensuring access to more than 60,000 tons per annum of specialty chemicals from leading local producers.

By positioning themselves as strategic partners on a global scale, chemical distributors can not only achieve competitive differentiation, but also create higher value for themselves as well as for their suppliers and customers. The key success factor here is a comprehensive, global state-of-the-art distribution, sourcing, and market intelligence network covering all major key markets for chemicals and related services.

By consolidating their relationships to a limited number of exclusive partners, producers are, for example, able to deliver their products to geographically dispersed customers more cost-efficiently. Specific benefits can be realized by increased operating efficiency in the areas of sales, logistics, and administration. Taking into account the general migration of chemical production to low-cost regions like Asia, the number of chemical products exported to the rest of the world is expected to increase continuously. Although China, for example, will remain a net importer of certain basic chemicals for many years to come (see also Chapter 32), it has already gained a clear competitive edge as a producer of many specialty chemicals, and commands more than 50 percent of the world's total production of vitamin C (L. Hepner & Associates). Since many migrated suppliers want to optimize the local logistics and sales support required to serve and sell to smaller customers, the efficiency gain will be even higher.

In addition, a focus on strategic distributors assures a coordinated approach to customers and markets throughout the world, with international distributors being able to ensure an adequate level of value-added services.

Distributors, in turn, will benefit from increasing the utilization of their comprehensive network based on additional business. Furthermore, as distributors' warehouses are generally located close to the point of final delivery, they are able to operate with lower delivery costs and shorter distribution times than the producers.

On the customer side, advantages can be expected from this fairly close proximity, in particular in the case of smaller and mid-sized customers which producers cannot serve at reasonable cost and service levels. Moreover, the "one-stop shop" concept offers significant savings potential through aggregated orders and coordinated purchasing. Finally, larger international distributors will offer their local customers all the benefits of a worldwide state-of-the-art sourcing organization, resulting in product and service offers at a superior price and quality.

12.3.2
Partners Ensuring Compliance with Rising QSHE Standards

As a result of more demanding environmental, health, and safety laws and regulations all over the world, adequate storage, handling, and transport of chemicals will increasingly drive the success of chemical distributors.

For example, once implemented, the European Commission's new chemicals policy REACH (Registration, Evaluation and Authorization of Chemicals) – endorsed in 2003 – would require manufacturers and importers of over one ton per year of any chemical substance to gather information on the properties, hazards, and use of that substance, and submit the data to the European Chemicals Agency. The Agency would then determine if the chemical substance needed to be further regulated, restricted, or banned from use. It is estimated that about 30,000 existing chemicals produced in or imported into Europe would be subject to REACH. Key objectives of the European Commission's initiative on the reformulation of chemicals policy are to raise human and environmental safety levels as well as to simplify and standardize the current legislation.

Although chemical distributors are not likely to experience a significant direct impact from REACH, they may indirectly be affected by REACH's impact on the entire chemical industry, such as cost increases, product availability, and restrictions on global trade.

Distributors could, on the one hand, benefit from opportunities to differentiate themselves as reliable business partners with a strong commitment to these higher standards, but on the other hand they will be forced to invest in new technologies. In many cases, the resources required for compliance – including the development of internal expertise, controls, and procedures – are limiting the growth options for smaller competitors.

Suppliers can also benefit from stronger cooperation with QSHE-experienced distributors in sales and technical support. Many specialty product lines are highly complex, requiring significant investments to train sales personnel adequately. Cooperating with distributors which have sales and technical support personnel locally available to follow up, answer questions, and provide technical expertise helps suppliers to deliver better service and foster long-term customer relationships. Since distributors are able to fund and amortize the investment in training sales people over larger volumes, they can operate at lower cost than the suppliers, particularly in remote emerging markets.

Chemical distributors are also providing customers, particularly the smaller players, with more and more direct service and guidance in the area of safety and environmental issues.

12.3.3
Partners along the Value Chain

In the future, the outsourcing trend based on chemical producers' strategic focus on core businesses and efficiency considerations is expected to continue in line with the increasing geographical coverage and availability of professional distribution services.

This trend has also been accelerated by the fact that producers have, for instance, steadily increased the minimum order size they are willing to ship directly, seeking to provide direct delivery focused on large and strategic customers and to outsource distribution for all others.

In this area in particular, distributors have opportunities for competitive differentiation by working in close cooperation with producers. We see this as the next industry S-curve: to establish win-win situations and create joint value through the efficient assignment of resources along the value chain.

Once more, distributors can benefit here from additional business and a higher degree of asset and personnel utilization. However, the major advantages for the chemical producers are less obvious: outsourcing their business with smaller customers can allow them to save costs, reduce complexity, and achieve additional sales through their distributor's network and closer customer proximity (Fig. 12.3).

The outsourcing of C and D customers, for example, allows producers to save sales, logistics, and administration costs. Furthermore, a focus on their core customers allows them to reduce their cost of complexity significantly and at the same time enhance the profitability of their overall customer portfolio. Further important added value arises from the distributor's in-depth knowledge of regional markets and ability to design tailor-made packages. As a result of their broader market access, distributors are often able to increase the sales of dedicated product lines outsourced by suppliers. New customers may be attracted and switch to products offered via distributors, as they normally cover a broader product portfolio and offer customers the possibility of one-stop shopping.

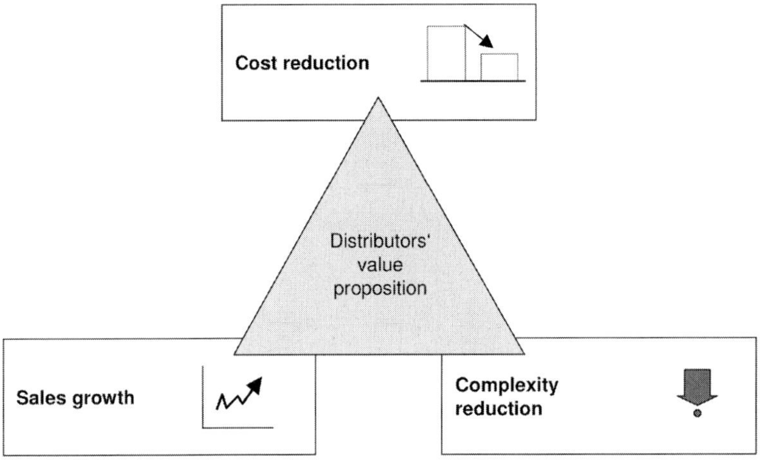

Fig. 12.3 Value proposition of chemical distributors for outsourced business.

12.4
The Future Outlook

As described, fundamental trends in the chemical industry particularly favor leading distributors like Brenntag, which can enter into partnerships with chemical producers on the basis of sufficient geographical coverage and the required financial strength to make investments in environmental standards, as well as adequate technical sales capabilities and services.

The chemical distribution industry has always been subject to ongoing, intensive M&A activities. Recently, private equity investors have also demonstrated an increased interest in distributors (Fermont, M.).

Their motivation is based on several key investment considerations: like the chemical industry, the chemical distribution market is huge, diverse, and shows sustainable long-term growth opportunities for well-positioned market leaders which can proactively capitalize on the consolidation and outsourcing trends. Due to the much lower capital intensity of the business, the earnings of well-performing distributors are by far less volatile over the cycle than those of most chemical producers. Finally, their very high diversification in terms of geography, products, and customer industries limits any downside risks and renders chemical distribution an attractive long-term investment.

Brenntag's success in growing its business in recent years and becoming the global market leader in chemical distribution is fundamentally based on a long-term strategy oriented towards suppliers and customers, as described above. Brenntag sees itself as a leader in identifying solutions when innovative business models such as preferred partnerships are to be realized.

12.5
Summary

- Over the course of time, chemical distributors have developed in the complex small-volume, rapid-response area of chemical distribution into playing an important role in the supply chain from chemical producers to consumers. In addition to fundamental delivery capabilities, distributors offer value-added services, skills in complying with environmental and safety standards, and economy of delivery to remote and small customers.
- Three interrelated trends of globalization among chemical producers and customers, increasing QHSE (quality, health, safety, and environment) standards, and outsourcing by the producers and customers are currently shaping the chemical distribution industry, offering distributors considerable opportunities for competitive differentiation.
- To tap these opportunities and create future value, chemical distributors have to strive for excellence as strategic business partners by delivering on the traditional local success factors and, building on these, the "partnership-specific" ones as well.
- In the changing environment, future value creation for the chemical distributors appears to lie in partnerships of three main kinds:
  - Supplying global distribution services to global chemical producers and global customers
  - Helping chemical industry players to ensure rising QSHE standards in the distribution of their products and providing expertise in this complex area to chemical consumers
  - Helping chemical producers and customers to focus on their core competences by offering more efficient and effective distribution, marketing and sales, or purchasing services.

Distributors which have sufficient resources to cover the geographical scope and skill requirements and to create new win-win business relationships with suppliers and customers are the most likely to benefit from the changing market conditions.

References

American Chemistry Council (ACC), *Guide to the Business of Chemistry*, **2004**

Fermont, M., Changing rules, *European Chemical News*, **2004**, July, Newsletter 13

L. Hepner & Associates, cited according to L. Jarvis, Vitamin C Prices Hit New Lows on Chinese Imports, *Chemical Market Reporter*, **2004**, June 14

13
Systematically Revitalizing Innovation in the Chemical Industry

Birgit König, Gary Farha, and Thomas Weskamp

Over the past century and a half, the chemical industry has changed the face of the world by applying boundless ingenuity and entrepreneurial spirit to solving many of the problems vexing humankind. With highly innovative product ideas from plastics to fertilizers, from textile dyes to explosives and medicines, chemistry transformed itself from alchemy into a value-capturing industry creating products with a high relevance for society. Chemistry as a discipline has continued to be a driver of innovation for many companies in the last decade (Fig. 13.1).

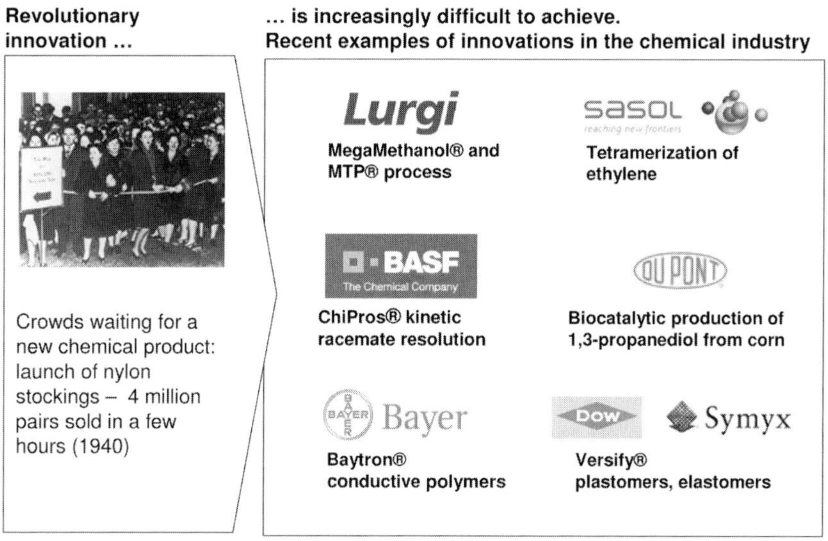

Fig. 13.1 Examples of innovations in the chemical industry.

However, over time the classic chemical industry of breakthroughs and high value seems to have given way to other industries or newcomers that are capturing value in areas very close to the chemical industry's home turf. Many other disciplines now work at the molecular level, clouding the mission of chemical innovation.

Value Creation: Strategies for the Chemical Industry. 2nd Edition. F. Budde, U.-H. Felcht, H. Frankemölle (Eds.)

ISBN: 3-527-31266-8

As a result, the chemical industry is in a situation where, as George M. Whitesides, head of the Chemistry Department at Harvard University, put it: "We think of ourselves as experts in quarrying blocks from granite; we have not thought it our job to build cathedrals from them." Even worse: many of industry's most famous chemical "cathedrals", such as fertilizers, pharmaceuticals, plastics, rubber, and agrochemicals, are also disintegrating more and more rapidly into commoditized blocks of granite, often making core competences to core rigidities for the future. Despite all ongoing efforts in cost reduction, the globalization of companies, and frontline improvements, the chemical industry has to revitalize the drivers of innovation in order to proactively generate sustainable competitive advantage (Fig. 13.2).

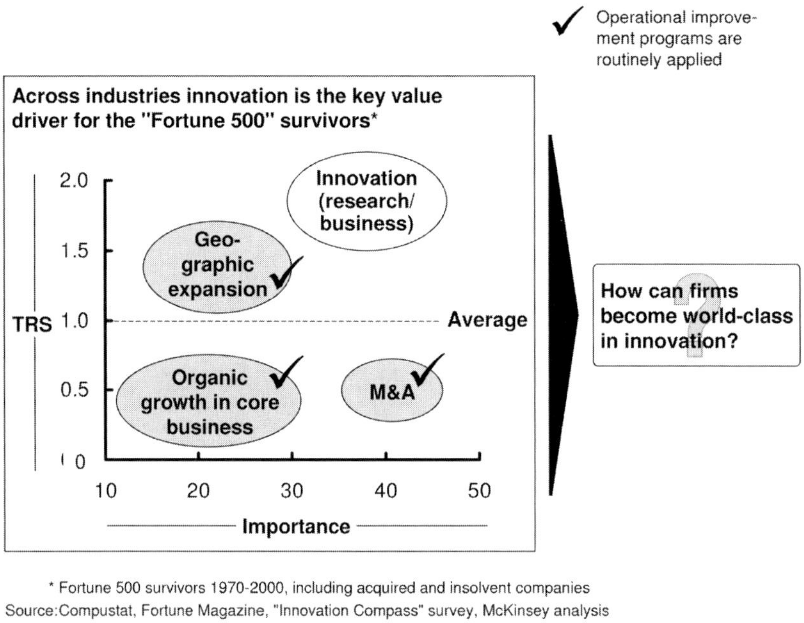

Fig. 13.2 Innovation: addressing an important white spot.

13.1
Drivers of Innovation

What is innovation? Breakthroughs can happen in a variety of dimensions. These often consist of new compounds, applications, or improved technical processes, but new services and new business models can be just as innovative and value generating (Fig. 13.3).

	Examples	
New products	syngenta Amistar	The agrochemical industry – probably more than other areas of the current chemical industry – is still largely driven by product innovations
New processes	DuPont Sorona polymer	Besides process improvements as part of efficiency programs, more and more companies are considering completely new approaches, e.g., biotechnology
New applications	makrolon the high-tech material	New applications for polymers, e.g., replacing other materials such as steel, is a major topic for polymer manufacturers
New services	DuPont Safety Resources	Providing services is a field of increasing importance, often initially forced by a switch from a cost to a profit center mentality
New business models	BASF Coatings AG	The idea of selling solutions rather than just products to achieve a price premium has become a goal for many fine and specialty chemicals companies

Fig. 13.3 Dimensions of innovation – examples from the chemical industry.

In our belief and experience, innovation is driven by a combination of deep knowledge, creative new approaches, and the perseverance to follow through. In this section, we focus on how these three elements can be used as the key motifs for making the revitalization of innovation a more tangible and operational goal.

13.1.1 Knowledge

> *"The strategic direction of industry's research departments changes much too often – usually when the VP changes. Accordingly, it is difficult to acquire a deep knowledge of the relevant topics and scientific questions that are crucial for the company to grow".*
> *Professor Henry N. C. Wong, The Chinese University of Hong Kong*

One indisputable ingredient of innovation is knowledge, tacit and explicit. This is required in two areas: the "state of the art" in science and technology and the jobs the customer would like to get done.

While stressing knowledge as an ingredient of innovation may seem trivial, accumulating it is far from easy. Most companies are organized as a collection of functional silos in which members of various parts of the organization, for example, researchers, rarely see the end customer and accordingly can only develop an abstract understanding of the problem to be tackled. Multiple filters separate the researcher from the real needs of customers or end users, making it virtually impossible to identify the most promising innovation targets. If it is difficult to find the right question, how much harder must it be to find the right answer? When looking for new solutions, most companies are overly reliant on internal expertise, even though most new thinking plainly occurs outside any given company.

We believe that tacit, lively insight into unresolved questions and unmet demand depends on regular, direct interaction between scientists and end customers. When scientists observe customers as they use current products, what they see typically unleashes a broad array of new ideas. It takes a scientifically trained mind and keen observation to see product opportunities in these customer problems. Reading market data collected by sales reps is rarely sufficient for this purpose. Visiting customers or end users and discussing technical challenges with them is certainly better, but real insights are often only created by working together with the customer for a couple of days or even longer to experience real life challenges and problems, an approach that some chemical companies have implemented successfully. To stimulate innovative thinking, researchers naturally will not stop after collecting customer insights. They need to translate customer needs into product features, derive scientific questions, and then resolve them. In our experience, regular interaction between researchers and end customers is one of the most powerful innovation boosters – and it is very easy to organize.

Innovations addressing end users' needs

1) Driving despite a flat tire
Having to change tires on a highway is surely something that many people would like to avoid, and being able to avoid it is probably one of the few things they would even pay for in a tire – a real commodity. Michelin has developed a new tire system – the PAX system – that depends on an internal polyurethane support developed by Dow Chemical in a solution-to-customer oriented partnership to allow the tire to run flat for 200 km at around 80 km/h. For Dow, the PAX system represents an opportunity to supply the auto industry with an advanced finished component, stepping beyond its traditional role of supplying raw materials to manufacture.
2) Avoiding cleaning rather than cleaning more efficiently
Creating value through product innovations in cleaning agents has become harder and harder for chemical companies. Thus, why not work towards the ultimate solution, i.e., "not cleaning at all" or – more scientifically – self-cleaning surfaces? Degussa has taken the first steps in this direction of water-repellent "super hydrophobic" polymer surfaces by using the so-called Lotus or similar effects for a variety of applications.
3) Putting out fires without flooding the inventories
Putting out fires in offices, computer rooms, or museums without destroying the entire inventory has been a challenge after the phase-out of halons. 3M has now come up with fluorinated ketone, which has been included in Time Magazine's "most amazing inventions of 2004".

Innovation 1: Michelin's PAX tire system contains a polyurethane support developed by Dow Chemical.

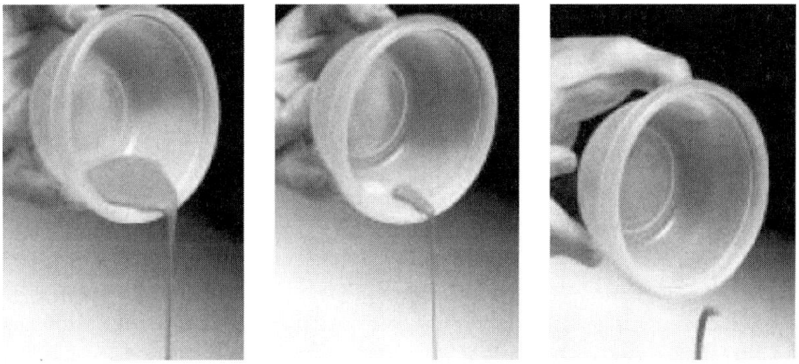

Innovation 2: With Degussa's Aeroxide LE 1, containers are completely emptied.

Innovation 3: 3M's Novec 1230 Fire Protection Fluid does not get things wet.

Once scientific questions are clearly formulated, the competitive advantage is driven by access to *all* relevant knowledge in the world. The sheer number of scientific publications makes this a daunting effort, quite apart from the fact that state-of-the-art knowledge is rarely available in published form (Fig. 13.4).

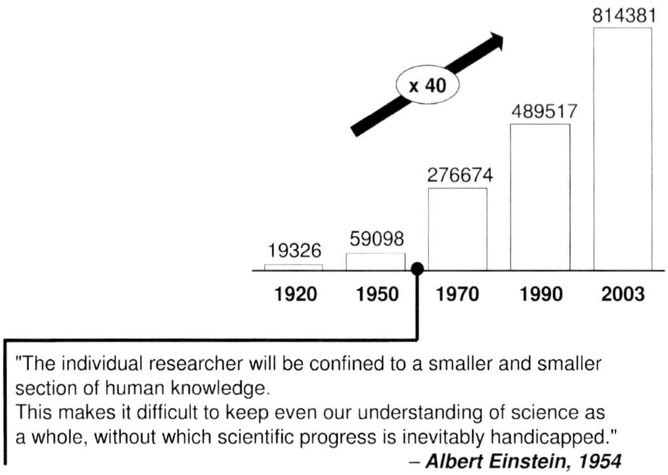

Source: Chemical Abstract Service

Fig. 13.4 Annual number of chemistry-related papers and patents.

Nearly all research-driven companies try to solve this dilemma by employing external scientific advisors, typically authoritative scholars in the given field. But how should these advisors be selected? Here, again, the key is knowledge – not subject knowledge, but knowledge of who the experts are. The scientific community clusters around common centers of interest. People who work at the border crossings between different fields serve as bridges between such clusters. Assuming that resource constraints limit the number of external advisors who can be engaged, companies should seek out the best-connected scientists, not necessarily those with the highest number of citations. A well-connected scientist can quickly refer researchers to relevant academic groups. The graphical representation of a scientific network in Figure 13.5 clearly shows the preferred collaboration partners.

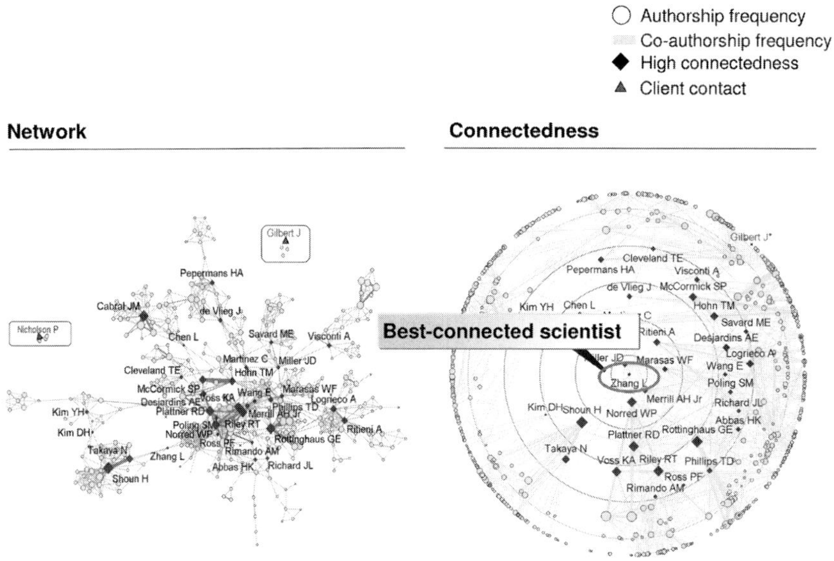

Source: McKinsey, Visone, SciFinder

Fig. 13.5 Calculated scientific network.

13.1.2
Creativity

> *"Creativity blossoms in small interdisciplinary units, where individuals experience that they make a difference – often simply through the survival of their company".*
> *Professor François Diederich, Swiss Federal Institute of Technology, Zurich*

Creativity takes place at the intersections between fields of knowledge, e.g., when marketing and sales transfer a business model that works in other industries, when researchers apply an approach to a different field, or when scientists with different backgrounds and experience work together at the same problem (Fig. 13.6). Accordingly, a large part of creativity is about breaking old connections and building new ones.

Why is it so difficult to tap into the full creativity of an organization? In the quest for efficient processes, many jobs are increasingly specialized. This high level of specialization can stifle breakthrough creativity: many researchers feel uncomfortable about discussing questions outside their immediate area of expertise. Accordingly, scientific problem-solving takes place inside a function, rarely between different scientific disciplines, where many new ideas are born.

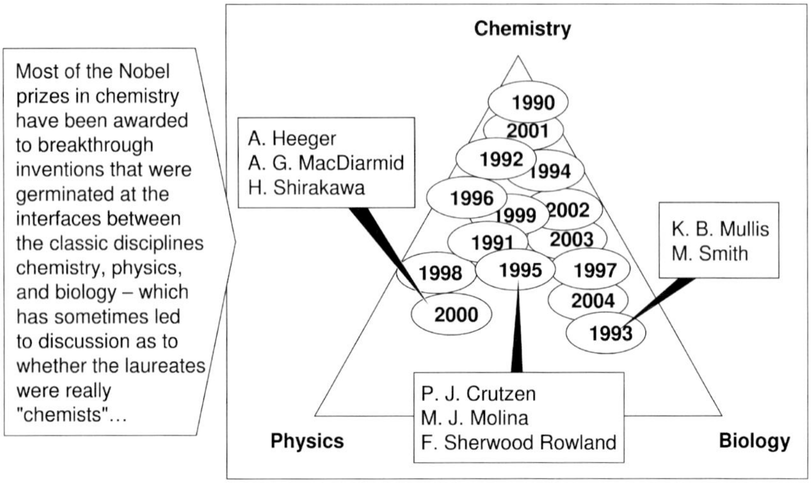

Source: McKinsey, Google

Fig. 13.6 Interdisciplinarity of Nobel prizes in chemistry.

Several companies have introduced cross-functional task teams to circumvent the specialization problem. Unfortunately, mixed composition alone does not change much, as many project teams know. We believe there are two success factors in creative interdisciplinary problem solving: 1) a clear task must be set that can only be completed if the team really works together, and 2) a long-term time scale needs to be provided, to give the team the chance to build mutual trust. Only these considerations make cross-functional teams really distinctive.

Do researchers in your company feel that they can make a difference? That their creativity matters? In many situations, creative ideas are never voiced, because no one further up the hierarchy ever asks for them. When we conduct scientific idea competitions in research organizations, they typically surface several hundred new ideas, many of which are truly fresh and innovative. The difference is that researchers submit their abstracts on these occasions because they know that every idea will be carefully evaluated and that senior management is personally involved in the process. They know that their creativity can make a difference.

Merging chemistry and engineering – a lesson in cross-functionality taken from the past

Scientific Design Co. was founded in 1946 by Ralph Landau, Harry Rehnberg, and Robert Egbert and later became Halcon International. This company has clearly been among the most enterprising and innovative chemical firms. It demonstrated that a sustainable competitive advantage can be achieved through being innovative. Or, in the language of the pharmaceutical industry, its pipeline was pretty full for several decades compared with all competitors (Fig. 13.7). Halcon's success was based on the insight that after the Second World War the manufacture of chemicals from petroleum would be a major growth area and that neither the traditional chemical companies nor the oil companies had fully acquired the needed skills. Building an organization that effectively merged chemistry and engineering undoubtedly became one of the key success factors for Halcon, as underlined by a quotation from Ralph Landau: "We are working on creative arrangements to bridge disciplines and encourage a new breed of sophisticated would-be entrepreneur. [...] We have chemical engineers right in the laboratory, working with our chemists at every stage of process development." In 1980, Arco bought out Halcon's interests and consolidated the venture within its Arco Chemical Co.

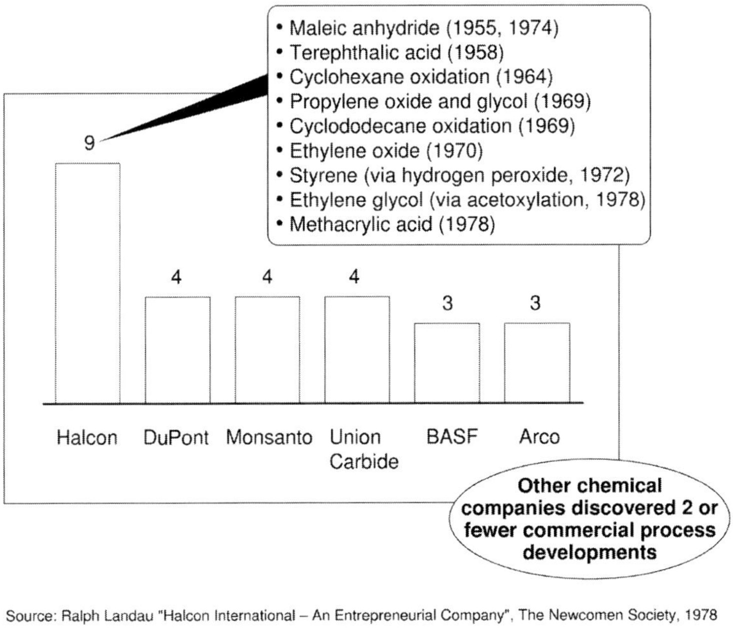

Fig. 13.7 Major commercial process developments, 1955–1980.

13.1.3 Perseverance

> *"Chemistry is like exploring unknown terrain. If you only walk in one direction you will most likely reach an insurmountable obstacle. Then you have to turn around and find a new path. Only patient explorers are successful, all others had better wait until they can buy a map".*
> *Professor Henning Hopf, President, German Chemical Society (GDCh)*

Perseverance is all about Edison's famous "99 percent perspiration." Implementing a solution often requires more creative effort than coming up with the solution in the first place. Only a few people are directly involved in the idea generation; everyone else who has to support the implementation will have to be convinced.

Quite often, companies abandon projects before they have exploited their full potential. In some cases, competitors demonstrate greater patience and bring a closely related product successfully to the market, leaving the original inventors with the sad knowledge that "they had it at their fingertips". In other situations, projects have the proverbial "seven lives" and become successes only because the inventor secretly keeps them alive, despite several no-go decisions.

Why is it so difficult to muster just the right amount of perseverance in a corporate environment? We believe perseverance is foremost a decision-making problem. To meet performance criteria, project management needs discipline. Typically, at the start of a project the project leader develops a milestone plan as guidance for future project evaluations. When milestones are not met after the originally estimated time, management typically decides to withhold further funding because it has lost faith in the project's likelihood of success.

Unfortunately, in many milestone plans, especially in research, we see huge overconfidence in the original estimate of time requirements. This initial overconfidence can be attributed to two factors: time pressure and personal frame of reference. If the talk around the water cooler claims that no project with a total timeline of more than two years will be funded, no milestone plans longer than that will be submitted, even if more time is realistically necessary. If the typical time a bench scientist needs for promotion to a management role is five years on average, milestone plans will not extend beyond two to three years. Accordingly, if all research steps have to fit this short time span, project planners will decide to set shorter individual milestones than are actually required. Little wonder that projects are abandoned prematurely or project overruns compared to originally expected timelines are so common.

Another decision factor that makes sufficiently deep exploration of an idea's potential difficult is rooted in a belief in serendipity: this is the underlying assumption that the likelihood of success increases with the number of different compounds and approaches the company screens. Of course, this will never be true if compounds are always explored at top speed and superficially, as can be seen from the often disappointing results of high-throughput approaches in the area of new compound discovery.

A final obstacle to perseverance lies in the way companies make go/no-go decisions. In many instances, these are consensus-based group decisions, and the consensus approach generally leads to low risk/fast success projects, because the immediately practical ones without problems are favored as "no regret" moves. More far-sighted projects or those with higher uncertainties always seem to find a majority vote against them.

Overcoming decision-making problems with smart project planning is one of the keys to earning enough time for true innovation. We recommend two main actions: first, compare the project in question with similar experiences within your own company and elsewhere to avoid the overconfidence trap as far as possible, especially in terms of timing, and second, sequence experiments selected for a project's specific risk profile in a "stage-gate" process to reduce the probability of late-stage failures.

The chiral switch of (S)-Metolachlor: 14 years of perseverance

Syngenta's Metolachlor is the active ingredient of Dual, one of the most important grass herbicides for use in maize and a number of other crops. Dual was introduced to the market in 1976 containing all four stereoisomers of Metolachlor. As early as 1982, Ciba-Geigy – one of Syngenta's legacy companies – reported that 95 percent of the herbicidal activity of Metolachlor lay in the two (S)-diastereomers. The application of these diastereomers only would reduce use rates significantly, giving the product an additional boost on the market and protecting it against generics. The quest for a viable catalyst for enantioselective manufacture had started – and it ended in 1996 with a full scale launch operation. (S)-Metolachlor is still one of the blockbusters of the agrochemical industry, with sales of around USD 400 million per annum (Fig. 13.8).

Another remedy has been implemented successfully by several companies: top management sets aside a certain amount of resources to pursue "wild cards" and invest in more long-term core research areas. The result is a portfolio that is equally well balanced in terms of both timing and risk.

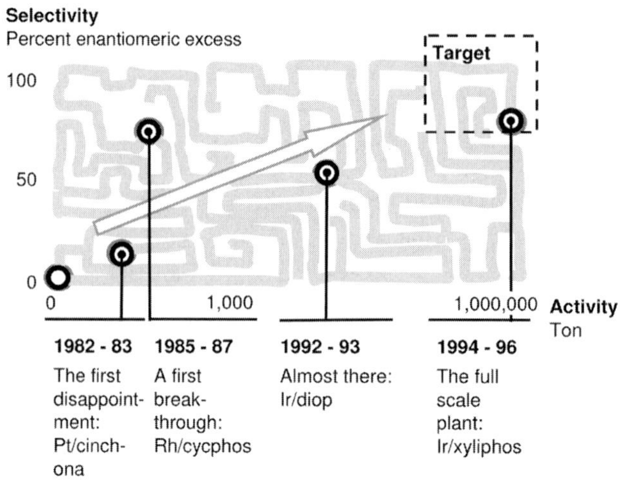

Source: Advanced Synthesis & Catalysis, H.-U. Blaser (Solvias): personal communication

Fig. 13.8 Metolachlor story – activity-selectivity labyrinth.

13.2 How to Organize for Innovation

In practical terms, how can you build an organization that fosters knowledge development, stimulates creativity, and is persistent enough to produce breakthrough innovations?

In the last decade, increasing commoditization of chemical products made cost reduction the industry mantra. Accordingly, most organizations embraced efficiency measures, taking out idle time, reducing interfaces between functions, and introducing throughput KPIs. As the track record shows, these measures can very effectively stifle creativity and thus innovation. However, the solution is not a return to the black-box approach. Instead, companies should combine modern management methods with five specific measures that stimulate innovation (Fig. 13.9).

1. *Redefine management aspirations.* Even today, many chemical companies are far from making innovation part of their core mission. Until top management wholeheartedly believes that innovation is crucial for the long-term health of the company, until management defines clear performance indicators for innovation and sets aside sufficient funds, innovation is not likely to happen.
2. *Perform a fact-based diagnosis.* Innovation can feel rather elusive, but in reality there is nothing magical about it. The innovation process can be assessed, like every other core process. Typically, core beliefs about sources of innovation in the company would be identified and challenged with facts and industry benchmarks. Equal attention should be given to the

internal environment (as outlined in the knowledge, creativity, and persistence sections above) and to the output parameters (e.g., pipeline strength, technology gaps, and competitive position). A targeted diagnosis reveals key areas for improvement, gives a good indication of innovation headroom, and helps top managers prioritize their activities.

3. *Clearly articulate the innovation strategy.* Knowledge-building is a quintessentially personal investment: scientists (and other innovators) will commit to it only if they believe that the investment will pay off in the long term. Accordingly, companies need to communicate their core innovation targets and areas, and not only communicate them, but stick to them.
4. *Redesign internal structures and processes.* To accommodate more knowledge generation, more creative problem solving, and better project planning, management must address some seemingly very mundane problems besides the actions discussed above: how much time can the typical scientist spend in undisturbed reading and thinking? Can he play around with new ideas? Does she have the freedom to initiate scientific collaborations or visit conferences without bureaucracy? Working constantly under pressure to meet timelines and budgets has become the norm for most researchers. Internal reporting and participating in task teams adds to the burden, so that the remaining time for undisturbed scientific thinking is quite short. A critical review of the company's meeting schedule, often organically grown and in need of pruning, is one important step. In many situations, however, it is also necessary to prioritize and streamline task team activities right down to the nitty-gritty level of personal time management.
5. *Implement.* Many aspects of implementation require organizational change or step-by-step learning of new decision-making processes. A company-wide effort can be unnervingly slow here. The success factor in implementation, therefore, is to run "controlled experiments" in a few critical units to create pull in the rest of the organization. In all organizations, you can find fast adopters who will readily change their conduct to a new paradigm. They are the ideal pilot leaders. And in all organizations you have the pessimists, who believe that nothing will ever get better. Ideally, you should keep them out of the way until the first visible successes can be celebrated throughout the organization.

A parting word of caution: rebuilding innovation can never be "just another project". It is a major restructuring task that requires full top management support, beyond just R&D. Innovators need encouragement to walk new ground; they need to know that company leaders expect novelty and innovation and understand the intrinsic risks. Authentic interest and support can go a long way towards stimulating innovation. Questions innovators may be asking themselves include, for example: how many full-day R&D meetings has the CEO recently attended? How much space does he give to the topic of innovation in his speeches? Is the board keeping abreast of current technical developments in the company? Is the incentive system designed to motivate innovators?

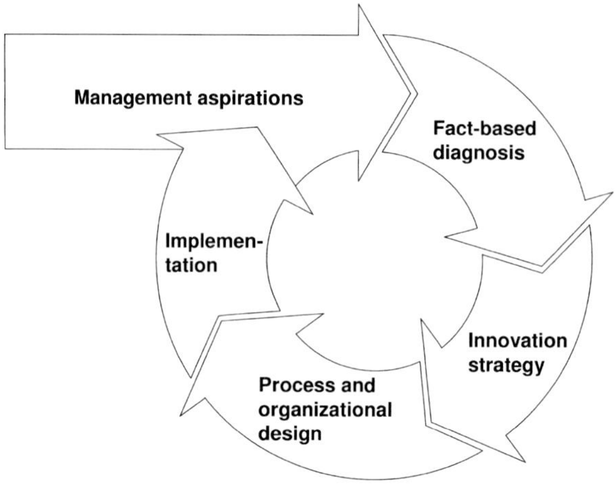

Fig. 13.9 Organizing for innovation: key modules.

13.3 Summary

- The chemical industry has to revitalize the drivers of innovation to proactively generate sustainable competitive advantage. Innovative breakthroughs can happen in a variety of dimensions, including not only new compounds, applications, and processes, but also new services and business models.
- Innovation is mainly driven by a combination of deep knowledge, creative new approaches, and the perseverance to follow through.
  - Knowledge has to be brought together from two areas: the "state of the art" in science and technology and the jobs the customer or end user wants to get done.
  - Creativity takes place at the intersections between fields of knowledge. Cross-functional teams with a clear task and a long-term time scale are generally the option of choice for creative interdisciplinary problem-solving.
  - Finally, the right level of perseverance is mainly a decision-making problem: optimizing the sequence of experiments according to a project's specific risk, "wild cards", and a conscious decision in favor of long-term research areas are means to come up with a well-balanced portfolio.
- To achieve a step-change in innovation, a transformation program has to include five steps: redefining management aspirations, performing a fact-based diagnosis, clearly articulating an innovation strategy, redesigning internal structures and processes, and finally, implementing the program.

14
Innovation for Growth

Thomas M. Connelly

There is still plenty of room for innovation in the chemical industry; that is if new and more robust strategies and tactics are used to transform and grow our businesses. Given today's increasingly competitive global environment with slow growth and limited returns in traditional chemical-based businesses, DuPont is looking to generate value with knowledge rather than capital intensive products and services, drawing on a wide range of sources. We see this as the way to grow, transform the science base for our businesses, and create significant value for DuPont and our customers.

14.1
The Ever-Present Challenges for the Innovation Process

Started in 1802 as a small manufacturer of explosives on the banks of the Brandywine River in Wilmington, Delaware, USA, DuPont today is a USD 27 billion company with operations in over 70 countries and about 59,000 employees worldwide. Our customers are primarily other manufacturing companies served by our five businesses, called "growth platforms" within the company: Safety and Protection, Electronic and Communication Technologies, Coatings and Color Technologies, Performance Materials, and Agriculture and Nutrition.

Our skill and capabilities as an explosives manufacturer dominated our first century, when we grew and prospered with the encouragement of the US government, supplying explosives for the military, construction, and land development. In our second century, we became a manufacturer of chemicals, paints, and polymers such as neoprene, nylon, Kevlar®, Teflon®, Nomex®, and Lycra®. This transformation was accelerated through our collaborations with universities and other industries. We have now begun our third century recognizing the need for a whole new transformation of our businesses, to be based on emerging sciences and technologies including biotechnology, nanotechnology, and microelectronics (Fig. 14.1).

Value Creation: Strategies for the Chemical Industry. 2nd Edition. F. Budde, U.-H. Felcht, H. Frankemölle (Eds.)

ISBN: 3-527-31266-8

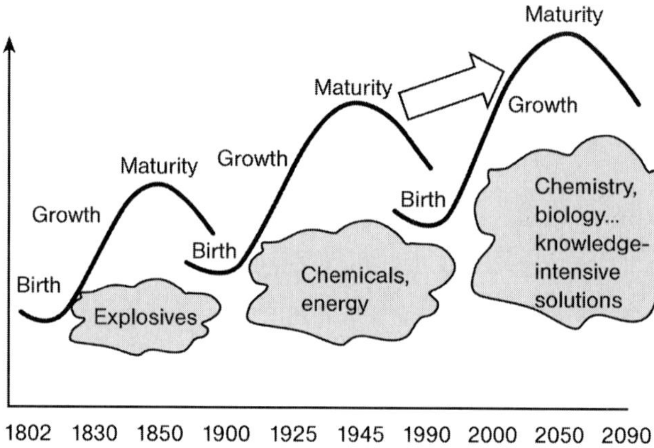

Fig. 14.1 DuPont business life cycles over time.

There are many influences acting upon us today; we are seeing intense competition from all parts of the world. Costs for petroleum-based raw materials continue to escalate, while manufacturing remains capital intensive, often with a high environmental profile, and in many cases promising only limited returns. How can we find more profitable opportunities for products and services from renewable resources with lower investment and a favorable environmental profile?

Additionally, our competitive position will not only be determined by the technologies developed but also by how and where they will be deployed, and how they will be integrated into and across our businesses. How can we involve the entire enterprise in pursuing innovation for growth?

A host of scientific domains are ready to be exploited in the research lab to develop these technologies – genomics, photonics, nanotechnology to name just a few. But which opportunities should be pursued first with the resources and talents at hand?

Revolutions are taking place in the way in which research is done – in computation, in bioinformatics, in analytical methods, and in information management. Stanford's High Wire Press (http://highwire.stanford.edu), for example, lets the individual researcher explore the most recent discoveries on line via topic maps or citation trails. How can our researchers take full advantage of these revolutions?

Information and technology transfer are done rapidly around the globe and technological competences are increasing every day in China, Korea, India, and the expanded EU. How effectively are we networked into the critical flows of knowledge?

The marketplace is making increased demands on product functionality and quality in highly segmented emerging markets that must be sought out and served, even if they are still in someone's dreams. How smart are we at recognizing the significant latent customer needs and anticipating the markets of the future?

Development costs continue to increase – the more successful research is, the more development work there is to be done, and the name of the game is speed. How ready are we to capitalize on our research successes and make the needed investments?

In such an environment, we have had to reshape the innovation "games" in which we play and develop a more robust innovation framework in which to carry out our growth strategies. The framework describing the five stages of the DuPont Innovation Process looks just like other depictions in countless organizations around the world (Fig. 14.2).

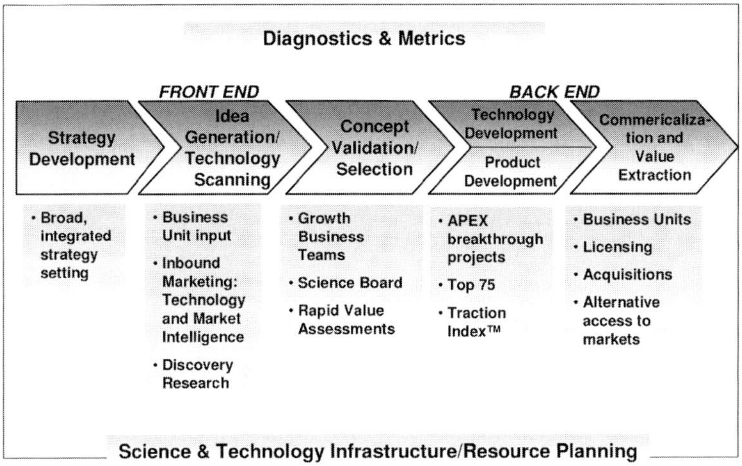

Fig. 14.2 DuPont innovation process.

The similarities end here, since significant differences exist in the specific tools and processes used in the execution and tactics for each stage, providing discipline while avoiding bureaucracy.

Strategy Development: Strategy setting with broad participation across the corporation tightly links project selection with the implementation of strategies at the business level and their successes and failures. "Learning Histories" help us understand what works and what does not. With a thorough analysis of earlier projects and initiatives, we have been able to find several critical keys for success such as an adequate understanding of markets, customers, and competitors, adequate team make-up, and the use of certain "best practices" to manage each project.

Idea Generation/Technology Scanning: In "inbound marketing", teams of marketing professionals and scientists develop technical hypotheses for potentially valuable market opportunities – merging competitive market and technology intelligence and asking "If only we could solve this technical problem, could we create value in this market, for this customer or future customers?" Over 50 percent of our current breakthrough or "APEX" research projects owe their beginnings to this effort.

Dave Hallahan, Research Biologist with DuPont, defines this innovative approach as "having a prepared mind combined with the technical skill and marketing experience to recognize or imagine a marketplace need". This requires both receptors that scour the scientific literature within a framework of the outside world and filters to understand the boundaries of the marketplace and the company's internal business structure. He and a small team of scientists, business managers, and marketing specialists are producing a yet to be commercialized replacement for the synthetic insect repellent, DEET (N,N-diethyl-m-toluamide). The effort is founded in his earlier scientific work in plant biology, which provided insight into specific biological chemicals that were naturally repellent and were exuded from "glandular" plants to ward off pests.

Another feeder stream for research proposals comes from formal and informal discovery research, in which scientists do the initial work of exploring the promise of a new technology or research finding. For example, in parallel with his formal research program, Hallahan pursued with some success a small discovery project in which he scouted areas of possible interest for several businesses in another plant by-product.

Concept Validation/Selection: Business and technology leaders together validate the projects that will be included in the portfolio and commit to supporting them. A formal structured method – "The Rapid Value Assessment Tool" (RVA) has proven invaluable in making this step more effective. RVA involves assessing the potential market size and "fit" of the opportunity with our capabilities, followed by in-depth analyses of both the market and geographic boundaries and the offering itself. Which organization in DuPont has the capability to develop the innovation to its fullest? Finally, the target value chain must be clearly delineated to enable an assessment of value creation opportunities: are there current offerings, what are their strengths and weaknesses, where is the value created, who are the participants along the chain, and where does the power lie?

Technology/Product Development: APEX projects – the breakthrough, high risk efforts – are aimed at large opportunities, conducted within corporate research, and funded by the corporation. These are managed by a staged or milestone process to move the successes toward rapid commercialization. Corporate oversight is also given to the "Top 75" projects in the business R&D units to assure adequate resources and skills, timely performance, and a high probability of success. A Traction Index™ provides a project by project analysis and portfolio prioritization for these projects using three criteria (each with five factors): market relevance; technical uniqueness; and quality of the commercialization plan (Fig. 14.3).

Market relevance

M1 Market need clearly defined
M2 Attributes required clearly understood
M3 Attractive market (size, growth trend)
M4 Attractive share potential (low competitive intensity)
M5 Impact of "environmental" factors (e.g., toxicity issues, regulatory requirements) low, manageable

Technical uniqueness

T1 Technology required known or can be developed
T2 Specifications cleraly defined
T3 Strong IP estate achievable
T4 Strong Voice of the Customer (VOC) input available throughout development process
T5 Technical skills required available

Commercialization feasability

C1 Marketing and sales resources available
C2 Total cost of good sold (COGS) leaves room for an attractive pre-tax margin
C3 Scale-up cost easily justified
C4 Rapid customer acceptance will be realized
C5 Value chain position is strong

Fig. 14.3 Traction Index™: factor definition.

Commercialization and Value Extraction: Here, a number of routes to the market are explored in addition to expanding ongoing DuPont businesses. An acquisition, a venture, licensing the technology, or a small stake in a startup all might be considered.

14.2
Three Lessons Guide Us

As we pursue innovation within this framework, we are seeking to profit from a number of lessons we have learned over the past decade.

14.2.1
Lesson 1 – Collaborate with both External and Internal Partners

We need to innovate in quite different ways than we did in the past – practicing what has been termed "Open Innovation". Innovative efforts must be structured differently and cannot be handled alone by DuPont's R&D community.

We need to collaborate with universities, small startups, government laboratories, other companies, and with other functions across the entire corporation.

The critical contributions in developing a new product, manufacturing process, service, unique business model, or business practice may well come from DuPont R&D, but may just as easily come from a myriad sources across the business or from external partners. Knowledge creation is occurring too fast, and the solutions to today's opportunities are too complex. We cannot succeed through our own efforts alone, and must become more adept at developing the right alliances and partnerships among government, industry, and academia.

Professor George Whitesides of Harvard has observed and told us that three models have emerged for governmental/industrial/academic partnerships. The microelectronics industry, for example, has successfully pursued a triumvirate model, relying on all three partners. Similarly, the biotechnology industry has expanded on this model, incorporating medical schools and venture capitalists. Historically, the chemical industry has had a more limited, linear approach, working with academia or with government but not melding the two into a single focused effort. That is now changing.

Our early interactions with academia can best be described as philanthropic, providing corporate aid with the hope of hiring more graduates and building relationships with the best universities and professors as consultants. However, academia today can provide opportunities to drive technology into the marketplace more quickly and more cost-effectively.

14.2.1.1 Polymerizations in Carbon Dioxide

In 2002, we started a major new development facility for manufacturing fluoropolymers, a direct result of unique technology developed by Professor Joseph DeSimone at the University of North Carolina. Professor DeSimone, who leads the National Science Foundation Science and Technology Center for Environmentally Responsible Solvents and Processes, discovered a revolutionary way to use supercritical carbon dioxide in place of conventional organic solvents for environmentally responsible manufacturing, cleaning, and processing. DuPont understood the uniqueness and relevance of his work and contacted DeSimone in 1996, offering to help review the patents and defray the patenting costs in exchange for an exclusive license of the technology. The company also offered to sponsor further research in his laboratories. DuPont then turned its efforts towards scaling and commercializing the new technology. This alliance has resulted in an entirely new process that lowers the cost of manufacture and the plant investment, decreases our environmental footprint, and creates more flexibility in our products and product offerings. Furthermore, we have now hired several of DeSimone's students, one of whom is acting as a liaison between DuPont and his labs as we prepare to commercialize additional developments.

14.2.1.2 Bio-based Technology with MIT

In January 2000, we formed a unique long-term alliance with the Massachusetts Institute of Technology stretching across multiple departments. The collaboration is focused on research and education in bio-based technology, including bioelectronics, biosensors, biomedical materials, bioengineering, and biopolymers. DuPont is responsible for providing the market relevance, development, and commercialization, and MIT is responsible for creating unique technologies.

An important aspect is that this alliance is highly collaborative. Proposals may be initiated by either MIT or DuPont, every project has a DuPont liaison and an MIT partner, and projects are reviewed at regular intervals and are continually evolving. Although this collaborative effort is still quite young, two technologies have already been transferred from MIT to DuPont for implementation.

14.2.1.3 **Nanotechnology for the Soldier**

We are also now partnering with MIT, the US Army, Raytheon Co., Massachusetts General Hospital, and the Brigham and Women's Hospital to form the Institute for Soldier Nanotechnology (ISN). We seek to enhance the protection and survival of the infantry soldier using nanotechnology to create revolutionary materials and devices to improve threat detection and neutralization, concealment, real time automated medical treatment, and reduction in the size of logistical support.

14.2.1.4 **Additional Collaborations with the US Government**

Similarly to our dealings with academia, our interactions with the US government are also evolving building upon past collaborations. Between 1985 and 1994, DuPont spent USD 400 million on R&D and capital to replace the entire family of Freon® chlorofluorocarbons (CFCs), following the Montreal Protocol and the impending ban and phase out of CFCs. We were dealing with familiar markets but quite unfamiliar technology, and needed help. The US government was an important partner then in a consortium that tested and developed refrigerants that were safe, environmentally superior, and required only minimal changes to equipment. The National Institute of Standards and Technology (NIST) and the US Department of Energy (DOE) modeled which compounds were viable candidates, while universities helped identify the best compounds. Success gave seven new manufacturing plants, 220 patents, a leadership award from the EPA, a remade billion dollar business, and an end to the production and consumption of ozone-depleting substances sooner than required by the Montreal Protocol – while not disrupting the benefits to society derived from CFC products.

Most recently, we entered into a science-based consortium funded by DOE for research leading to an innovative "biorefinery" – capable of producing fuel and high-value chemicals from renewable resources such as biomass and corn. The consortium includes DuPont, Diversa Corporation, the National Renewable Energy Laboratory, Michigan State University, and Deere & Company, and is a good blend of complementary technical expertise. DuPont is contributing expertise in applied chemistry, biotechnology, chemical engineering, and project management. Diversa brings its proprietary genomic technologies for the rapid discovery, development, and optimization of novel proteins and microorganisms. The National Renewable Energy Lab's participation is anchored in its expertise in biomass conversion, and a biomass expert from Michigan State University will address life cycle analysis. Deere & Company offers world-class expertise in agronomic systems analysis and materials processing.

We now understand that collaborative efforts with external partners work best when a strong, shared purpose is identified as valuable to all of the parties in the alliance. In addition, we have found that the route to the most effective partnership lies in building relationships at all levels of the interaction, from the highest management down to the scientist-to-scientist interactions. Alliances must be well planned; selecting the right partners certainly affects success. Each partner fulfills a separate and specific role in determining the relevance of the project, creating unique technologies, or providing commercialization capabilities. Finally, a successful collaboration must commit to continuity: the work must be of strategic importance to the collaborators so that the alliance receives the continued intellectual and financial support required to succeed.

14.2.1.5 **Partnerships are just as Important In-house**

We continue to learn that partnerships can be very powerful tools that can be applied to our in-house innovation as we collaborate across functions and businesses inside DuPont. Most recently we did this when we faced a challenge because patent protection was ending for a very profitable family of herbicides.

In 1975 DuPont researcher George Levitt discovered sulfonylureas, potent chemicals that interfered with plant growth enzymes but were non-toxic to mammals. In the early 1980s DuPont introduced its Glean® herbicide to wheat farmers, and then developed in total thirteen sulfonylurea herbicides for every major food crop in the world. By 2001 DuPont manufactured 40 products and maintained operations in 40 countries around the world. Success was due to product performance, low mammalian toxicity, exceedingly low use rates, and extensive and tight patent protection for the ingredient and for the dry formulation as granules. However, the first patent out of the thirteen was scheduled to end in 1999, with all others following over a period of a decade, leaving the business open to generic counterparts. A team from marketing, operations, and R&D met the threat with each function playing a critical role in developing a successful innovative new offering – novel, patent-protected sulfonylurea formulations, dry blends, and delivery systems taking product performance to a new level. Sulfonylurea herbicides are usually suspensions that are sprayed onto fields. Suspensions still have some very small particles that cling to the walls of mixing devices and spraying tanks. The new formulations eliminated these problems by being totally soluble in water, with no particles to stick to any walls. At the same time, these new formulations were shaped in a different way through extrusion and were sized in a unique patented manner. Multiple herbicides, shaped and sized with the same technology, could now be blended in wide ratios and would not segregate as happens with granules of different size and weight, and blends could be customized for each grower's needs. Crop profiling could be generated, analyzed by geography, conditions, and specific seed type to create a customized "sharpshooter" approach. This new technology of non-segregating, fully homogenous blends could be offered in bulk distribution machines, developed by DuPont, allowing farmers to be served by their dealers with the exact amount of product they need. The bulk container could be taken back and refilled by DuPont, reducing packaging waste and allowing farmers to buy the exact amount of product they needed. Three unique new technologies were thus integrated into new and improved offerings for farmers, leaving even less of a footprint in the environment.

Success in stopping losses from generic market entrants and creating new sustainable revenue growth for a matured portfolio was achieved with the combined effort of marketing (providing a clear understanding of the challenges of the distributors and growers), manufacturing (providing technology, infrastructure, and logistics for newly formulated blends in returnable bulk distribution systems), and R&D (integrating an understanding of formulation, extrusion engineering, particle science, and crop protection biochemistry). By such efforts, helping customers increase productivity and deliver healthier, better tasting food to consumers, DuPont Agriculture and Nutrition has now achieved a double-digit compound annual growth rate for the past three consecutive years.

14.2.2 Lesson 2 – Draw upon Many Sciences

Once we start practicing open innovation we can find significant opportunities with an "integrated science" approach, drawing upon a multitude of scientific disciplines and exploring the interfaces with the chemical sciences.

14.2.2.1 A New Business in Pathogen Screening

Over the past decade as we developed systems to track bacterial contamination in microbe-prone products such as pharmaceuticals, cosmetics, and food, leader of the effort Vinay Chowdhry called upon microbiologists, geneticists, molecular biologists, chemists, chemical engineers, electrical engineers, and software experts. He then turned to business skills to find funding, as his team developed an entirely new business. DuPont Qualicon is now an important player in the USD 27 billion global diagnostics and USD 1.7 billion food diagnostics industries, providing advanced molecular technologies and systems for pathogen screening. The BAX® system, for example, an innovative DNA-based technology for screening food and environmental samples for pathogens or other organisms, provides polymerase chain reaction (PCR) assays to screen food and other samples for such pathogens as Salmonella. The automated system, which looks just like a desktop computer, is now used by governments, food companies, and laboratories around the world. The RiboPrinter® system, the world's only automated DNA fingerprinting instrument, rapidly pinpoints sources of bacteria using genetic information to provide an automated snapshot or RiboPrint™ pattern of any bacterium in less than eight hours. With more than 6,400 patterns in the identification database, electronic data security, and characterization sensitivity below the species level, the system surpasses US FDA guidelines for preventing contamination and assuring consistency in aseptic processing. Such a development would never have been possible without the power of integrated science.

14.2.2.2 Biology Provides More Economical Routes

DuPont is now integrating biology into polymer manufacturing, producing Sorona® polymer. Identified in the 1940s, the copolymer based on 1,3-propanediol (PDO) and terephthalic acid was never launched commercially because of the high cost of the PDO monomer. However, an astute DuPont research manager noticed the structural similarity between PDO and certain biological chemicals, and pursued the notion that a biological approach might be economical.

Combining metabolic engineering with polymer science, researchers introduced a microbial process that used a sustainable, renewable resource – glucose. Scientists and engineers from DuPont and Genencor International, Inc. redesigned a living microbe to produce 1,3-propanediol. Inserting biosynthetic pathways from several microorganisms into an industrial host cell line allows the direct conversion of glucose to 1,3-propanediol, a route not previously available in a single microorganism. Modifications maximize the ability of the organism to convert glucose to 1,3-propanediol, while minimizing its ability to produce biomass and unwanted by-products. The catalytic efficiency of the engineered microorganism allows replacement of petroleum feedstock, reducing the amount of energy needed in manufacturing steps and improving process safety. The microbial process is environmentally green, less expensive, and more productive than the chemical operations it replaces.

14.2.2.3 A Top Innovation in Nanotechnology

Recent developments in bioelectronics involved molecular biology, biochemistry, engineering, chemistry, material science, and analytical chemistry. Single wall carbon nanotubes (SWCNT) contain many varieties of CNTs, both metallic and semiconductors with a wide range of bandgaps. Such SWCNTs will only be useful in electronic devices if populations with highly controlled electronic properties can be isolated. Earlier research by others failed to find ways to selectively separate such CNTs. DuPont was successful with phage display, a technology used by a young professor at the University of Texas sponsored by DuPont. She was able to develop affinity between peptides and inorganic materials such as CdS. DuPont researchers identified a family of peptides with affinity to SWCNTs and used computer simulations to gain some insights into the fact that single-stranded DNA of the correct composition might have selective affinity for SWCNTs. Experiments verified this. In fact, the DNA formed a very stable complex, wrapping around the CNT in a helical fashion that was surprisingly stable as a hybrid up to 100 degrees Celsius. The hybrids could then be separated by "standard" chromatographic techniques used by biochemists to separate various proteins. Though the commercialization and sale of an offering are distant, following the publication of the results in Science, Forbes selected this development as one of the top five innovations in nanotechnology for 2003 – made possible by such a diverse environment.

14.2.3
Lesson 3 – Technologies Can Find Applications in Many Other Markets

The next step in practicing open innovation with integrated science allows us to extend technology platforms into new market spaces. This requires an increased understanding and further development of the underlying technologies which help us see the opportunities in new applications and in entirely new markets.

The extension of the sulfonylurea technology platform was one example of how "platform building" is an important business activity. Another example is found with our digital ink technology, where applications have just begun.

In the early 1990s, a DuPont customer was developing a new ink-jet printing system and had been meeting with our marketing people. We understood that the customer's new system would require colored inks that met specifications far beyond the capabilities of available technologies. The inks needed to be stable under high shear, dry instantly, provide high-definition images, come in a wide range of brilliant colors, and hold up to moisture, abrasion, and ultraviolet radiation. Our customer's search for such inks had been fruitless and frustrating, and at the time we viewed this customer narrowly, as one that purchased special coated paper for engineering drawings. But when our paper product offering was rejected again and again, the account manager with extreme persistence sought out technologies to meet the customer's need. He found his way to researchers in automotive finishes. Thus began the difficult and complex process of marrying a number of new and old technologies. Dozen-year-old "group transfer polymerization", for example, provided unique polymer architecture for the pigment binder, giving the needed stability, flow, and resistance to UV radiation, abrasion, and moisture. The development effort, which was nearly terminated during aggressive cost-cutting in the late 1990s, survived to make DuPont the world's largest digital ink developer and manufacturer, with half the market in desktop printing ink. Now, this fundamental knowledge of digital inks is being extended to the development of a digital fabric printer, Artistri®, that can print 16 million colors by combining a handful of inks on silk, cotton, linen, stretch fabrics, and other synthetics. This increases the sale of inks and helps us realize that extension into even more markets with other substrates appears to be in our future.

14.3
What Will the Future Bring?

We will continue platform extensions, taking Nafion® fluoropolymer membranes from chloralkali cells and catalysts to advanced fuel cells, and Kapton® polyimide film from flat electronic computer circuits to complex flexible circuits in cell phones. We are learning to build on our competences in laboratory and plant safety to expand the DuPont Safety and Protection business, and to follow the pathway to sustainable growth by generating value with knowledge rather than with capital-intensive products and services. We understand the importance of bringing diverse skills into our R&D community as we maintain a worldwide quest to hire or access the best technical talent – the true basis of gaining value from knowledge.

We continue to keep score using metrics to stay focused in the uncertain endeavor of innovation. We are seeing a positive trajectory in results. Since 2000, the technical effort in the businesses on growth initiatives has increased from 40 to 62 percent: patent applications increased from 950 to 1700, and revenue from new products increased from 22 to 29 percent of total sales, nearly half being non-replacements. We expect that figure to reach 33 percent by the end of 2005 based on products in the pipeline. More needs to be done, but we know we are learning, moving in the right direction, and laying the foundations for further growth.

14.4 Summary

- Innovation in the chemical industry now requires some different strategies and tactics, built upon increased knowledge of what works and what does not in today's highly competitive world.
- We needed to build a disciplined framework for innovation with specific tools and approaches that fit our culture and the nature of our businesses.
- We realize that is critical to practice "open innovation" with extensive collaboration with both external and internal partners.
- That is not enough: we must draw upon many scientific disciplines with an integrative approach if we are to transform the technological base of each business.
- Furthermore, we need to practice open innovation and integrated science to extend our technologies and products into entirely new markets and new applications.
- Finally, we need to seek other lessons from our successes and failures in implementing the innovation process as it fuels growth for DuPont.

15
The Four Pillars of Sustainable Purchasing Transformations

Helge Jordan, Nicolas Reinecke, and Khosro Ezaz-Nikpay

Improving purchasing performance remains one of the most important levers for chemical companies to boost profitability: according to the German Chemical Industry Association (VCI), in 2002, chemical companies in Germany spent more than 50 percent of their gross production value at third party suppliers: 31 percent went to purchase raw materials, 13 percent was spent on re-sale items, three percent on energy, and another three percent on maintenance services.

For an average chemical company or business unit with sales of EUR one billion, an EBITDA margin of ten percent, and a purchasing volume of EUR 500 million, reducing its overall spend by ten to 15 percent will result in an EBITDA margin improvement of five to seven and a half percentage points, if sustainable.

In our experience, savings opportunities in double-digit figures can be identified through purchasing improvement programs. Typically, 30 to 50 percent of opportunities identified can be captured within the first year, and 80 percent within the second year following implementation.

It goes without saying that savings in individual spend categories range from fairly low percentage numbers for commodity chemicals (e.g., polyethylene, titanium dioxide) to high double digit figures for indirect spend items like maintenance, repair, and operations materials (MRO) or service categories (e.g., facility management).

Sustaining the initial momentum is a common problem in purchasing improvement programs, which often lead to widely varying results: while most companies start out by launching multiple cost reduction initiatives, investing significant time and resources in skill building, and focusing on quick delivery of savings, the wheat soon separates from the chaff.

For the unsuccessful, savings do not turn into visible bottom-line impact, the project teams quickly return to business as usual, and frustration, distrust, and skepticism about the validity of the savings claimed creep into the organization. The successful, however, deliver sustained and acknowledged performance improvements (with savings figures that are trusted by the organization), new, cross-functional ways of collaboration persist, and organizational energy and the capability to move ahead remain high.

Value Creation: Strategies for the Chemical Industry. 2nd Edition. F. Budde, U.-H. Felcht, H. Frankemölle (Eds.)

ISBN: 3-527-31266-8

Traditionally, the purchasing function alone is given the task of proving that the savings claimed are actually being delivered to the bottom line. R&D, production, finance, and sales & marketing may sit at the same table during the performance review sessions, but the attitude towards the purchasing function can often be described as "guilty until proven innocent."

Tracking purchasing-driven improvements can truly be a daunting task – once an organization has adopted a Total Cost of Ownership (TCO) or even a Total Value approach, and moves beyond simply reporting purchasing price variances (PPV), the trouble begins: in direct materials, changes in production schedules and sales forecasts drive changes in the overall amount of raw materials and the mix of specifications used, and different inventory practices mean that new materials suffer delays in moving into production. In indirect materials and services, frame agreements negotiated by the central purchasing group are often only reluctantly adopted, thus leading to significant leakage of volume leverage and renegade buying.

Shifting successfully away from a narrow and short-lived approach, however, opens up the potential for purchasing to become a valued and trusted partner for the other functions. A sustainable purchasing performance transformation initiates change that penetrates deep into the heart and mind of the organization. To be successful in the long term, it has to rest on four strong pillars:

1. A well-planned and well-orchestrated *transformation program* (based on a thorough diagnostic of current performance)
2. *Behavioral change* anchored in the organization
3. A proven *value creation process*
4. A tailor-made *performance measurement and management* system.

15.1
Designing a Performance Transformation Program

An old axiom says that in order to plan a route, you have to know both where you are starting from and where you want to arrive. Performance transformation therefore needs to follow a well-planned and well-orchestrated program, based on a thorough diagnostic of the company's current purchasing practices, and a clear vision of its desired future state.

The diagnostic should address four key dimensions:

1. *Performance*: world-class purchasing organizations have a system in place that links their own aspirations with regard to cost savings, efficiency improvements, and additional value added to the business unit's strategies and tactical plans. Purchasing has a "steering dashboard" in place so that it is not flying blind (e.g., raw material requirements are linked to sales forecasts through correct bills of materials), but at any given point in time knows – and is able to communicate – what its contribution is to mastering the business challenges.

2. *Value creation system*: a well-set-up function has commodity strategies in place which are periodically reviewed and updated. In addition, streamlined processes for both strategic sourcing and transactional procurement as well as state-of-the-art tools like spend analysis, supplier cost modeling, electronic RFQs, and bidding tools are in place and adhered to. Furthermore, all of the above are supported by good information management systems which help the function to capture, analyze, and create key insights from information.
3. *Management infrastructure*: the way in which purchasing resources are managed within the company through formal structures and processes, and how purchasing interacts with other company functions, can make a crucial difference. In the worst case, functional separation of tasks between purchasing and business units can mean that purchasing is not interested in the needs of its customers, and customers are unaware that they can reach out to purchasers for advice on what and where to buy. In addition, in many cases purchasing performance is not measured or targets set. Purchasing's role as a function that only carries out orders also has the knock-on effect that it may not attract top talent. In world class organizations, however, purchasing and supply management staff collaborate with the business units in multi-functional teams. With mechanisms in place to set targets and monitor, manage, and reward performance, the purchasing team will be more motivated and the function will become more of a development arena for top talent.
4. *Mindsets, capabilities, and behaviors*: many traditional purchasing functions are purely reactive, there to carry out orders and expected to wait for the customer to call, and with the unpleasant duty of screwing the last cent out of their suppliers. If, instead, they have a sense that they are welcome members of multi-functional teams, collaborating with their customers and the suppliers to produce savings and innovation, success will feed on itself and purchasing staff will increasingly become productive members of the corporate team. They will own their part of the business instead of being owned by all the rest, drive change as well as being driven. This can best be achieved by setting aspirations for purchasing performance (in collaboration between top management and the purchasing function itself), and aligning the function's values and overall direction with that of the company as a whole.

Once the actual and target states have been defined and analyzed, the best way to articulate the changes required is to draw up a mission statement laying out what purchasing's role and remit will be (Fig. 15.1 gives an example), and to exemplify the desired changes with a schedule of "From… to" statements (Fig. 15.2).

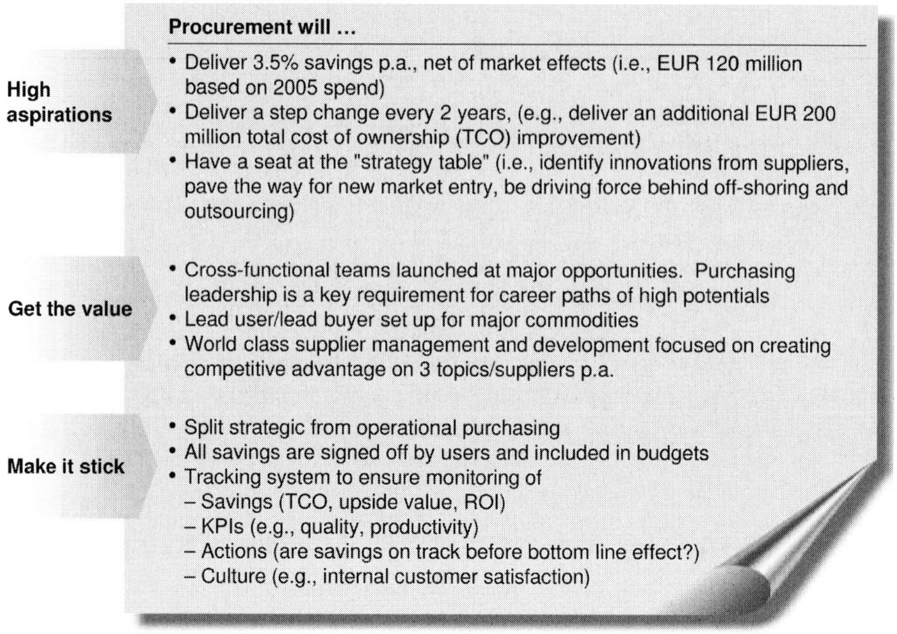

Fig. 15.1 Articulate a purchasing mission statement – example.

15.2 Initiate and Anchor Behavioral Change in the Organization

Ensuring organizational buy-in starts at the top: all members of the senior management team need to be brought on board with the initiative and understand what the program will contribute to achieving the company's strategic objectives. Purchasing can, for example, make a major difference to such corporate goals as becoming the industry cost leader by having the cheapest sources of raw materials, out-smarting the competition by acquiring intellectual capital and leveraging ideas from outside, or building up an overseas supply base from scratch to prepare for a major shift of the company's global manufacturing "footprint". Every senior leader also needs to understand what time and resources his/her organization needs to contribute to make the program successful.

It is also key for top management to decide ahead of time what additional cost savings will be used for. For example, site budget holders may be allowed to use the money to cover additional high NPV investment needs, the savings could be used to boost bottom line profit or dividends to shareholders, or they could be channeled into new marketing campaigns to help accelerate growth.

Once the call for "all hands on deck" is heard, purchasing project teams will be launched. Multi-functional sourcing teams have been common practice for quite some time – at least for the more advanced purchasing functions. However, even

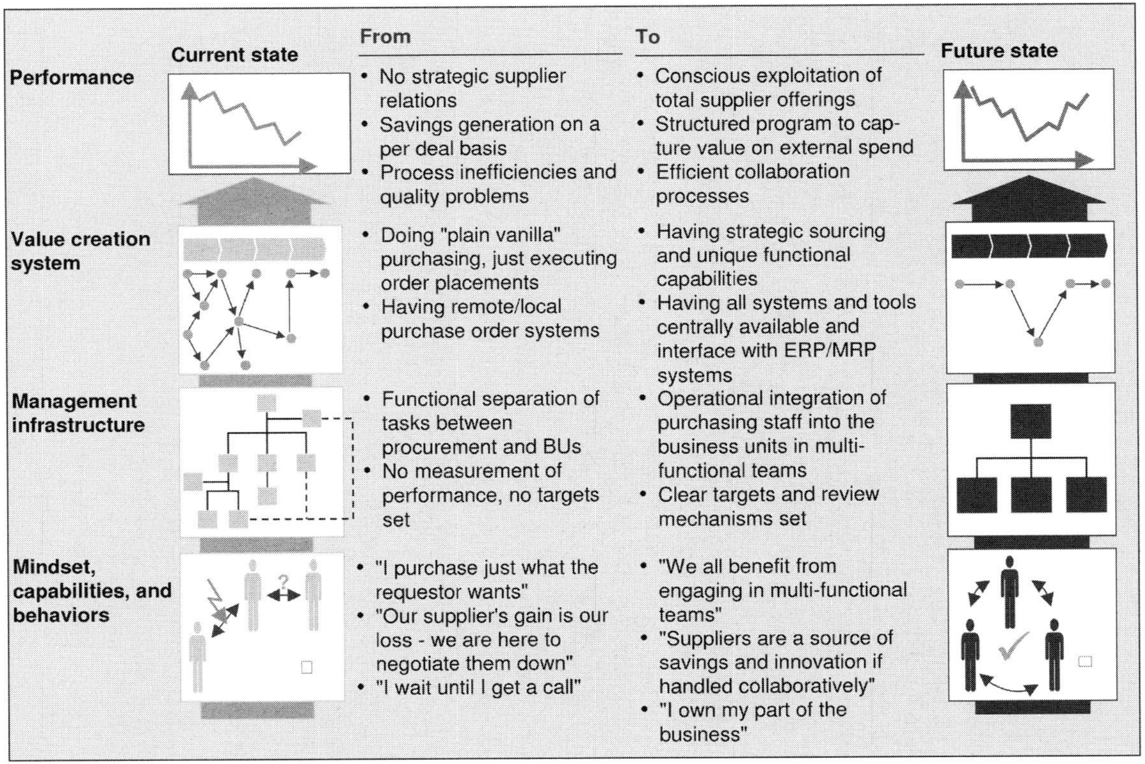

Fig. 15.2 Describe the desired change in "From ... To" statements.

these teams often fail to deliver sustainable results despite the function-spanning collaboration of experienced subject matter experts. What are the reasons? The teams often lack full-fledged senior management backing, start to dissolve immediately once the apparent project tasks are fulfilled, and do not have the organizational clout to see complex changes through to completion. At one advanced materials company, for example, savings depended critically on the approval of new raw materials from new suppliers. Top management intervention was required to ensure that the necessary time-consuming lab and production trials were approved and completed. This was supported by a fundamental organizational change where, for the first time, the R&D and purchasing functions reported to the same senior VP, thus making it possible to align the goals and incentives of the two and to push through an aggressive new raw materials approval program.

Lasting behavioral change depends on three key success factors that will generate attention and commitment to the program:

1. *Full-fledged management endorsement and backing from the senior management team, business unit leaders, and site leadership teams*: the leadership teams need to endorse the purchasing initiatives and "walk the talk". If standardization of specifications is identified as one of the major improvement levers – a significant lever in chemical feedstock or engineered equipment, for example – and is backed up by the appropriate convincing business plans, site management needs to make sure that these changes are embraced and implemented by operations. A "pull" attitude needs to be created in the user community, instead of purchasing having to "push" every single change.
2. *Significant time commitment from purchasing and other subject matter experts after the initial project is finished, to see initiatives through to completion*: most changes in raw materials are not mere drop-in replacements. R&D and operations have to invest significant time in making sure the new supplier delivers the appropriate high-quality material, sales & marketing may need to seek approval from end customers to make switches, and purchasing needs to secure a seamless supply chain. It is imperative that each function involved pulls its weight in the approval process, and is held accountable for delivering its piece.
3. *Alignment of individual and team performance goals and codification of these goals in performance contracts*: often performance goals are aligned neither between individuals and the projects to which they are assigned, nor between different functions. While purchasing is incentivized to deliver cost savings, R&D and Engineering look out for the best technical solution regardless of cost, e.g., if a new production plant is being designed, and operations seeks the highest overall operational effectiveness (OEE). In the course of a purchasing transformation, these seemingly conflicting goals need to be aligned and incentivized, and in the best cases are mirrored one-to-one in the suppliers' scorecards as well.

15.3
Generate Impact through a Proven Value Creation Process

One key success factor is that the purchasing function uses a formal value creation process (Fig. 15.3). This cannot be a mere "three bids and a buy" exercise. We have found that in order to be effective, the process has to:

- Move through formal stages – supported by analytical tools, templates, decision support documents, etc.
- Focus heavily on hypothesis-driven problem solving to prevent teams from getting side-tracked or even lost in low-impact activities
- Engage stakeholders early on to test the initial improvement hypotheses, and then continuously to ensure timely feed-back and buy-in for the proposed changes
- Use explicit alignment/quality gates which need to be passed to ensure impact and sustainability. It is better to stop the project temporarily if certain quality requirements are not met or roadblocks need to be removed than to carry on regardless (Fig. 15.4).

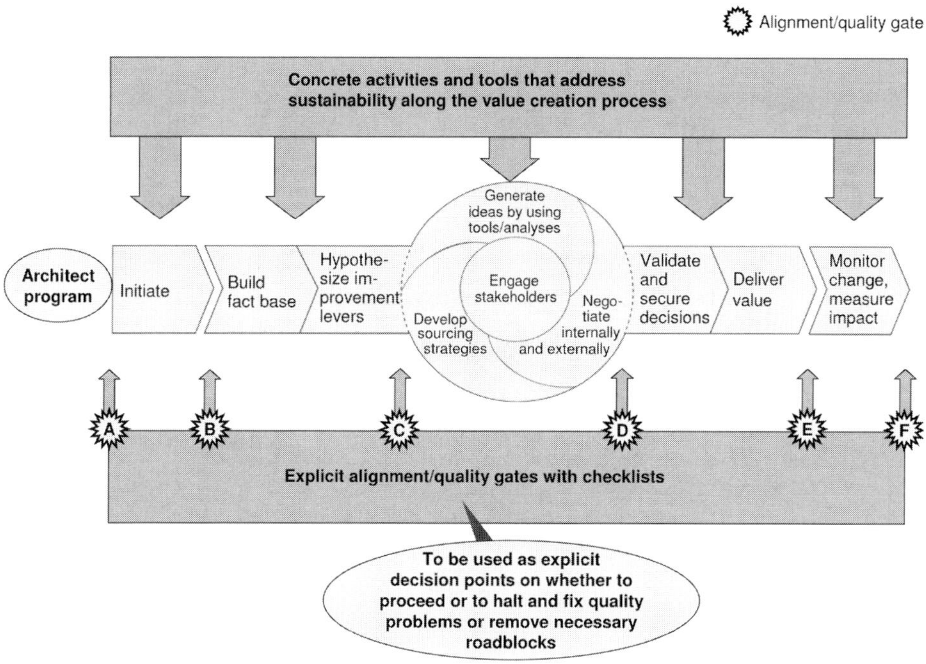

Fig. 15.3 In the formalized value creation process, specific steps need to be followed and alignment/quality gates passed to ensure sustainability.

Above all, the value creation process has to focus on the Total Cost of Ownership (TCO), not just the purchase price. The principal levers in optimizing TCO

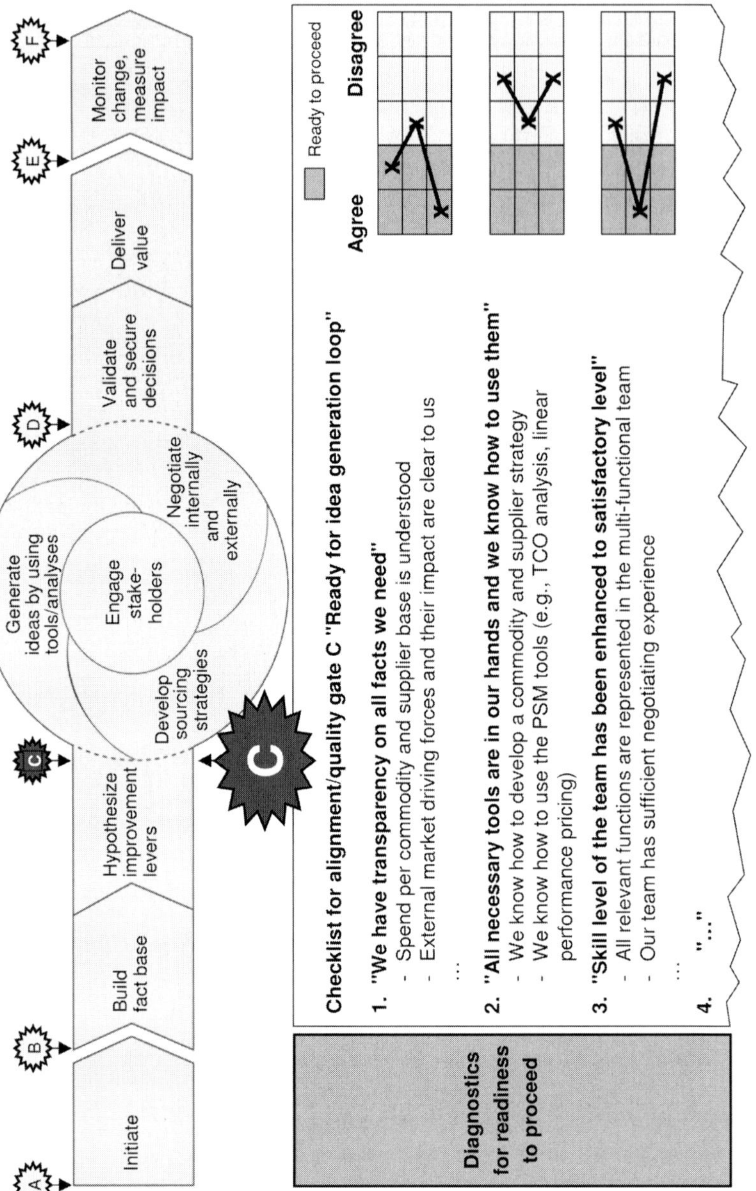

Fig. 15.4 Example of alignment/quality gates in the value creation process.

are: exploiting purchasing power and commercial levers (including sourcing from low-cost countries; changing out products and optimizing specifications; and improving production, supply chain, and management processes. They can be pulled individually or in combination. The following case examples exemplify various aspects of these levers and demonstrate their impact.

Case example 1: How company A was able to reduce its purchase prices by developing a deep understanding of suppliers' economics and leveraging new suppliers (Fig. 15.5).

Company A was facing price increases for a number of specialty polymer products. The purchasing department teamed up with a number of senior scientists from the R&D group to model the suppliers' production costs and thus understand the profit margins for the different products. When it obtained additional price quotations from low-cost countries, it became apparent that the incumbent supplier was trying to push through unreasonable margins. Using the detailed cost structures and the threat of shifting volumes to "wildcard" suppliers, the company achieved net savings of seven and a half percent for this product category. This was only possible through joint, cross-functional collaboration between the purchasing and R&D departments, using a deeply fact-based negotiation approach.

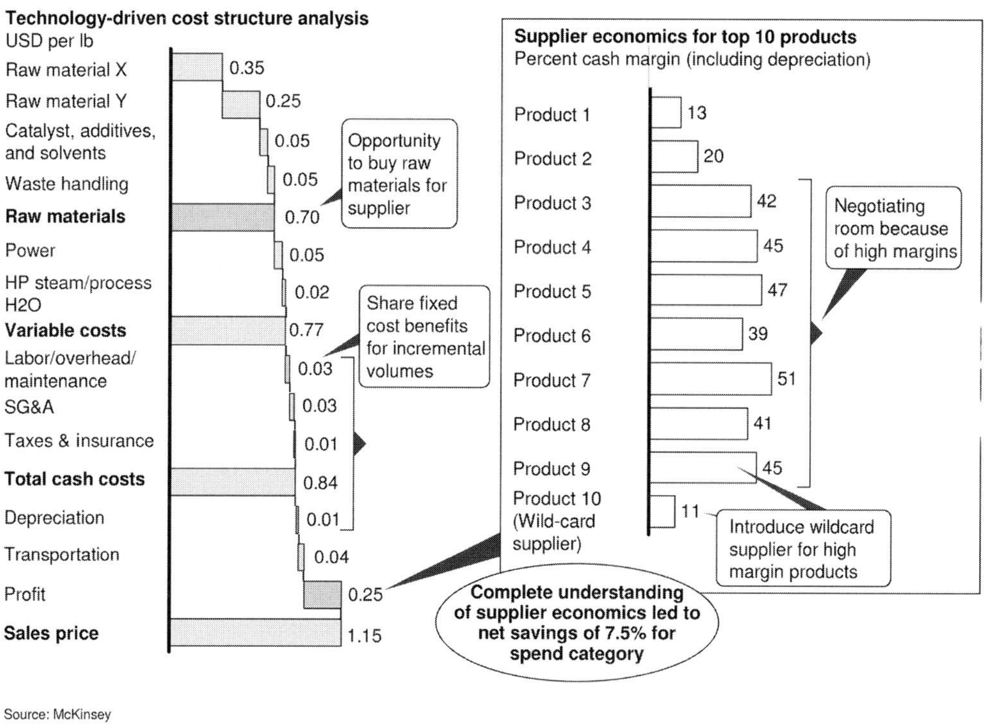

Fig. 15.5 Using deep understanding of supplier's economics to reduce purchase price.

Case example 2: How Company B reduced the Total Cost of Ownership for caustic soda (Fig. 15.6).

Caustic soda is primarily produced by either the membrane or the diaphragm process, which differ in their production costs. Also, different physical forms (beads, microprills, solutions) and various concentration levels are available, and packaging/transport modes range from bags to bulk barging. After thoroughly reviewing all options and calculating the lowest TCO, company B chose to buy 50 percent solutions produced by the membrane process and supplied via barge shipments. Overall spend was reduced by 25 percent, and the number of specifications from ten to four. The number of suppliers dropped from ten to three.

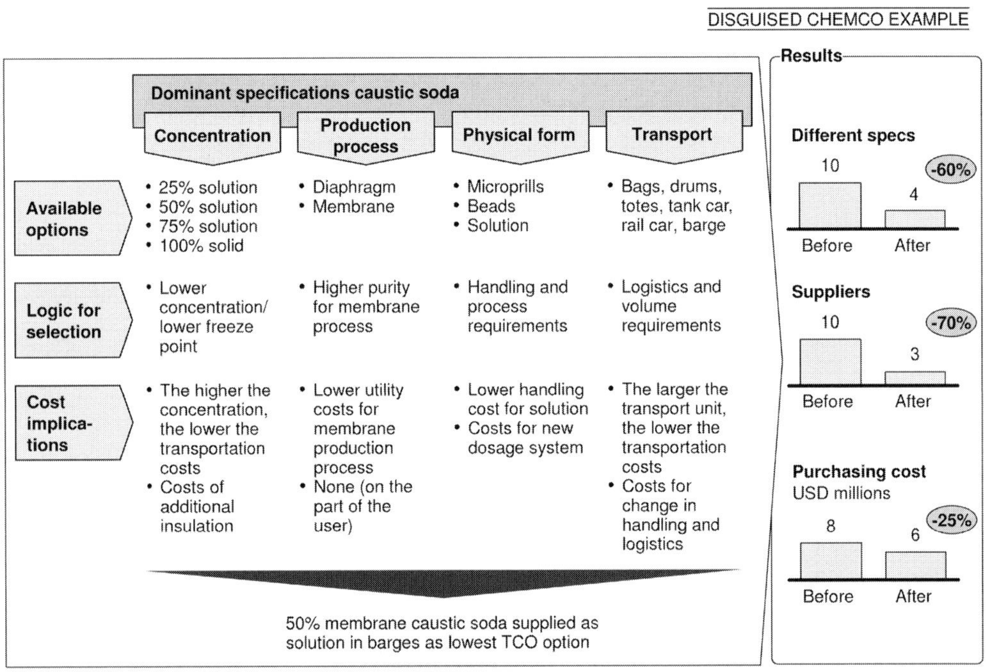

Source: McKinsey

Fig. 15.6 Reducing TCO for caustic soda.

Case example 3: How Company C consolidated spend along the product value chain (Fig. 15.7).

One of the key characteristics of the chemical industry is the complexity of the value chain from basic feedstocks to finished products, and the varying degrees to which the different players are integrated backwards and forwards in their own markets, leading in many cases to complex and expensive business processes. Company C conducted a thorough review of its existing and potential supply base, and the opportunities to consolidate spend by using fewer suppliers. By reducing the number of suppliers from 30 to six, the company achieved savings of almost ten percent. Savings resulted from consolidation of feedstock purchases, reduced fixed costs for volumes produced, and increased negotiation leverage.

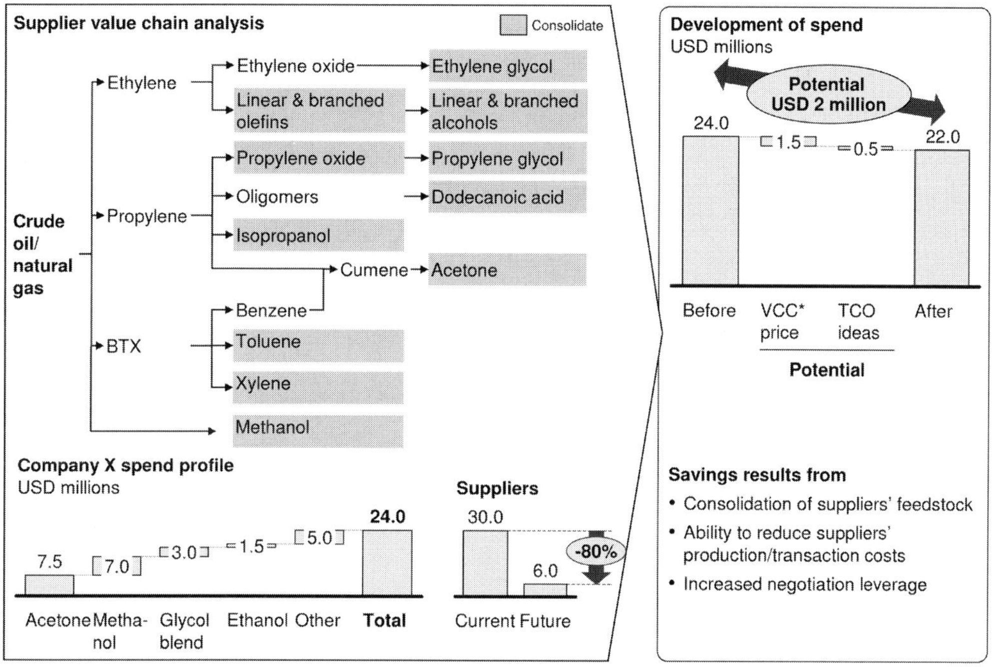

* Value chain consolidation
Source: McKinsey

Fig. 15.7 Consolidating spend along the product value chain.

15.4 Create a Tailor-made Performance Measurement and Management System

While the backbone elements of a solid tracking system are straightforward (e.g., action tracking, financial impact, and improvement of non-financial indicators), the following principles need to be closely followed to ensure lasting success:

- *Tailor performance information to the needs of the different stakeholders (e.g., CEO, CFO, CPO, BU heads, plant managers, department heads, risk management department, investors, etc.)*: naturally, information needs differ for every stakeholder group within and outside of the organization. On the highest strategic level, senior management needs to understand:
  - What bottom-line impact purchasing does and should deliver and how the resulting value is spent;
  - What strategic moves can be prepared and exploited with purchasing's help (e.g., building a local supply base in countries where new production sites are being built);
  - The kind of risks to which the company is exposed by entering into agreements with suppliers (and how these risks are being mitigated);
  - What well-structured supply agreements look like, what strategic significance they have, and how ill-structured agreements can wipe out competitive advantages very quickly;
  - Whether important suppliers are also key customers or joint venture partners, and how these complex relationships can best be strategically shaped and managed.

  On the tactical level, a plant manager will want to know how his or her site's or unit's operating costs compare against industry benchmarks, how he or she can influence these costs through superior purchasing, and how improvements are actually flowing down to the bottom line. Plant managers particularly need to have a differentiated understanding of the performance of all purchasing-related efforts compared with other improvement initiatives such as sales stimulation, productivity improvements, or overhead cost reduction. We often find that there is too little transparency on the performance of each initiative, and that over-delivery in one area covers for low performance in another.

  Purchasing needs to understand these varying needs, and will have to put a data/information collection and reporting system in place that provides the different stakeholders with tailored insights to support their respective management decisions.
- *Establish specific controlling principles for every product and service segment*: chemical companies buy a wide variety of goods and services, ranging from petrochemical feedstocks and local maintenance services to capital projects (including design & engineer-

ing, parts procurement, and the physical construction of new assets). Broadly, these purchases can be classified into three major segments: (a) general business expenses, (b) directly production-/output-related expenses, and (c) infrastructure-related spend. Different cost drivers exist in each of these segments which need to be understood and subsequently pro-actively influenced. Therefore, purchasing has to apply a variety of measurement principles and controlling mechanisms, ranging from basic PPV reports to the monitoring of compliance with frame contracts or the utilization of external market indices to benchmark performance. In this context, finding the right parameters that help define a real or proxy market index is key to ensuring year-on-year vigilance on performance improvement. Figures 15.8 and 15.9 show an example of a "performance dashboard" which enables purchasing to monitor and communicate its performance.

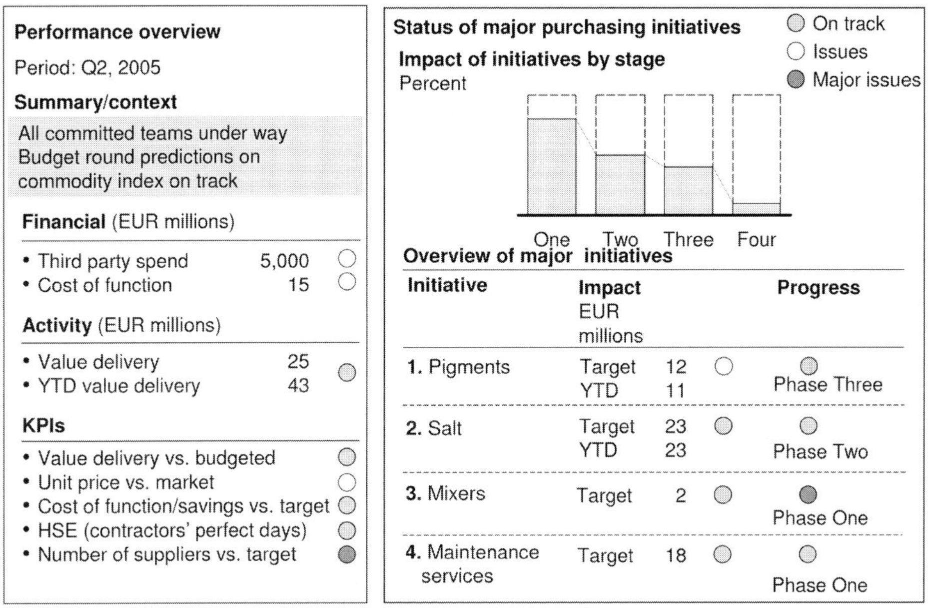

Fig. 15.8 Tailor-made measuring system – example.

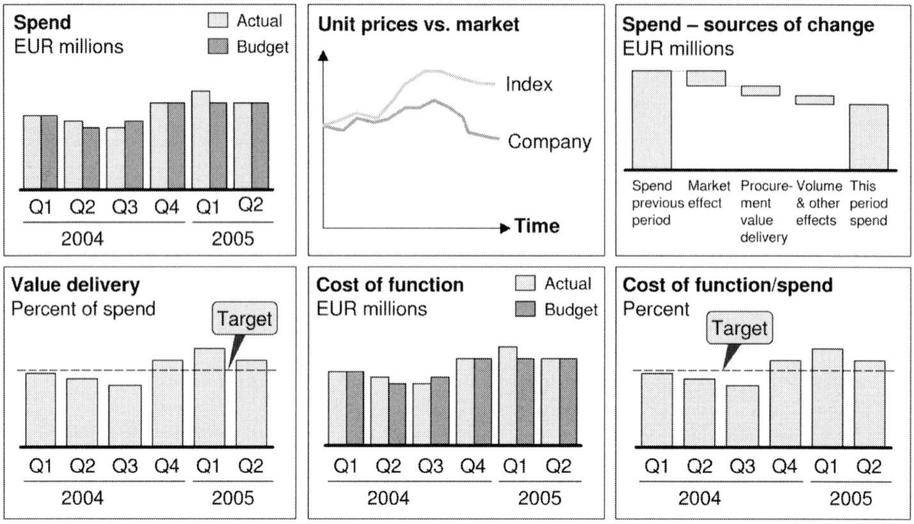

Fig. 15.9 Tailor-made measuring system – selected metrics.

- *Hold performance review meetings with "real teeth" where corrective actions are taken in the case of under-performance*: as mentioned earlier, performance review meetings must never be forums for "bashing" the purchasing function for not delivering the expected impact. They have to be productive sessions utilizing the agreed controlling principles and following the right steps to arrive at a set of corrective actions if initial targets are missed. If these performance reviews are run correctly, our experience shows that comments like "I don't see the money on my bottom line" often turn into "Now I understand where the value leakage in my business is".
- *Restrict the authority to commit to third party providers to the purchasing function alone*: it needs to have overall control over the company's purchasing, in order to receive the best prices through scale and gain the overview of problem areas. In many companies today, however, the purchasing function does not have the sole authority to commit spend to external suppliers. Site and operations managers keep control over their budgets and barely involve purchasing in the selection process for critical contractors like maintenance services, while departments responsible for non-traditional spend areas like marketing, IT, or legal services often deny purchasing any say whatsoever in the sourcing process for these commodities/services.
- *"Hard-wire" savings and other value added into budgets*: performance improvement targets need to be hard-wired into the busi-

ness units' budgets, and individual budget holders must be held accountable for delivering these results (whether budget adjustments are made directly or "shadow savings budgets" are used is a secondary issue).

- *Reduce renegade buying by removing unapproved vendors from the accounts payable list*: users are typically resistant to changing vendors they have been working with for years. It is far too easy to call up the supplier next door to get the parts or services delivered right away rather than complying with corporate frame contracts. Finance and purchasing need to work hand in hand to remove obsolete or unapproved vendors from Accounts Payable so that no payments can – or will – be made even if orders are placed with these vendors. Renegade buying will die very soon afterwards.

15.5 Summary

Well-designed and well-executed purchasing transformation programs enable chemical companies to sustain cost reductions and create additional value. Linking purchasing's objectives to the business units' goals, truly changing the mindsets and behaviors of the people involved, and tightly managing performance using a set of key metrics build the foundation for long-term success, and position purchasing as an essential driver of competitive advantage.

Once these elements have been put in place, the purchasing function will have earned the credibility to have a seat at the strategy table. The broader organization will then look to purchasing, for example, to help source innovation from the supply base (e.g., new technologies, products, value-added services), help develop new markets (e.g., by leveraging the supply base to support market entry), reconfigure the value chain (e.g., insource/outsource, optimize the global production footprint), or contribute to the risk management of key cost drivers such as major raw materials, energy, or capital expenditure.

16
Feedstock Price Volatility and How to Deal with It

Scott Andre and Sunil Sanghvi

The decimation of US nitrogen fertilizer and methanol production since late 2000 due to rising North American natural gas prices is the most dramatic consequence to date of the trend towards higher feedstock prices and higher pricing volatility, a trend that is affecting the chemical industry globally. Producers of both commodity and specialty chemicals have seen major increases in feedstock prices and price volatility since 2000, and have in too many cases been unable to pass their higher costs through to customers, which has damaged returns. These developments have put chemical makers on notice that some important shifts are under way in the relative attractiveness of different types of feedstock as well as in the competitive positions of different geographic regions.

The changes are unfolding just as industry leaders are confronting important choices on their companies' future growth plans: first, they need to decide their strategies in emerging markets, notably in China, and second, they need to decide what – if any – investments they should make in many product lines where profits are high and supply tight after an extended period of low capital spending. In this chapter we outline the long- and short-term drivers of price volatility in different feedstock and chemical companies' exposure to them, and describe the steps companies can take to minimize the impact on their profitability. At the same time, we advise companies to be on the lookout for new opportunities arising from changes in the feedstock landscape.

16.1
What is Driving Feedstock Price Increases and Volatility?

Chemical producers have witnessed a dramatic run-up in their feedstock prices in the last four years, whether natural gas, crude oil derivatives such as naphtha, or petrochemicals such as benzene. The US chemical industry's feedstock bill in 2004, for example, was up USD 31 billion compared with 2002, representing a 57 percent increase in the net sales value of these basic chemicals. This dramatic overall surge conceals a number of different drivers, each with its own dynamics (Fig. 16.1). Basic hydrocarbon costs were up 72 percent, led by crude oil and nat-

Value Creation: Strategies for the Chemical Industry. 2nd Edition. F. Budde, U.-H. Felcht, H. Frankemölle (Eds.)

ISBN: 3-527-31266-8

ural gas price increases. These accounted for about two-thirds of the total feedstock price hike. The remainder came from pricing and margin spikes in several building block chemicals such as ethylene and aromatics. What were the causes?

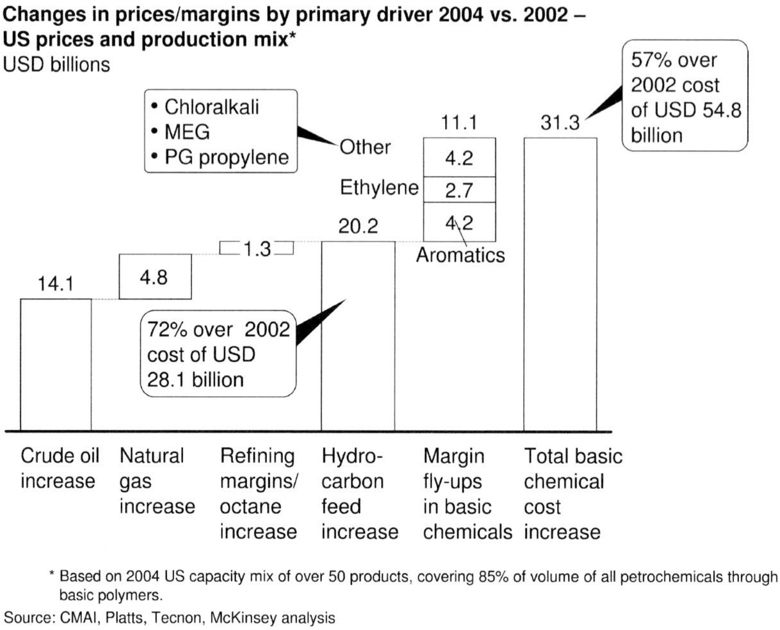

Fig. 16.1 The large increases in basic chemicals costs and prices were due to a few very big factors.

16.1.1
Crude Oil Surprises

Crude oil prices have been pushed by healthy demand led by China, which has outstripped new supply. Political concerns about the stability of the Middle East have combined with this tight supply-demand balance to enable the OPEC cartel to keep control of world oil pricing at a higher level than forecasted as recently as 2002. While a full examination of crude market prospects is beyond the scope of this chapter, the long-term outlook is for OPEC to continue to hold crude oil prices above USD 30/bbl due to the revenue needs of major OPEC member countries. At the same time, the expectation is for oil prices to continue to exhibit considerable volatility around that price.

16.1.2
Natural Gas Price Contrasts

Natural gas pricing varies from one production location to another, in contrast to crude prices which have an even-handed effect on the chemical industry worldwide. The dramatic shift since 2000 has been the more than doubling of US natural gas prices, as supply tightness has resulted in a shift to higher pricing mechanisms. US natural gas prices have always been capped by the price levels of the alternative fuels that power plant operators can burn, but in the past decade, the pricing trend has moved up as latest-generation power plants have started to use higher-priced distillate instead of fuel oil as an alternative fuel (Fig. 16.2). Severe gas price spikes occurred in the winter of 2000/2001 and early 2002/2003, when demand could not be met by substitute fuels, and gas prices increased to levels that forced a number of users to shut down since their operations were no longer competitive.

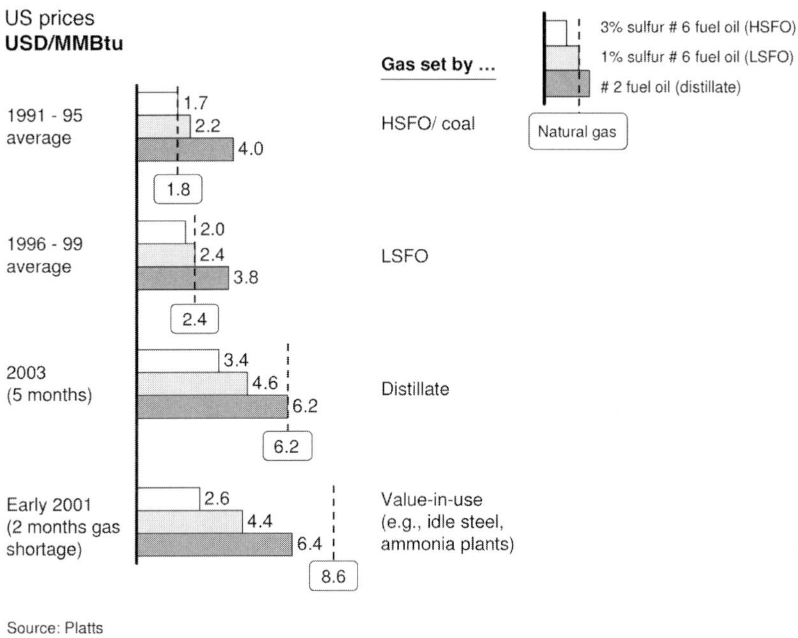

Fig. 16.2 North American natural gas prices link to the alternative fuels needed to meet energy demand.

Prices for the major natural gas fractions – methane, ethane, and propane – settled after the spikes at much higher levels than those seen in the 1980s and 1990s. The result has been a change in worldwide competitive economics: the unseating of the giant US petrochemical industry from its former advantaged-feedstock position as swing producer of numerous products, and its forced pull-back from export markets. US exports of styrene, for example, fell more than 50 percent in 2001 after the natural gas price spike at the beginning of that year.

There seems little likelihood of the USA regaining its advantaged low-price position on natural gas without government intervention, given the growing demand for natural gas as a power plant fuel. At the same time, US natural gas prices are likely to continue to show extreme volatility during peak demand periods such as extreme cold or hot weather, swinging between several levels such as fuel oil and distillate equivalence and possibly further influenced by liquefied natural gas (LNG) and Canadian gas flows to the USA.

In direct contrast to North American chemical companies, the chemical makers already sourcing their natural gas feedstock in the Middle East or in other low-priced locations such as Trinidad have seen their advantage multiply as crude oil prices and North American natural gas prices rose (Fig. 16.3).

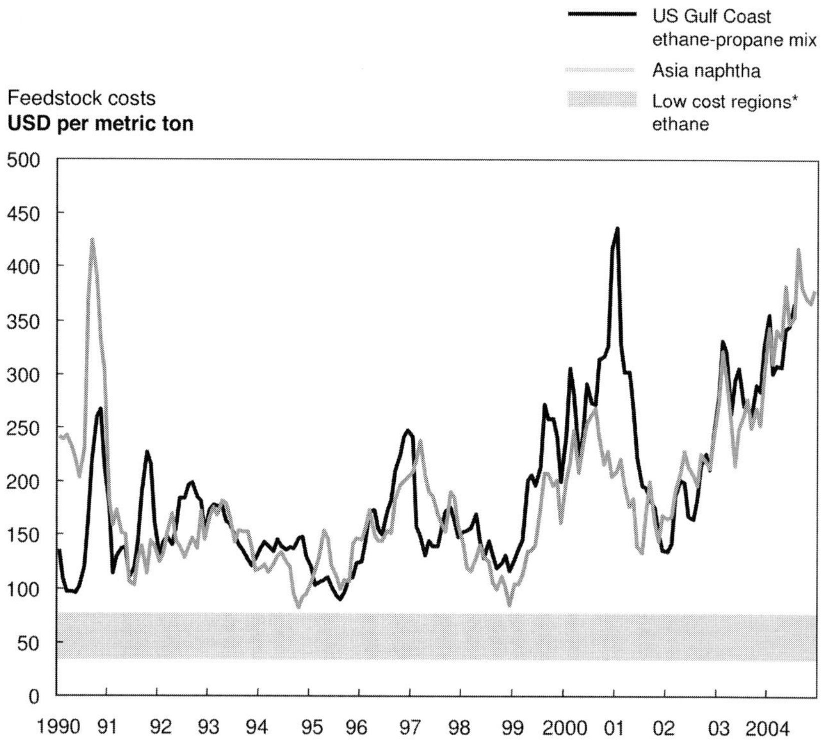

* Includes the Middle East
Source: Oil and Gas Journal

Fig. 16.3 The feedstock cost differential between major markets and low cost producers is widening.

16.1.3
Petrochemicals – Layers of Volatility

The price spike in the benzene market during 2004 highlights the complex layered volatility effects that can touch feedstock prices, in addition to the underlying push from higher crude oil or natural gas prices. Benzene prices spiked up in the first half of 2004 to USD 4/gal, triple the prior five-year average, during a period when crude oil prices rose 68 percent. Besides higher-priced crude, the key reasons for the benzene price rise were a combination of aromatics demand growth and tight capacity at a number of processing stages in refineries, which have been starved of investment due to poor returns during the past decade (Fig. 16.4).

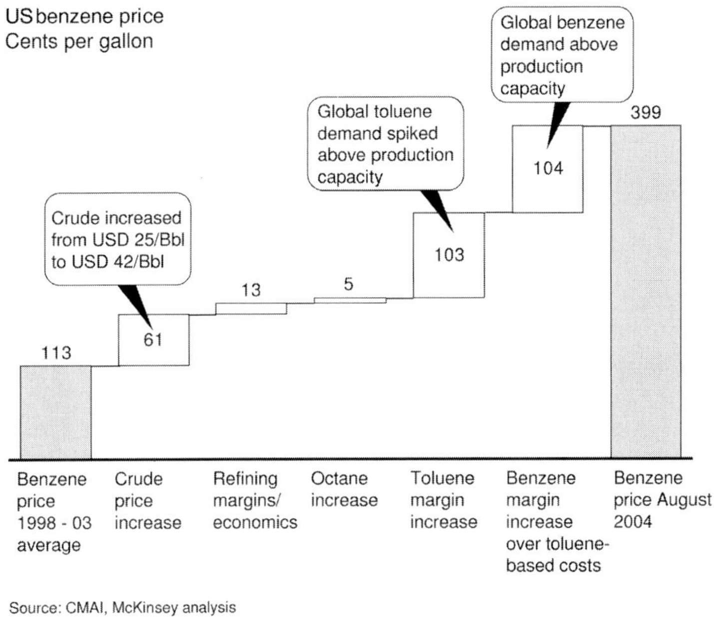

Fig. 16.4 Benzene's price spike resulted from pressure all along the chain.

Similar dynamics play in other non-refinery petrochemicals that moved to fly-up conditions in 2004, such as monoethylene glycol (MEG) and styrene. Petrochemical producers had held back investments during the previous several years due to poor returns. The underlying upward trend in demand tightened the supply-demand balance, and restored pricing power to producers' hands, enabling them to harvest exceptional margins at this point in the cycle.

Certain products experience multiple volatility effects, due to tightness at different points in their production chain that create a number of layers of volatility. Take the case of polystyrene and styrene monomer: the price of the benzene portion of the styrene molecule rose in 2004 due to a combination of higher crude oil prices and tight toluene as well as benzene capacity, while higher natural gas prices and tight ethylene capacity and plant outages pushed up the price of the ethylene portion.

Chinese demand has been the main driver of the world petrochemicals recovery since 2002, and while scenarios exist under which China's demand falters, the consensus view is that it will continue, and that petrochemical prices will be sustained at high levels until new capacity comes online. The products' elevated price levels and margins are likely to attract new investments in the next few years. At this point the products seem likely to resume the cycle of overinvestment leading to a price bust later in the decade, followed by a period of underinvestment leading to a new price spike several years later. While the drivers of the price movements contrast with those of crude and natural gas and are more inherently short-term, they ensure similar and continued price volatility.

In the case of petrochemicals, other factors will compound that volatility. China's purchasing rates can vary considerably, for instance cutting purchases when large inventories of downstream products build up, or accelerating purchases when price increases are expected, with consequences for international market pricing. In addition, there are a number of products that can quickly move in and out of export markets depending on margins, contributing to price and supply volatility. Finally, tight availability in the shipping market leads to periodically higher shipping costs, and increasing regional price differentials.

16.2
A Brave New World of Feedstock

The consequences of these feedstock price increases and volatility are being felt around the world. Natural gas price spikes have shut down much of the US's methanol and ammonia industries, and on occasion forced temporary closure of US ethane-based olefins capacity. The higher natural gas prices have eroded the historic feedstock cost and capital advantage that North American producers enjoyed over European, Japanese, and Korean competitors, whose petrochemical industries are primarily based on naphtha feedstock refined from crude oil, and crimped North American demand growth prospects. With North America's fall from grace, all the traditional established bases of petrochemical production are challenged on competitive feedstock access, and do not appear to provide attractive underpinnings for new investments. Methane and ethane supplies that are "stranded" in isolated gas and oil production locations are a more attractively priced source, and high oil prices are also reawakening interest in coal as a chemicals raw material as well as prompting companies to reconsider the possibilities that could lie in bio-based feedstock. However, while companies caught up in higher priced and more volatile feedstock need to consider all their options in the worldwide picture, these avenues are far from delivering the whole answer to their volatility problems.

16.2.1
Step on the Gas

Saudi Arabia has built a chemical industry since the late 1970s in partnership with Western oil companies to make chemicals from natural gas produced in association with crude oil that was previously flared. Other countries in the region have followed suit, including Kuwait, Iran, and Qatar. All these countries continue to expand production, usually in partnership with Western chemical companies anxious to secure low-priced feedstock. Other parts of the world with stranded gas are seeing increased investment: ExxonMobil, for example, is negotiating a petrochemical project in Jose, Venezuela, a location where methanol and ammonia are already produced.

New business models are also emerging. The launch of giant liquefied natural gas (LNG) and gas-to-liquids (GTL) projects in the Middle East is increasing ethane availability from incremental natural gas production. Also under evaluation is the production of a range of chemicals based on giant methanol plants sourced with low-cost stranded natural gas.

However, while stranded gas supplies are essentially unlimited, access to low-cost ethane for ethylene plants does have limits. Ethane supplies are developed only with natural gas or crude oil production: regional consumption and LNG demand will set the demand for natural gas, while OPEC will determine crude oil output. We estimate that the available ethane in the Middle East could only cover about 40 percent of all new worldwide ethylene demand through 2010, by which date the region's ethylene market share would reach 17 percent. Even if stranded natural gas and ethane in other locations are put to use, they will be far from covering all demand growth, and chemical producers will need to continue to rely on the established types of feedstock, while at the same time considering alternatives (Fig. 16.5).

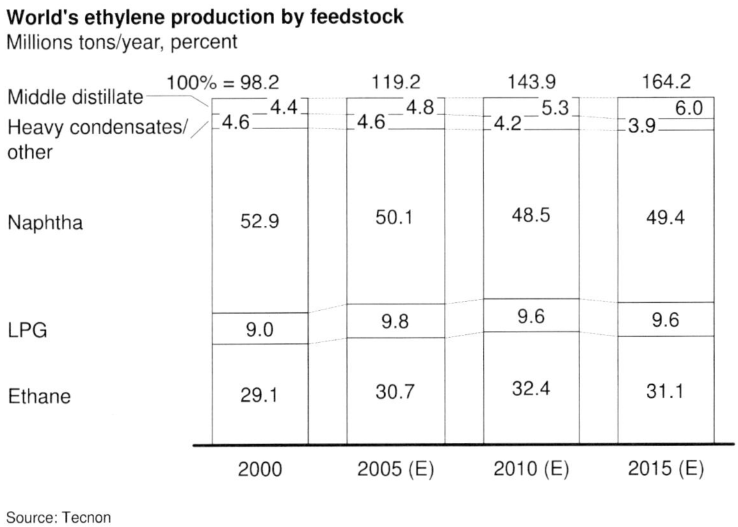

Fig. 16.5 Supplies of "stranded ethane" are limited and thus other cracking feeds will grow along with ethane.

16.2.2
Coal Gets a New Look

High oil prices have led to a revival of interest in the use of coal as a chemical feedstock. Coal had been the primary feedstock for the world's chemicals production until the mid-20th century, but its use was eclipsed by ready availability of oil and gas. The major exception was apartheid-era South Africa, where Sasol employed Fischer-Tropsch coal-to-gasoline production technology and made chemicals as well. Coal's relatively stable and low price compared with crude oil and natural gas are bringing the material back into discussion, particularly in the context of serving China's huge market growth potential since the country is a major coal producer. Coal gasification plants are very capital intensive, and the technology is still in need of process improvements, but, with the new interest, it is attracting research investments. A number of projects are progressing in China involving Chinese entities and Western partners, including Shell and Dow Chemical. Our analysis suggests that under a high crude-oil price scenario, coal-based polymer production in China would undercut naphtha-based plants and be competitive with Middle East imports – although bettered by methanol to olefins (MTO) based production (Fig. 16.6). No large-scale MTO plants have yet been put into operation.

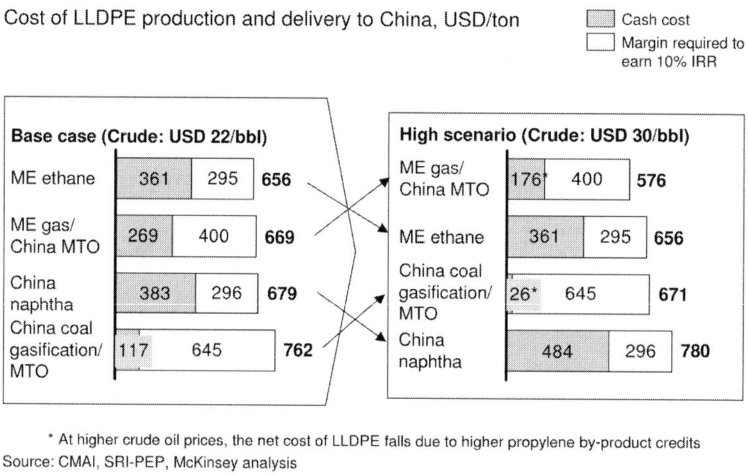

Fig. 16.6 Higher crude prices make additional gas- and coal-based routes to petrochemicals more competitive.

16.2.3
Bio-based Feedstock

The use of bioprocessing – the enzymatic conversion of natural raw materials such as corn or sugar cane to chemicals – is another feedstock route that a number of companies are exploring and which will become more competitive if prices of basic hydrocarbons increase greatly. DuPont is developing specialty polyester production based on a bioprocessing route, and a number of companies, including Cargill, are working on building block chemicals derived from bioprocessing, including lactic acid and succinic acid, that could be used to make specialty polymers and chemicals. There is currently widespread skepticism at chemical companies that bioprocessing can deliver production volumes to compete with hydrocarbon feedstock. However, the technology is advancing steadily.

16.3 Meeting the Challenge of Feedstock Volatility

As we have seen above, alternatives to high prices are hard to find, and producers still have to cope with volatility. What can they do in such a situation? The first step is for a producer to analyze its exposure to feedstock volatility across its complete value chain to avoid inadvertently overlooking a natural internal hedge. It should examine competitive conditions in the various markets in which it participates to analyze where it would be able to pass along price increases to protect its margins if feedstock prices rise, and where it would not.

Where a producer is not going to be in control of its pricing destiny – which is likely to be the majority of cases – it has a variety of approaches available to cope with feedstock volatility, which can be used separately or in combination: in the short term, it should adapt its contracting approach to the unstable feedstock pricing environment, and it should consider financial and physical hedges for its feedstock positions when costs cannot be passed through. In the medium term, the producer should make its operations more feedstock flexible and optimize them. Finally, in the long term, it should consider whether to restructure its asset and product portfolio to reduce its exposure to volatile feedstock, e.g., backward integrate directly or via alliance with an upstream producer, and build up feedstock-advantaged operations. Once a company has integrated back – directly or contractually – to basic hydrocarbons that are traded on exchanges, financial markets can be used to hedge the volatility of those hydrocarbons and thus eliminate the chain's remaining exposure. For accounting purposes, a producer may want its suppliers to do the hedging and provide fixed or indexed priced contracts on the physical sales to the company. Producers should also be constantly alert to occasions when feedstock price volatility gives them an advantage over competitors.

All of these actions require a thorough understanding of the current and potential cost and price drivers along the entire chain. Downstream players more focused on customer needs and product performance often lack this knowledge and may need, for example, to hire upstream experts to achieve the skills needed to dissect these exposures and identify cost-effective hedging solutions.

16.3.1
Optimizing Contracts

As companies move into the era of volatility, they need to be aware that they are bringing with them a portfolio of legacy contract arrangements that may put them at a severe disadvantage in the new circumstances. Producers need to analyze their portfolio, and identify contract changes that will protect them against losses while maximizing gains. Any producer encountering feedstock price volatility needs to keep as free a hand as possible to renegotiate its sales prices to reflect higher feedstock costs, and many will need to develop skills at rapidly pushing price increases through. Producers in a number of markets have started to respond to these changed circumstances, for example switching contracts to monthly instead of quarterly settlement, eliminating "price protection" clauses which delay price increases, and adopting escalation clauses tied to changes in prices for key inputs such as feedstock and energy.

As well as incorporating standard terms and conditions, such as building a margin floor into a contract to assure a minimum margin or reducing payment terms, producers should consider sometimes-overlooked changes such as requiring that a customer accept an equivalent grade, and eliminating truck sales in favor of cheaper-to-handle railcar sales. Our experience suggests that such overlooked changes can add nearly 5 cts/lb of value to a company's sales contracts, compared to less than 1 ct/lb for standard changes.

16.3.2
Hedging – Financial and Physical

The most efficient way of dealing with volatile feedstock prices is to be able to pass them straight on to customers, but producers unable to do so should consider a hedging approach that includes getting physical access to feedstock as well as financial hedges. Before embarking on a hedging strategy, a company should consider how this fits in with its corporate policies on risk tolerance and make sure that one division's hedge fits with those of other divisions. It also needs to consider the wide variety of hedging vehicles available, such as hedging against peaks versus the entire exposure. But it should beware of the usually high costs associated with hedging vehicles offered by third parties: better deal structures – and easier accounting treatments – are usually available from negotiating directly with suppliers.

In the case of natural gas, physical hedges could include investing in natural gas production or setting up a direct supply relationship with a natural gas producer; investing in LNG imports as Dow Chemical is doing in Texas; and investing to get access to cheaper hydrocarbon sources, such as coal. The same model can be applied further downstream by customers who purchase petrochemicals as feedstock. Over the years, major consumers have invested in their supplier's operation, and continue to do so: according to press reports, a number of major benzene consumers are considering investment in benzene production – either directly or through cost-based supply contracts – following the 2004/2005 price spike. These companies need to assess the best timing for such investments, considering how much new capacity is already coming on-line in response to the fly-up, which will impact future benzene pricing.

Consumers can also negotiate with feedstock suppliers on upfront payments or payment terms under which they pay a higher price than the lowest market price at the trough, but pay lower prices when product prices spike. An interesting application of this is the potential for an ethane cracker operator to convert the economics of its cracker to those of a "virtual" naphtha cracker, by paying an integrated gas producer-processor a price for ethane indexed to naphtha-based ethylene production costs.

Chemical companies that are part of integrated oil companies such as ExxonMobil, Shell, or BP have their own built-in hedges in the sense that the upstream business reaps profits from high feedstock costs. BASF's successful gas and petroleum business has similarly served as a hedge to its hydrocarbons-exposed chemical operations.

16.3.3
Fleximizing

Producers can also consider increasing the flexibility of all the aspects of their production operations, in order to increase the impact of optimization. We call this approach "fleximizing". In the new environment of volatile prices, the ability to switch between different types of feedstock for olefins production, for example, has become more valuable than ever. Chemical companies should also fleximize their production to maximize output of the highest margin products, if necessary making investments to enable them to be more nimble in switching between co-product yields where the process generates multiple products.

Producers can also source across a mix of plants, and maximize their production at plants offering the best economics at a given moment. Leading methanol producer Methanex exemplifies this approach, optimizing production across its global network. A number of players in the fertilizer industry are also developing this approach, including Koch Industries and Yara.

For a company to be able to switch its output to the plant offering the best short-term economics, it must have the capability to rapidly stop and start production – a difficult undertaking with large petrochemical facilities. But in the new era of volatile feedstock, producers should consider investing in R&D to improve their performance in this area.

Where a producer is likely to shut down production for months at a time when feedstock prices are too high for economic operations, it should investigate converting as many of its fixed costs as possible into variable costs, including not only outsourced IT and maintenance but also a part-time workforce. Companies can also fleximize their procurement operations, for example increasing their sourcing flexibility by having multiple feedstock contracts with different pricing terms, and making sure the contracts offer considerable leeway in terms of the maximum and minimum quantities the buyer can take under the contract.

16.3.4
Looking for Advantage

Producers can use their in-depth understanding of their feedstock exposures relative to those of their competitors to take advantage of their positions of strength. They can capture market share from competing materials, or maximize margins, when confronting a competitor with higher feedstock costs. For example, at times of high natural gas prices relative to crude oil, polypropylene producers – whose costs are tied to crude oil-derived gasoline – might be able to capture certain plastic-processing applications from polyethylene makers whose costs are tied to gas. Under the same scenario, ethylene producers using crude oil-derived naphtha would enjoy lower costs than ethane-based competitors, which they could either exploit by trying to win business at a lower price, or simply enjoy by banking their higher margin (Fig. 16.7).

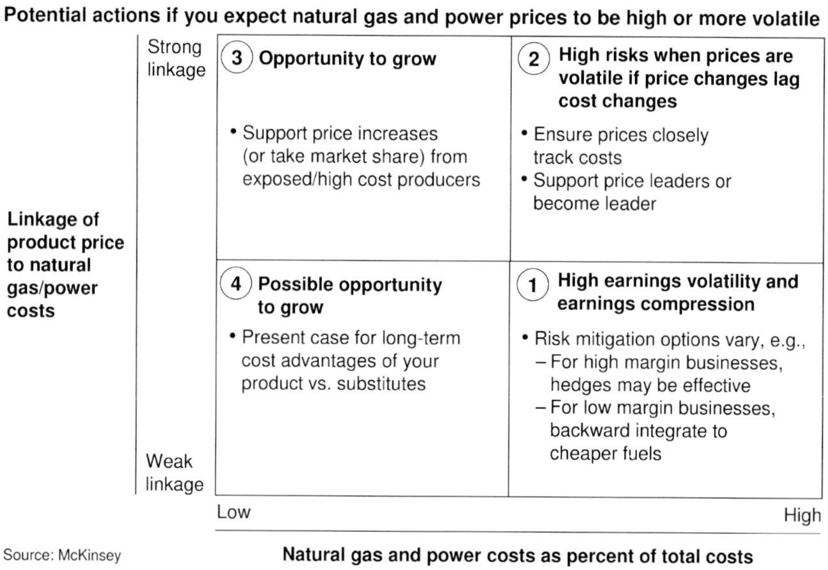

Fig. 16.7 Actions to cope with high and volatile gas and power costs.

16.3.5
Restructuring Assets and Taking the Offensive

Producers also need to consider the attractiveness of restructuring their asset base in order to reduce their exposure to high feedstock prices and feedstock volatility, and to get access to better opportunities. Many companies with global operations are already well down this path. Methanol producer Methanex, mentioned above as an expert optimizer, has permanently shut down production in areas with high natural gas costs such as the USA and expanded production in low-priced gas locations such as Chile. Similarly, the launch of world-scale acetic acid production in Saudi Arabia to take advantage of low-cost natural gas for syngas and methanol supply is likely to displace production in higher-cost locations such as Europe over the medium term.

Producers of commodity plastics and petrochemicals such as MEG have followed a similar course of repositioning their asset base to best take advantage of future opportunities. Although they have not closed significant capacity in their established home markets, they are dividing their new investments between the Middle East, where they can take advantage of low-priced gas feedstock, and China, the major growth market. Some companies such as BASF are focusing their new investments primarily on China, others such as Dow Chemical are focusing on building in the Middle East to serve Asia, while others, including ExxonMobil, Shell, and Total, are pursuing both approaches.

Producers also need to look out for ways to take advantage of new technology developments that will enable them to access low-cost feedstock. As mentioned above, these include MTO technology using "stranded" natural gas, the revived interest in coal-based production, and bioprocessing.

16.4 Conclusion

Few drivers of profitability in the chemical industry have such dramatic effects as movements in feedstock pricing. In volatile feedstock price climates, chemical company executives need to make sure that their companies are taking active roles in managing their feedstock exposures. While few producers enjoy the luxury of being able to pass higher feedstock costs through unchallenged, managers can nevertheless access an array of approaches, resources, and tools to moderate their companies' vulnerabilities and exposures, as outlined in this chapter.

16.5 Summary

- Chemical producers have entered a period of higher and more volatile feedstock costs, whether they consume natural gas, crude oil derivatives, or basic petrochemical building blocks.
- Prices for oil are likely to remain at elevated levels and to continue volatile due to OPEC's influence on the oil market, while a tight supply-demand balance will hold natural gas prices up in North America, maintaining the region's disadvantage to competitors using stranded gas. High profits may encourage petrochemical

capacity additions that will bring prices down later in the decade and provide relief to buyers, but prices will continue to be volatile.

- Chemical producers should assess their exposure to feedstock volatility. In cases where they will face difficulty passing higher feedstock prices along to their customers, they can consider a variety of approaches to mitigate their impact.
- In the short term, producers should adapt their contract arrangements to the unstable environment, and consider financial and physical hedges.
- In the medium term, producers should ensure their operations have feedstock and production flexibility to enable them to ride out price volatility.
- In the long term, producers should evaluate restructuring their portfolios to reduce exposure to volatile feedstock, and build up feedstock-advantaged operations based on low-priced gas or – if competitive technology is developed – methanol-to-olefins, coal, and bioprocessing.

17
Taking a Leap in Purchasing

Gregory Nelson

For the chemical industry, with raw material cost ranging anywhere from 45 to 60 percent of the cost of sales depending on a company's place in the industry segment or value chain, profitability is highly susceptible to variations in feedstock and raw material prices. Reducing the cost of sales is therefore as important to the profitability of the enterprise as sales growth. The old adage rings true – a penny saved is a penny earned.

Over the last decade, a small but growing number of chemical companies (for example Dow Chemical, DuPont, ExxonMobil, but also a number of smaller players) have discovered procurement as a virtually untapped source of value creation, and have leveraged it from within their organizations to meet the profitability challenges facing their industry.

These companies' experience shows that cash can be delivered to the bottom line by leveraging the procurement function: the total cost of procuring feedstock but also other goods and services such as maintenance, repair, and operations (MRO), logistics, and corporate services can be significantly reduced, making a real, immediate and long-term positive impact on profitability.

Furthermore, the purchasing organization is in a unique position to initiate and drive important strategic opportunities to expand the company's degrees of freedom through its vast supplier network: for example, the product and service range can be extended by leveraging suppliers, and new regions and markets can be developed through global sourcing.

17.1
Maximizing Value Creation

Successful chemical companies have discovered that multiple levers need to be pulled to achieve functional excellence and create a sustainable competitive advantage through procurement: standardizing products and aggregating volumes to leverage purchasing power, changing to more cost-effective input materials and simplifying products to optimize their manufacturing, and improving processes. In addition, world class companies have all the facts about markets, prices, cost

Value Creation: Strategies for the Chemical Industry. 2nd Edition. F. Budde, U.-H. Felcht, H. Frankemölle (Eds.)

ISBN: 3-527-31266-8

structures, etc. at their finger-tips, and utilize their knowledge to prepare well for negotiations.

17.1.1
Using Purchasing Power

Often, chemical companies develop a culture that leads to the use of (over-) specialized raw materials for products and unique equipment to build and maintain their manufacturing facilities: the "not invented here" syndrome. This is sometimes taken to extremes, with different production units in the same location developing different designs or specifying different grades of raw material to produce the same product for their customers. Little or no commonality among the raw materials or other products purchased means that the purchasing function is unable to leverage volume in the marketplace.

By developing common standards and specifications for raw materials, packaging, equipment, etc. companies can pool their volumes across locations, business units, and geographies, and instead of approaching suppliers with small lots of differentiated materials, exert significant volume-based power in the market. This packs a real punch when the organization is also able to rationalize the number of its suppliers, pushing up the volume, and leveraging its clout with suppliers even further. It can be lucrative for the chosen suppliers, too, as they can simplify their own business processes and cover fixed costs better.

For smaller-volume materials and services, collaboration with third parties to achieve critical mass is an effective means of increasing buying power. Ways to do this include partnering with suppliers, forming consortia, or buying for customers and strategic partners. For small and large companies alike, this extension of the volume game outside their own boundaries can prove a powerful tool.

With globalization increasing in all industries and the international trade in chemical products growing faster than global chemical production, companies are being forced to locate production capacities closer to the sources of key raw materials and customer markets. It is therefore critical to pro-actively broaden the strategic focus by sourcing globally and rapidly qualifying new, low-cost suppliers in conjunction with optimizing the production network.

For many raw materials, viable alternative suppliers are developing in most low-cost regions, and over the past ten years there have been significant improvements in the quality and reliability of materials, especially those from India, China, and Eastern Europe. These advantages combined with the lower prices make it an imperative of global competitiveness for all chemical companies to be able to access global sources (Fig. 17.1). It is essential to include low-cost suppliers in most requests for quotations (RFQs), and secure the impact quickly by committing to them and ensuring that they are given as much business as possible if price, quality, and delivery are right. Commitment is necessary in order to build credibility: this should not be an exercise in getting pricing information to use as leverage with existing suppliers, but a real business opportunity for new suppliers.

However, global sourcing is an excellent way of understanding and leveraging the world market. By understanding the cost structures of global players, real improvements can be achieved with existing suppliers. When procurement volumes are given to the best global sources, local players are forced to improve their productivity to become more competitive.

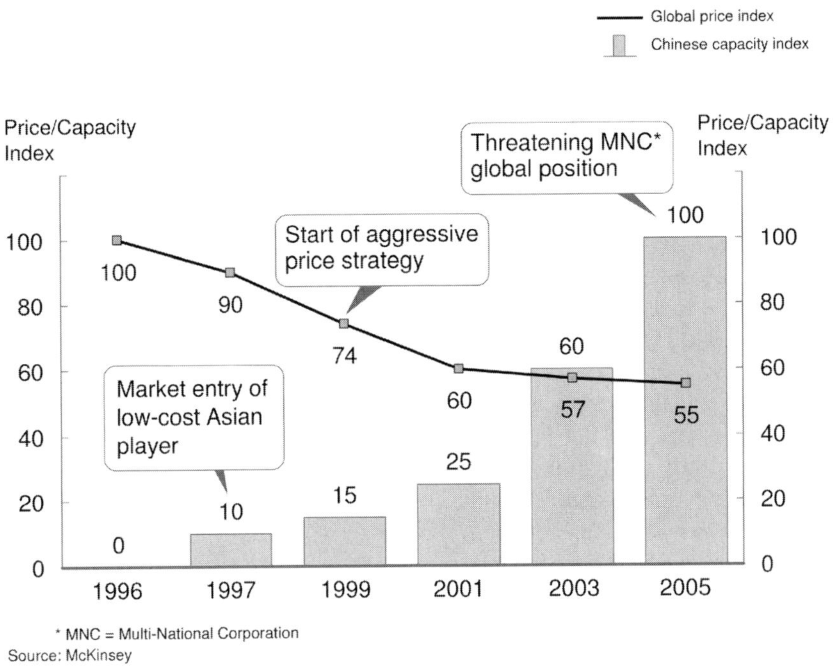

Fig. 17.1 Example of price development after market entry of low-cost Asian player.

A particularly powerful application of global sourcing is the introduction of global suppliers to break up oligopolies. Such cases require a deep commitment to developing the capacity and capability of new sources of supply, perhaps by training and educating global suppliers to take on new products from further up or down from their current place in the value chain.

17.1.2 Making Aggressive Input Changes

Changing the qualifications of input materials to make them simpler, more effective, and cheaper is another important lever. Major savings can be achieved by partnering with the R&D, manufacturing, maintenance, and marketing functions to define clear customer specifications and to determine the requirements on "fitness for purpose" or quality that will prevent over-engineering and enable purchasing to source the most cost-effective raw material for the application.

It is sensible to take the attitude that it must be possible to simplify or substitute inputs. The purchasing function as a whole just has to be determined enough in benchmarking across business units and suppliers. Selecting a few products and comparing them across business units typically uncovers substantial value.

Here, too, suppliers in low-cost countries have proven to be a particularly important source of improvements in creating innovative and highly effective products at low cost, leveraging low cost input and capabilities to develop attractive products for local markets.

17.1.3
Scrutinizing Processes to Cut out Costs

To optimize processes, total cost and value have to be examined along the entire value chain, from the suppliers' suppliers to customers' customers, to understand where cost can be eliminated, or where suppliers can be leveraged to take over certain parts of the supply chain (e.g., logistics, quality testing). The administrative system also has to be scrutinized to find ways to remove unnecessary steps. Logistics costs can be significantly improved by shipping larger truckloads (instead of smaller packages), by grouping a large number of product groups in a single shipment, and by implementing a vendor-managed inventory system. Establishing the right level of supply chain decoupling, with committed lead times and safety stock across the value chain, can accomplish much lower inventory levels even for critical products, without the company compromising supply reliability or becoming dependent on suppliers.

17.1.4
Getting the Most out of Negotiations

Excellent negotiating skills are necessary to create competitive advantage, but they are no longer sufficient on their own. In the current environment, good negotiation skills have to be re-enforced with a superior fact base, tools, and well thought through negotiation strategies.

First, in order to bring major competitive pressure to bear on the supply market a company has to create the maximum transparency about its suppliers. It has to gain an understanding of cost structures and supply market dynamics, with the aim of having a better appreciation of the market than suppliers do themselves. One of the most important tools here is aggressive entitlement pricing, i.e., rapid-fire commercial negotiations based on an understanding of supplier cost structures, redirecting volumes to the suppliers with the most favorable cost structures, and understanding win-win options.

When purchasing power is leveraged, suppliers tend to insist that the lowest price is reached because of the volumes purchased. They will also often say that their preferred customers already receive the lowest prices (Fig. 17.2). To validate this and ensure that the company is getting the maximum value, negotiators should "titrate" the situation by finding the price at which the supplier is willing to walk away from the business.

"For God's sake, think! Why is he being so nice to you?"

Fig. 17.2

Finally, all negotiations need to be prepared and set up in a way that leverages cross-functional knowledge, and ensures a large degree of freedom for the negotiator. Companies should always have the right suppliers participating (for example, including wild cards from low cost countries), have the right cross-functional mix of people with the right skills (e.g., analytics, product knowledge, negotiation experience) lined up, and make sure that purchasing has a mandate to shift volumes or abandon existing suppliers if required.

17.2 Creating Sustainable Value for the Business

To unlock and capture the value within purchasing, a high-performing organization is required. Transforming purchasing performance means embarking on a major journey of improvement, in which the purchasing function has to engage the rest of the organization by painting a picture of the strategic possibilities of the next horizon and what they mean for the company. It has to deliver significant bottom-line savings as well as significant value to business units. And it has to ensure that capabilities are built and performance measurement systems are put in place to make sure savings are sustainable.

17.2.1
Carrying Out an Improvement Program

Right at the outset, senior management needs to state explicitly that the purchasing program is not a one-off effort to make temporary savings, but the basis for a permanent transformation. Furthermore, top executives themselves need to be strongly committed to change, and express a clear and unambiguous willingness to look at purchasing in a new light, to find ways to leverage the full cash-generating and strategic possibilities. In essence, purchasing has to be empowered to influence the company's strategy.

The program is as much about doing things in the right sequence as about doing the right things. By conducting a high-level diagnostic, purchasing gets the rest of the company's attention and understands barriers to performance. By drilling down to uncover savings opportunities in pilot studies, it will reveal where the biggest opportunities are. And by ensuring that everyone in the team has his or her mind set on aggressively chasing quick wins as they occur during the pilot stage, it demonstrates the ability to capture value in well-defined, challenging areas.

Using the initial pilots and other available benchmarks from the chemical industry and other industries known for purchasing excellence (e.g., automotive, retail), purchasing leadership must draw up an aggressive business plan that lays out clear goals and a focused strategy for the purchasing effort – tackling all levers (purchasing power, product changes, and optimized processes), influencing total cost of ownership rather than price, building top-notch skills and a new company-wide performance culture. To create the focus and drive needed to carry out the program, it is particularly important to push targets above benchmarks in the industry, and make them truly aspirational.

Purchasing has to recruit the rest of the organization to help deliver and maintain the full impact, and must take every measure available to ensure full collaboration with the users. To achieve this, getting the right talent on board (e.g., analytical strength, technical knowledge, leadership qualities) involving the relevant decision-makers (e.g., leaders of business units, top management), setting up clear roles and responsibilities within purchasing and across organizational boundaries, and putting in place the right mechanisms (e.g., creating the right governance, incentives, and buy-in), involving external expertise, and institutionalizing cross-functional teams are more important than setting up the formal organizational structure.

The most critical of these actions, in order both to generate the best ideas and to get them implemented, is ensuring the full participation of cross-functional resources on teams, e.g., manufacturing, maintenance, R&D, and marketing. When all key areas are represented, buy-in and implementation questions can be addressed in a continuous and iterative manner. In addition, once an idea has made it through discussions within cross-functional teams it stands a much better chance of getting through the implementation phase.

It is also fundamental to maintain clear and concise communication with the broader organization throughout the program, focusing on what is going to change as a result of it: to provide transparency to all key leaders on the impact on the largest cost blocks and on the way purchasing and users work together, and to broadcast achievements as they appear. This can generate a great deal of internal goodwill and momentum, and create clarity and security throughout the company about planned and realized achievements.

17.2.2 Building Capabilities

Key for implementing the savings and making them sustainable is to put in place the right leadership and expectations and build appropriate capabilities throughout the company. In fact, showing the initial savings is less of a challenge than ensuring sustainability by truly embedding a change in behavior. In essence, for chemical companies to deliver world-class performance in purchasing, they need to do things in a completely different way, which in turn means that people must build an improved, and to some extent entirely new, skill set.

It is important to rapidly put in place a number of key capabilities: analytical skills for understanding supplier and market economics; strategy development capabilities for all major product categories (raw materials as well as other products and services); leadership skills to manage processes and stakeholders; capabilities in tracking implementation; and new knowledge about supplier qualification, including in-depth regional know-how – China, India, Eastern Europe. This full set of capabilities needs to be developed and enhanced to secure a competitive advantage. Any missing element will leave significant value on the table, delay the realization of the savings, and make it harder to make them stick.

In addition to capabilities, a core set of superior problem-solving and operating practices and tools (e.g., cross-functional teams, design-to-cost, e-tools) that meet the specific needs of the category strategy should be understood and applied to deliver value across the entire organization.

The best way to build capabilities and roll out tools is to make sure that people are exposed to them on the job. Likewise, getting the organization to train itself ensures that existing expertise is further developed and skills are cascaded throughout the organization. This includes, for example, taking a 'train-the-trainer' approach, holding focused learning sessions to disseminate new knowledge, and fostering a strong performance and feedback culture.

An excellent means to make sure knowledge and capabilities are spread and leveraged is to set up a Center of Expertise to collect information and know-how and make it easily accessible. Such a Center of Expertise should harbor product and supplier intelligence from across the chemical supply industries, organizational experience and knowledge (e.g., case studies and 'yellow pages' of experts), training materials and tools, and tracking systems for group and individual performance. It should function as a means to develop and disseminate new knowledge, share information and insights, continuously improve capabilities, and manage performance.

17.2.3
Tracking Performance

One of the reasons why the purchasing function has not been considered a strategic business function is because it has not systematically measured and tracked savings rigorously and demonstrated that they reach the bottom line. Traditionally, purchasing groups have included cost avoidance as a saving and excluded price increases from the savings figures.

In addition, budget holders are often reluctant to make the savings transparent, as this will inevitably create massive pressure from top management to exclude these savings in the next budget round.

A professional purchasing function needs to develop a process to track real savings generated by the purchasing function, and have them validated by the finance department to demonstrate how these savings are impacting the profitability of the company. A net-net approach including price increases and excluding cost avoidance should be used to track savings right down to the bottom line. What happens to the savings also needs to be made clear (e.g., are they used to invest in high NPV opportunities, or for arbitraging low performance in other areas?).

In addition to measuring financial aspects, it is important to set up a system to track other key elements ensuring full value delivery, to continuously follow up both on key performance indicators (for example, number of suppliers, quality of performance, etc.) and on whether actions have been followed through and milestones fulfilled.

Finally, tracking mechanisms should be directly linked to budgets and incentive schemes, so that people care about developing new ideas and challenging the status quo. Creating transparency between teams within and beyond the purchasing organization makes it possible to give clear rewards for high performance, and ensures an immediate reaction to any issues that arise.

17.2.4
Putting a New Organization in Place

Once the processes and mechanisms are tested and evaluated, it is important to put the right structure in place for the purchasing function.

Many companies still use a decentralized approach, with purchasing personnel reporting to the business units. This structure is supported by a small corporate strategic sourcing team that provides functional expertise for the rest of the organization. The advantage of this approach is that purchasing is close to the businesses and can respond optimally to user needs, and the businesses' leaders have a sense of control over their entire operations. However, this approach makes it difficult to maximize the benefits across the entire company, and hard for purchasing to challenge the status quo.

A center-led purchasing function that uses cross-functional mechanisms to provide stakeholder involvement and optimize demand and total value (e.g., by having raw material sourcing managers participating in the business units' strategic planning teams) is the most effective way to provide functional excellence and deliver the maximum savings (Fig. 17.3). Such a group looks across organizational silos, challenges assumptions, and transfers best practices, and can better enable purchasing managers to commit specific volumes to suppliers. The cross-functional setup ensures that the benefits of the decentralized model are not jeopardized, and at the same time the advantages of a more centralized approach can be captured.

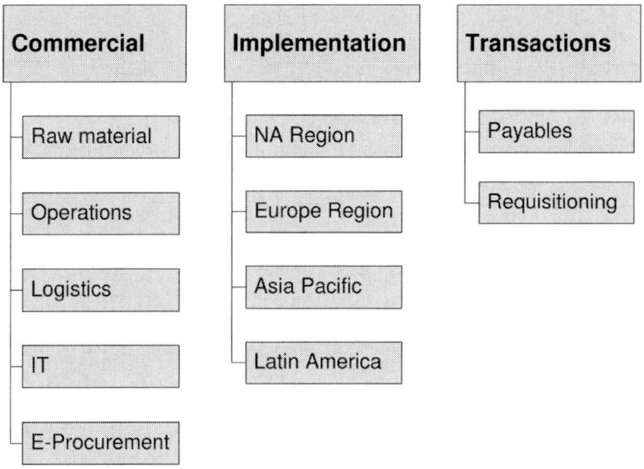

Fig. 17.3 Example of a center-led purchasing organization.

Building a world class purchasing organization involves setting up a small cadre of strategic sourcing managers leading cross-functional teams which are tasked with designing and implementing global strategies for cross-company categories (e.g., raw materials, MRO, IT, corporate services). The local buying organizations should be kept in place for materials management duties and to enable adaptation to necessary local technical and regulatory conditions, but with a clear definition and communication of supply management roles and responsibilities based upon economic, not organizational criteria.

In setting up the organization it is also crucial to access skills and capabilities in order to be able to put the right people in the right places, get rid of underperformers, and establish cross-functionality as a permanent way of working, including rotation between business units.

The geographic dimension requires particular attention when building the new organization. Coordinated operation across continents and countries with very different conditions and ways of doing things makes communication and alignment particularly demanding. The best way to address these challenges is, once more, to drive common aspirations, strategies, and methodologies rigorously across the global network. For example, global training programs and workshops should be held with the full purchasing group, global purchasing teams must meet in person several times during the course of a year and talk on the phone almost daily, and such tools as spend databases, RFIs/RFQs, and e-tools must be launched and used on a global basis.

17.3 The Next Horizon

Once a world class purchasing organization is up and running, it is possible to broaden the focus and take even more ambitious steps to further increase savings and create value. The next logical progression for an excellent purchasing function in a chemical company is to tackle: make versus buy decisions; turning cross-functional projects into a broader change management force; pushing the strategic agenda in globalization; and exploring how products and services can be improved by extracting new ideas from the supplier network.

All products should be evaluated in order to decide whether they should be made in house or externally. A key factor is the consideration of total landed cost, i.e., making sure all costs associated with an out- or insourcing decision are included (e.g., sunk costs, skill transfer, delivery and quality considerations), not just the potential gains, in order to ensure fact-based decisions that are viable in the long term.

The cross-functional mechanisms set up to optimize the purchasing function make purchasing a natural integrator to drive large-scale change throughout the company. For example, purchasing could be used as a test ground for new ideas, with the implementation of new practices starting in purchasing and working its way through the organization, or high-performing purchasing project managers could be employed to drive other cross-functional topics (e.g., strategic expansion, product development) in the organization.

Purchasing organizations' vast supplier networks provide an opportunity to drive the strategic agenda in globalization. By utilizing purchasing as the first step into a new market, major value can be added by transferring superior market knowledge, market contacts, and local knowledge to the rest of the company before production or marketing and sales are moved to that market.

Purchasing can also leverage its supplier network to identify new ideas for products and services, make improvements to existing products and processes, and initiate new strategic partnerships. By leveraging the supply base for new methods, benchmarks, technology, and competitive intelligence, purchasing can become a key source of intellectual capital, breakthrough innovation, and true advantage through cost efficiency and clear customer value orientation.

In the long term, purchasing will not only have to become more sophisticated in its core value proposition of optimizing demand and total value and influencing supplier markets, but will increasingly have to become one of the major catalysts of business optimization.

17.4 Summary

- Purchasing is a virtually untapped source of value creation to meet the profitability challenges facing the chemical industry by enabling a real, immediate, and long-term positive impact on profitability.
- Multiple levers need to be pulled to achieve functional excellence and create a sustainable competitive advantage: standardizing products and aggregating volumes to leverage purchasing power, changing to more cost-effective input materials and simplifying products to optimize their manufacturing, and improving processes.
- To capture savings and value within purchasing, a high performing organization is required. This is a transformation journey which involves engaging the rest of the organization to capture savings, deliver significant savings to the bottom-line, build capabilities, and manage performance.
- A center-led purchasing organization leveraging a cross-functional approach is superior to a de-centralized business-led approach, to maximize the benefits of all levers.
- Once in place, a world class purchasing organization can also be leveraged to take the next steps in purchasing excellence and make purchasing a major catalyst of business optimization.

18
Excellence in Operations – the Never-ending Journey Continues

Leonhard Birnbaum

Chemical companies may feel they have already wrung out all the value they can from operational improvements: it has, after all, rightly long been recognized as an area that demands serious management attention. The author takes issue with this stance – such programs have rarely resulted in real sustainable continuous improvements. As a result, improvement opportunities have in many cases still not been tapped to the full. This is good news in a sense, given the ongoing need for productivity increases in the chemical industry.

To achieve the sustainable continuous improvements of truly lean operations, companies need to execute holistic programs. They have to give their organizations clear visions for operational excellence and the ability to identify and execute on situation-specific targets, but also individual approaches to address the soft side of change – capability building and changes in mindsets and behaviors. The journey is difficult, but the rewards are still fast and very considerable.

18.1
Operational Improvement – the Bar Is Rising

In the 1980s and 1990s, the chemical industry embarked on a series of operational improvement programs, and made impressive gains – in many cases saving companies from plunging into the red, or dragging them out of a mire of losses.

Such programs also helped bolster the industry's defenses against new low-cost competitors by raising the barriers to entry for newcomers – even those that could undercut staffing costs dramatically. Innovation has also been critical in maintaining market share, but as the industry continues its trend of commoditization in specialties and a steady price-cost squeeze overall (see Chapters 5, 8), the spotlight is shifting once more to operations as the fundamental supporting structure for a successful chemicals business – except that it will become even more challenging in the future to realize sustainable savings.

Finally, operational excellence made the chemical industry competitive enough in some instances to enter new markets. The fact that the industry can now reach

Value Creation: Strategies for the Chemical Industry. 2nd Edition. F. Budde, U.-H. Felcht, H. Frankemölle (Eds.)

ISBN: 3-527-31266-8

beyond the traditional areas of Europe, North America, and Japan has played a major role in laying the groundwork for future profitability.

The McKinsey Global Institute has analyzed operational performance improvements in the industry over time, based on an investigation into labor productivity. Between1991 and 2003, the average annual growth for German chemical companies was 5.5 percent, in France it was 3.3 percent, 3.2 percent in the USA, and finally 2.6 percent in Japan.

Behind these broad numbers lies an interesting detail. With the exception of Japan the smallest growth in average labor productivity has basically occurred in the most recent time period (1998-2003). The difference is most pronounced in Germany, where the average productivity improvement was just 2.5 percent per year in 1998–2003 compared to the overall figure of 5.5 percent for the full twelve-year period. France also fell considerably, from 3.3 percent to just 1.5 percent (Fig. 18.1).

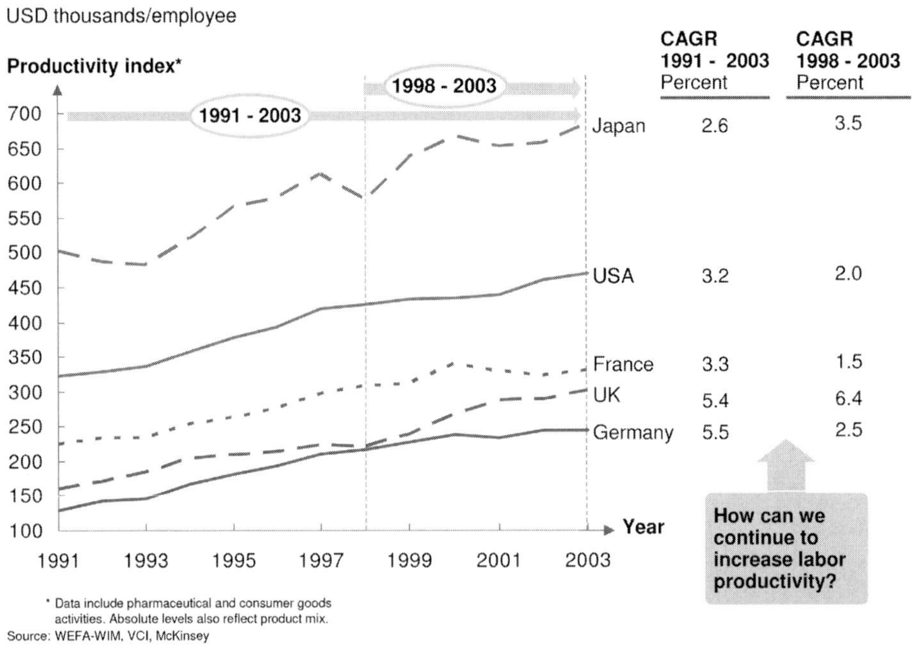

Fig. 18.1 Labor productivity in the chemical industry.

Thus, we can reasonably assume that companies have already picked off the quick wins, in which case, the figures indicate that the remaining value creation potential is declining and probably becoming more difficult to realize. This trend is certainly also supported by the fact that growth and the related capital investments are declining overall in Europe. It is always easier to realize productivity improvements in a growth environment.

In addition, the labor cost advantages of Chinese companies often compensate for their inferior overall productivity, clearly implying that the challenge of Chinese competition cannot be met sustainably by operational improvements alone. However, a series of recent operational diagnostics and large-scale improvement programs shows that operations remains a significant lever, and one that companies can ill afford to neglect. Figure 18.2 shows as one representative example the reduction of total site cost (including depreciation and amortization) realized in a global operational turnaround program during 2004. On average, return on sales (ROS) improved by approximately 3.9 percent – and the full benefit has not yet been realized.

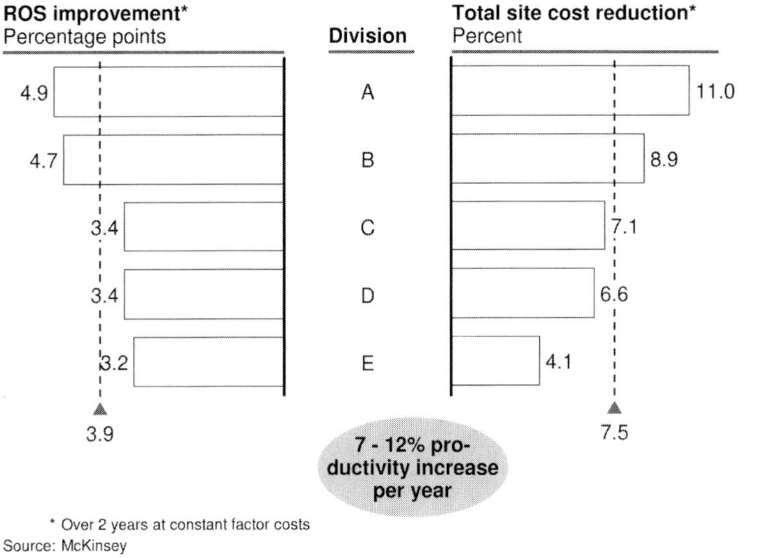

Fig. 18.2 Total site costs were reduced by between four and eleven percent in large networks.

All these issues have to be addressed in an increasingly difficult business environment. Executives have to set up operational programs not only in local (often integrated) sites, but increasingly in global production networks of locations of widely disparate size. In addition, capital markets and competitive pressure grant them less and less time for such change programs – speed of change is becoming a key factor for stabilizing or potentially expanding a business's position. However, driving sustainable change fast across all producion sites worldwide poses particular challenges. Exchange of best practices becomes more difficult due to logistics, language barriers, different mindsets, and different cultures. For example, in a turnaround program at a global producer of specialty products, it proved close to impossible to include the company's Chinese sites, since their perception of demonstrated best practices differed widely from European thinking. Furthermore, small sites in particular have limited management capacity beyond day-to-day needs and bringing in external or internal expertise to such sites is quite expensive in relation to the improvement potential. Including them in any type of overarching program is therefore difficult. On the other hand, large sites often need special attention and tailored approaches due to the complexity of having very different businesses, warranting different business systems, in one location. One company found it necessary to set up a completely separate program for its largest integrated site. In another large scale operational program, one and a half years were spent on improving the global production network of a leading chemical player, with sites in locations ranging from Europe to South America and Asia. The company found that a traditional approach addressing groups of different sites sequentially would have taken too long and would have been too costly. Instead of the "usual" structured measure-generating approach, the program therefore focused on building capabilities around operational excellence within a structured setting. This enabled the responsible production managers to execute the turnaround with much higher leverage.

It all boils down to one point – it is a huge challenge for executives to reach the whole production organization in order not only to achieve sustainable impact fast, but also to really initiate continuous improvements. It is becoming more and more crucial to avoid tradeoffs between speed and sustainability and to achieve both in parallel. The benefits to the successful organizations that figure out how to execute such programs are still very substantial.

18.2 Making Lean Operations Happen in Chemicals

There are obviously situations when simply increasing labor intensity and improving productivity with straight-forward cost-saving programs is still the best solution, in the short term at least. With a little luck, this approach will allow a good organization to maintain its performance, provided that it is running in the right direction and not yet stretched to its limits. However, the majority of players will need a more sophisticated approach. Assuming that they have already achieved the easy wins, companies will need to find completely new ways of working – doing things differently, not just more quickly – in order to achieve sustainable improvements.

Other industries indicate one possible path for chemical companies going forward. If correctly implemented, the principles of lean manufacturing, developed first in the automotive and assembly industries and then used in process industries (such as chemicals, metals, pulp & paper, and power) by companies like Alcoa, offer a promising approach. Major impact has been achieved by this method in large scale transformations over the last few years. In this chapter, we will focus on how to implement such a transformation rather than on the elements of lean manufacturing, since there are many publications on that subject.

Companies aiming to reap the benefits here need programs that not only address the optimization of the operating system, but also performance management as well as mindsets and behaviors across the organization. Only if organizations can address all three elements in parallel will they achieve true transformations that will take them far beyond today's operational standards – falling down on one means failure (see Fig. 18.3). In the following, three key questions that need to be answered in the transformation process are addressed:

- What defines a successful transformation in detail, i.e., which elements need to be addressed for sustainable impact?
- How can executives embark on a journey without making any trade-off between speed and sustainability at all, or only making this trade-off later?
- What do executives need to do to really change the performance culture towards a continuous improvement process?

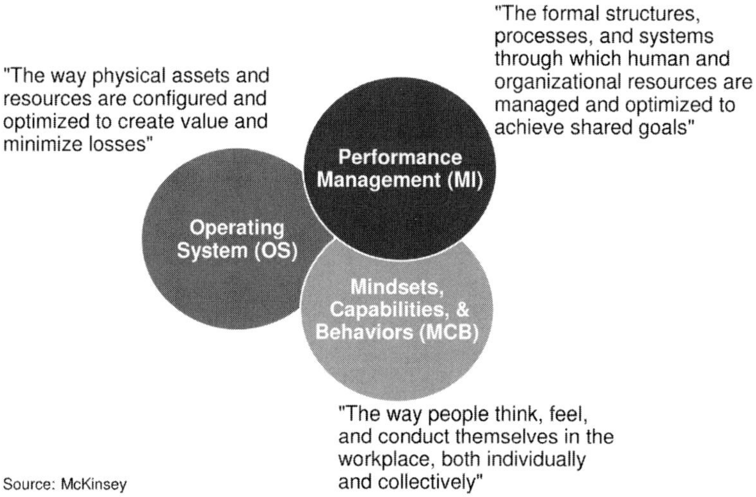

Fig. 18.3 Three elements of real transformation.

18.2.1
Defining Relevant Aspects of Successful and Sustainable Operational Transformations

As mentioned above, a successful transformation must occur along three dimensions if it is to be sustainable:

- The operating system – this is the way physical assets and resources are configured and optimized to create value and minimize losses;
- Performance management – this refers to the formal structures, processes, and systems through which human and organizational resources are managed and optimized to achieve shared goals; it requires a systematic approach to measuring and managing performance down to the front line;

- Mindsets, capabilities and behaviors – this is the way people think, feel, and conduct themselves in the workplace, both individually and collectively.

Many programs in the chemical industry have encountered the problem of unilateral focus on the future operating system. In many instances, little attention was paid to performance management or to the people side. As a result, programs often failed (sometimes completely), and improvement measures were not executed. Even if measures were implemented, they did not lead to sustainable change and continuous improvements afterwards. But the opposite can happen as well. For example, a large European industrial conglomerate decided to introduce the Six Sigma methodology by training champions across its business units. This was a promising idea in itself, since the problem-solving rigor enforced by the methodology can result in improved end product orientation and better problem-solving results. However, because the company failed to link this capability-building effort with performance management in particular, the opportunity to achieve real and sustainable change was threatened. The new champions felt that the time they spent on training and capability building was detrimental to their careers, since it pulled them away from their "real" task – day-to-day line operations. Thus, good though the Six Sigma methodology is, it is not sufficient on its own for a real transformation – at the very minimum, it needs to be linked with the management system and with the mindsets and behaviors of the employees.

However, change beyond the operating system (which engineers and chemists understand much better than the other aspects) does not come easy. It requires a

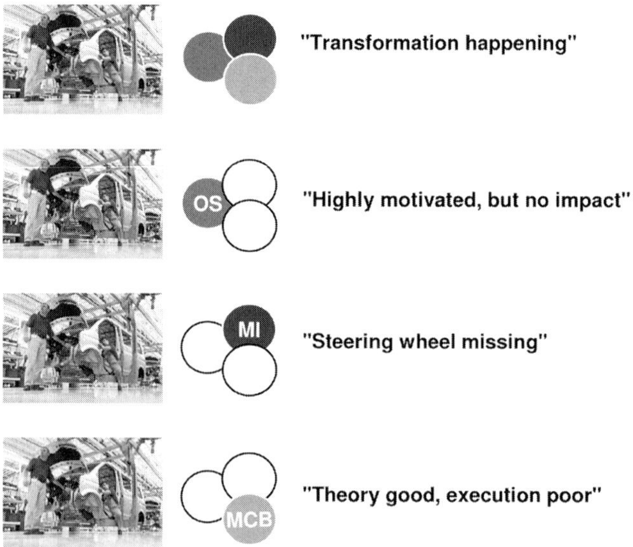

Source: McKinsey

Fig. 18.4 All aspects are crucial for success.

new type of leadership. Executives have to act as role models of the new behavior. Plant managers have to spend a significant share of time with the frontline people. They have to foster understanding and conviction by initiating a regular and open performance dialogue on operational business issues, so that employees clearly understand what is expected of them. Executives have to ensure that their employees feel they have the skills and confidence to behave in the new way, and finally, management has to reinforce the requested change in its formal structures, processes, and systems (see Fig. 18.4). One good example here was a major turnaround in a global production network, where the setup and execution of a large-scale training program reaching 80 percent of the employees was part of the cost reduction program. More often, training turns out to be something executives pay lip service to, but do not really regard as top priority.

All this is easily said, but so far all we have done is raise the bar for executives. Where should the company make a start? What does it take to make a holistic transformation of this kind happen – addressing all the elements described above? Which elements need to be in place to achieve both speed of change and sustainability?

18.2.2
Making Transformation Happen – Front-end Loading with Content

As a starting point for setting up operational programs, we believe it is imperative for executives to drive their operations managers to develop a clear blueprint of where the organization should be heading, in order to load the program with content right from the start, at the front end. In many programs, the desired outcome remains so open that executives find themselves unable to push through consistent change across the global network. An organizational blueprint does not just include structures. It represents a joint understanding within the organization on the execution of its core processes, and on the best practices the individual company needs to achieve in its specific situation.

- The organization needs to understand its core processes and should have a clear idea of what these should look like in detail. This includes not only activities, but also a shared understanding of how and where interaction happens and decisions are made. Organizational implications are usually obvious from this. Let us take maintenance as an illustrative example. Maintenance should focus on keeping operations running as efficiently as possible, and should not, for instance, have the scope to expand its role by defining changes in production equipment. Such changes are based all too often on weak economics, and do not create value. At one global fine chemical producer, strict gate-keeping, i.e., separation of project execution from task definition, followed by the application of clear technical and economic criteria, led to a significant reduction of maintenance activities. In the end it proved possible to reduce capital expenditure for maintenance-related projects by 30 percent. Ongoing spending on maintenance also went down by a fifth. This approach might sound hardly worth mentioning, but too often we see unsustainable opex and capex reductions driven by budgeting needs rather than a clear understanding of long-term business needs. Another example is the management of health, safety, and environmental regulations, which is one of the main tasks of a chemical company. Such functions should be separated from production to avoid goal conflicts.

- The organizational blueprint should be combined with worldwide (internal) best practices, so that the final setup is not only effective but also efficient. These practices should reflect the institutional experience of the global organization. They should also leverage all the well-established lean concepts, such as operational equipment effectiveness for bottleneck equipment. We have rarely seen a production manager who does not claim to have these data available. The acid test is whether the methodology is clear throughout the organization and the data are used for actual day-to-day operations, i.e., there is a regular (daily) dialogue with operators around this data. Despite all the successes of the past, many chemical companies still have a long way to go to arrive at such "best practice" setups. Compare this with the automotive industry, where production experts at the best companies have a global uniform understanding of what benchmark manufacturing assets should look like. We acknowledge, however, that it is much harder to identify any such benchmarks in chemicals, as the different production processes vary much more depending on the actual product and chosen technology, but also on factors like health, safety, and environmental regulation.
- We do not believe that there is one best practice organization that fits all needs. Companies will need to develop the optimum organization for themselves. The example of control room operations demonstrates this point. A company could decide to train employees to become multi-purpose operators (MPOs), or it could train specialized control room operators and specialized field operators. The MPO model ties in with the belief that a disparate, multi-tasking organization is better than a specialized, functional setup in terms of performance and employee motivation. The chemical industry, however, is too diverse for any such sweeping assertion. The MPO model might make sense in multi-purpose units for the production of fine chemicals or pharmaceutical products, with operations controlled directly at the equipment on the shop floor. In fact, real multi-purpose production buildings do not even have control rooms. However, in physically spread-out, complex continuous operations such as in petrochemicals it may well be far more sensible to have specialized employees,

> especially if highly qualified workers are scarce (a real issue, especially in developing countries). The discussion is very similar for the setup of maintenance and other functions. Companies therefore need to understand the pros and cons of a range of organizational structures, in order to deploy the best one for their specific situation. The optimal organization will depend on the company's product range and production technology, its employees' skill set, and its operational performance level.

How does a clear blueprint help? For a start, it provides a reference for comparing organizational performance levels, and thus forms the basis for deriving the targets and levers to be addressed in the improvement program. Such individual targets help to achieve credibility for yet another round of performance improvements. They are much better than the alternatives, such as imposing change targets based on financial imperatives or defining general cost reductions across the board.

Drawing up the blueprint also sets a common language and communicates the shared vision for the level of operational excellence required across the production network. It provides a clear idea of the company's future operating system. Company executives also benefit, as it sets their individual activities in an overarching context. Finally, a blueprint can be codified and expert groups can be established to help roll it out globally. However, caution must be exercised so that this does not lead to yet another corporate standard that is not appropriate for the local sites, and therefore is never realized.

We have seen in a number of situations that such perspectives allow short diagnostics (i.e., of around one week) to be conducted, which lead to clear targets and identification of the main improvement levers. This results in heavy front-end loading of operational programs and accelerates change.

18.2.3
Making Transformation Happen – the Soft Side of Change

The development of an organizational blueprint including corporate best practices and the definition of company-specific targets is only one element of setting up an operational program, however. Executives also need to build new capabilities and address mindsets and behaviors (Fig. 18.5; see also Chapter 24). Companies need to find ways to mobilize action across their production network and establish mutual learning experiences. We will focus on four key elements here – the setup of an expert community, training, performance dialogues and leadership, and, finally, program management. These elements are not exhaustive (e.g., incentive systems are not covered, much more could be said about details of operating systems, etc.), but they are the ones which are neglected more frequently than others, causing transformations to fail.

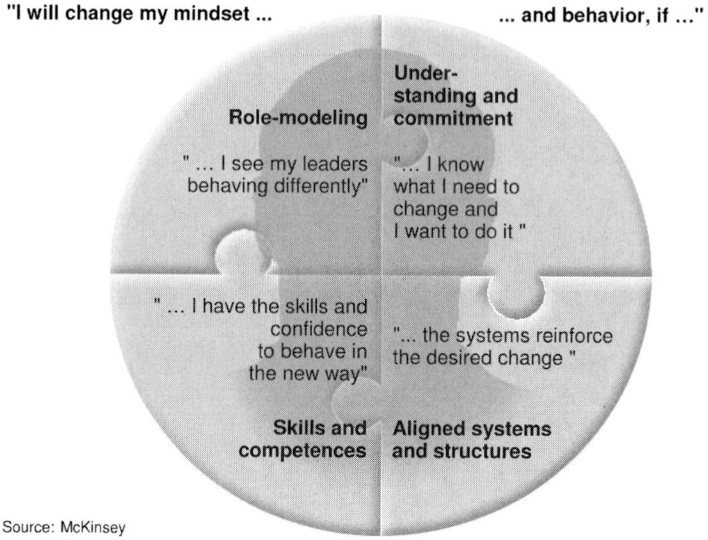

Source: McKinsey

Fig. 18.5 Successful transformations personalize change in order to change mindsets and behaviors.

18.2.3.1 An Expert Community

Our first key element is the setup of systematic skill building and development of a nucleus of experts within the organization. This provides the resources needed to reach out to the whole global organization fast, and prevents trade-offs between fast change and sustainability. Academy programs have proven a successful approach in various process industries here. The exact setup of each of these programs differs, but each of them required a strong effort. The organization will need expertise at different levels. All operational managers (the practitioners or "users") need to be able to apply the best practice blueprint to their specific situation. A smaller group of people (the "experts") is required with more in-depth expertise about specific aspects, e.g., maintenance processes and organizations, control room operations, etc. A still smaller group of experts (the "navigators") has to be able not only to apply every detail of the blueprint, but also to develop it further and design new change programs. These stages of expertise need to be defined individually by every company. Care should also be taken to prevent the academy from becoming a corporate staff function. Managers from operations need to rotate in and out of the expert group. Strong leadership has to be applied here. Belonging to such a group for a certain period of time must be perceived as a career booster, not a dead end. There needs to be a clear curriculum for theory and field work that will lead to the different levels of expertise.

Across process industries, we are seeing an increasing number of such skill-building efforts being undertaken with widely varying degrees of success. To do well, companies should combine the setup of such communities with genuine improvement programs. Capabilities are then built "on the job", and a successful program also lends credibility to the experts. A large special alloy producer achieved skill building by combining weekly one-day classroom sessions with parallel work in improvement programs. The duration of this parallel capability building differed between three and twelve months, depending on whether the employee was supposed to become a practitioner ("user") or a deep expert ("navigator"). The program for navigators was designed to last one full year and give participants exposure to improvement program work in at least three different production units. A total of one to two percent of production employees were trained in the production academy in order to reach critical mass. The management also made sure that external support focused on the skill-building activities rather than on the generation of individual improvement measures for short-term gains. This example shows the massive investment needed to create communities that are recognized as leading edge by their own organizations.

18.2.3.2 **Training**

In parallel, training really needs to be taken seriously, and time and money need to be invested in it. Not only do the different degrees of experts need to understand what is expected of them, but workers on all levels of the organization must also feel that they have the skills to deliver against expected performance. If expertise cannot eventually be leveraged into the front line, the sustainability of all improvements remains at risk. Serious training is also a morale booster for employees. It shows that they are really important to the company, and that the company wants to develop them. This has immediate positive side effects on productivity as well. The example above of the special alloy producer shows that training is closely linked with the formation of expert communities. But it clearly goes beyond that, since it is also necessary to reach the front line in all shifts and in all production units. At one chemical company, the local management of individual production assets – with the support of the expert community – changed the shift operator concept and maintenance philosophy, among other things. This demanded a new kind of behavior on the part of most of the production operators. It was only achieved after a large training effort that made clear to everybody what needed to change and what was expected of individuals.

18.2.3.3 **Performance Dialogues and Leadership**

Managers may also need to change their leadership style. They must establish a regular performance dialogue down to the shop floor. This requires them to create transparency around key performance indicators on the shop floor, ideally on a personal/shift level. Problems must be addressed directly and in a solution-oriented way, root causes of issues need to be identified. Such dialogues need to happen daily, not weeks or even months after the fact. This requires management to be more visible in operations. In many corporations, managers are tied up in administrative tasks away from their line jobs. Top management should address this by ensuring that only really value-adding meetings take place. Delegation of tasks should be encouraged. This sounds simple, and the concept is indeed simple – the difficult thing is to stick to that behavior. It can be helpful if peers review each other, this takes away the evaluation dimension of top-down reviews and increases the willingness to accept support from the outside. Asking for help should not be perceived as a sign of weakness, but rather as an indication of real efforts. Open feedback is an opportunity and not a criticism.

From all that we have observed, this element of change is probably the most difficult to implement. Training efforts, expert communities, definition of target operating systems, etc. are all elements that managers can push forward with conscious efforts without changing their own behavior. But the end product-oriented leadership required in real transformations is something that requires personal change by each and every manager. It can be useful to define agenda items and participants in regular meetings of production shifts, maintenance, etc. However, this definition only helps if people live by it.

18.2.3.4 **Program Management**

Finally, although program management may not sound very exciting, it remains a key element in embedding change in the corporate environment, with the familiar components of setting up controlling mechanisms for initiatives, checking for actual implementation and bottom-line impact, conducting regular reviews, and communicating information about the program and about progress throughout the organization. If these actions are not implemented, the program will lack credibility and "individual accountability of management" will remain a hollow phrase. In addition, top management will find it difficult to intervene in time if problems arise, and may only realize with hindsight that the program has failed.

Since benefit tracking is the crucial basis for successful program management, let us look at the example of a global market leader in the production of high performance materials. Program progress was controlled here on two different levels. First, local management checked the factual implementation of changes in processes, behavior, purchasing contracts, etc. This ensured that the right things were done. Second, the program office conducted benefit tracking, where the financial impact of the implemented measures was estimated, taking into account volume and price changes compared to the reference period (baseline) for the program. Both levels – implementation monitoring and benefit tracking – were essential to get the full picture at any given point in time. They also made it possible to adjust the rhythm and pace of the program to the progress made at any point in time and to initiate corrective actions where needed.

18.3 Summary

To sum up: although the easy wins are gone and driving operational change is becoming ever more of a challenge, operational improvement remains one of the keys to sustained value creation. Companies should invest in this area by setting up programs that address not only the operating system, but also the management systems of production facilities as well as changes in the mindsets, capabilities, and behaviors of employees. Programs will only be successful if all three elements are combined. Among the most crucial pragmatic issues for executives to bear in mind when embarking on operational programs are:

- Organizations should develop a company-specific understanding of how they want to operate in order to optimize value creation and minimize losses.
- They should front-end load their programs based on this joint understanding via individual target-setting and program design.
- They should create a clear blueprint of what production should look like over time – including a clear perspective on continuous improvement.
- They should consider the soft side of change with the same intensity as the definition of the future state.
- This involves setting up real expert communities, recognized as such by the rest of the organization.
- It also includes training throughout the organization.
- As the most difficult element of all, it also requires executives to change their leadership style.
- Finally, stringent program management is still key to making sure that implementation really happens.

The task is sometimes daunting, especially within global production networks, but in the end the rewards of success are still very substantial.

19
State-of-the-art Production Concepts in the Chemical Industry

Uwe Nickel

The chemical industry's environment has changed over the last decade, and production with it. We are aware that raw material prices and wage increases have raised costs, competition is taking place more than ever on a global level, and markets are maturing (see also Chapters 1, 3, and 8). The numbers of patent-protected products and of new chemical species have shrunk, leading to a significant reduction of ground-breaking technologies and to commoditization in many fields. Along with that, producers from low-cost countries like China and India have successfully developed in the market. All this has made cost optimization through production management more important than ever to a company's economic success.

But the task has not become any easier. A global production network adds complexity that needs to be managed. Interdependencies within this network require more sophisticated management processes and fluctuations of lead currencies call for risk limitation strategies. Furthermore, turning the rising flood of information into true knowledge for the company is a substantial challenge for production management.

As a result, the production function has to deliver excellent complexity management and also produce at even lower costs (Fig. 19.1). In order to manage the increased complexity, Clariant, one of the leading specialty chemicals companies, has established a dedicated network management function that balances capacities and production volumes, with dedicated supply chain management controlling the global flow of goods. Technology and knowledge management processes are in place to filter state-of-the-art processes and concepts and turn them into institutional knowledge. Cost reductions, on the other hand, are addressed through structural and operational levers. The structural levers consist of an optimized global production network and a focus on distinctive value creation. On the operational side, functional excellence is to be achieved by establishing a system for continuous performance improvement and continued plant optimization. This chapter describes the efforts that have been made at Clariant since the late 1990s, and the experience and learnings we gained.

Value Creation: Strategies for the Chemical Industry. 2nd Edition. F. Budde, U.-H. Felcht, H. Frankemölle (Eds.)

ISBN: 3-527-31266-8

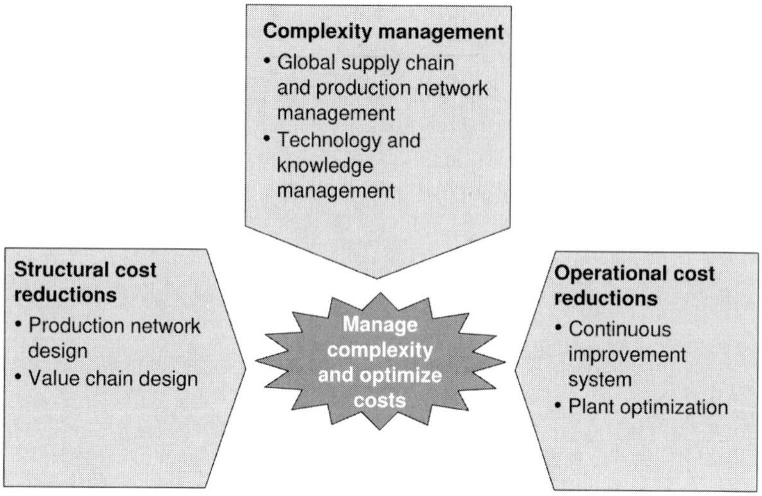

Source: Clariant

Fig. 19.1 Challenges in production and how to address them.

19.1 Operating in a Transformed Environment

Over the past decades, and especially over the last one, the environment of the chemical industry has undergone some fundamental changes: innovation has slowed down considerably, geographical barriers have largely vanished, and the skills of competitors in low-cost countries have improved.

19.1.1 Decelerating Innovation

Over the past decade, sales of products younger than three years have decreased. In 1999, only 29 percent of sales were generated from these new products, much less than in other areas. While other industries, especially in the consumer sector, have increased their share of new products, chemicals saw a drop of 19 percent between 1992 and 1999 (Fig. 19.2).

The pipeline of innovation, expressed in filed patents, has also changed. The Derwent World Patent Index shows that the number of patents in the world nearly tripled to 710,000/year between 1980 and 2004, whereas in the chemical industry it only doubled to 213,000/year. Since 1992, the number of patents in the chemical industry has been steady.

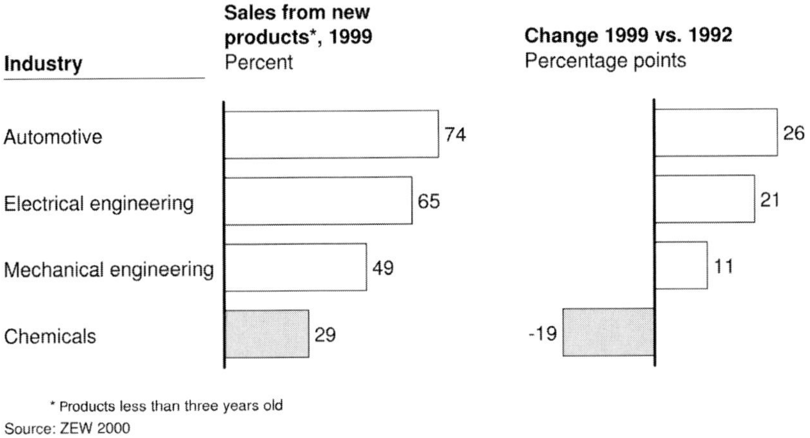

Fig. 19.2 Decreasing sales from new products in chemicals.

What is more, the innovations in chemicals were largely incremental around existing technologies and markets. New technologies (e.g., fuel cells) are growing quickly but will not have a major substitution impact on traditional areas. For example, the number of patents in pigments, where Clariant is a market leader with a turnover of almost EUR 1.2 billion in sales, is consistently above that of emerging technologies (Fig. 19.3). As a result, instead of breakthrough innovation and blockbuster products, the chemical market is characterized by marginal changes and is fundamentally a very stable environment from the product perspective.

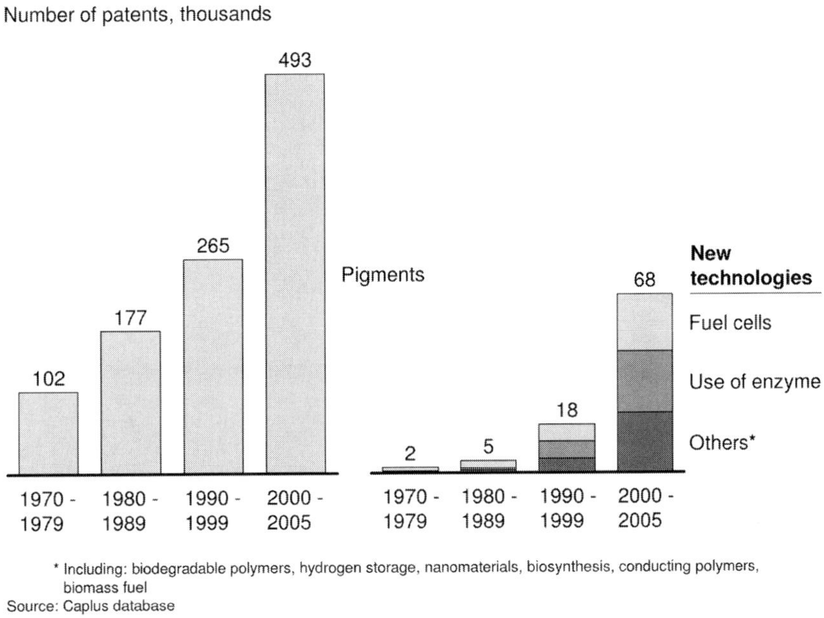

Fig. 19.3 New technologies in the specialty chemicals industry, 1970–2004.

In contrast to other industries, the introduction of new production processes faces some substantial barriers in chemicals. For one thing, customers are reluctant to accept price increases, even for innovative products. Even if product substitution is possible from a physical-chemical point of view in a wide range of applications, as a result of consumers' increasing focus on price rather than value, more and more specialty chemicals have turned into semi-specialties or even commodities without changing in themselves. Companies' product portfolios have undergone a creeping aging process. In many areas, competition for the better price rather than the better product is giving efficiency of production a pivotal role for a company's prosperity.

Second, process changes that require new production facilities – and the restructuring of old ones – often do not stand up to financial scrutiny. Only if the cost savings from the new process are very substantial do the investments have an acceptable pay-back. Furthermore, due to the growing importance of asset-driven financial indicators like return on invested capital (ROIC) or economic value added (EVA), tight management of assets has become more important than in the past. Investments into newly designed processes are being handled with increased caution. This has led to an increasing amount of investment into existing equipment rather than new. In the USA, for example, investments into existing assets rose from 15.2 percent of total capital spending in 1997 to 36.5 percent in 2003 according to the American Chemistry Council.

19.1.2
Chemicals Have Become More Global than Ever

Political and technological changes have reduced the barriers between countries and have created a truly global industry. Attempts at protectionism soon go by the board, and have a greatly diminished ability to safeguard a local economy and, still more importantly, to ensure permanent GDP growth. Customs procedures have been simplified, excise duties have been lowered, technical standards have been harmonized, and the Internet has created a completely new market transparency. All this has made it easier, cheaper, and more convenient than ever for the customers to compare offers and sources on a global level rather than look only for local suppliers. Operating margins have dropped from eight percent in 1990 to 6.7 percent in 2000, and competition is taking place on a global level for almost any product.

19.1.3
Competition from Low-cost Labor Countries Is Changing the Industry Landscape

Political and economic progress in Asia, especially in China, has allowed Asian companies to successfully compete on the global market (see also Chapters 7, 32, and 33). This has also brought opportunities for foreign companies to take advantage of the low labor costs and the proximity to a huge emerging market. Beginning in the early to mid eighties with just representative offices, many companies have now moved significant production capacity to this region (Fig. 19.4).

China has modernized its economy by opening up to foreign trade and investments, reducing bureaucracy, and fostering the education of skilled labor.

Furthermore, China and India have agreed to comply with intellectual property rights acts and global patent protection practices, which has caused a quantum leap in these areas: today, 55 percent of all chemical patents come from Asia. With wages in China only one thirtieth of those in Western European countries (EU of 15) and the country's infrastructure becoming increasingly acceptable, its comparative advantages for production are tremendous. As a result, the gross output of chemical products in Asia (excluding Japan) increased from eight percent of the global market in 1989 in real terms to 22 percent in 2004, and is expected to reach 26 percent by 2010 according to *Global Insight – World Industry Monitor*. One consequence of this rapid structural shift of capacities to Asia has been a drop in capacity utilization in the US chemical industry (including pharmaceuticals and consumer products) of 16 percentage points to an unfavorable 73 percent since 1990, according to the Federal Reserve Board. Today, a substantial number of plants in Europe and the USA are underutilized and require restructuring.

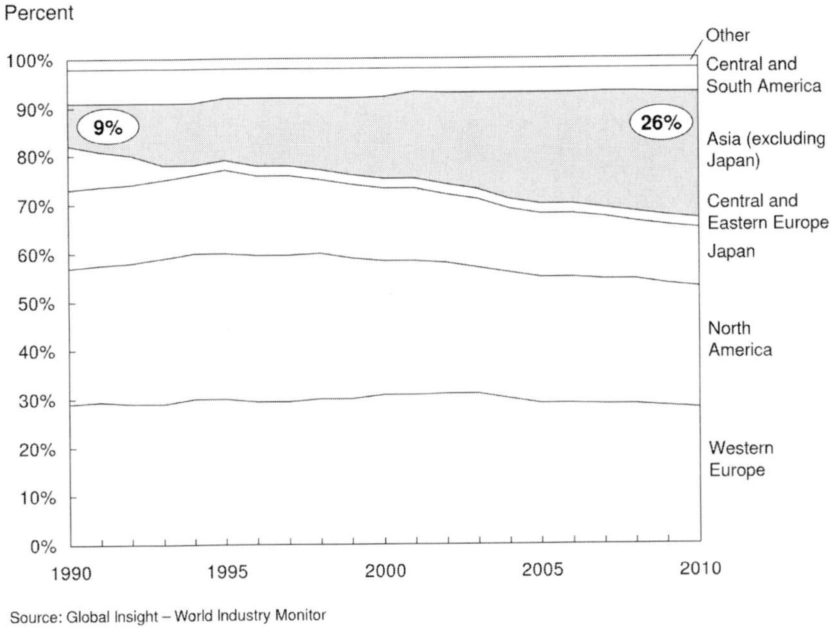

Fig. 19.4 Increase of production capacity in Asia.

19.2 Challenges and Responses

What are the implications of these trends for the chemical companies? To optimize costs and manage complexity, Clariant pulls structural as well as operational levers and manages the global supply chain and production network through dedicated functions. In addition, technology and knowledge management have to enhance institutional knowledge in order for the company to become best in class. The two structural levers that are systematically assessed are network optimization and value chain design. The two operational levers are a powerful system for continuous performance improvement and ongoing plant optimization. Finally, the increasing number of and distance between sources and sinks within the supply chain systems obviously requires global management, highly standardized processes, and a high degree of planning accuracy. Today, capacities and product allocations are managed within the global production network.

Leading companies will have to achieve an annual productivity increase of three to five percent with these and similar actions to avoid losing competitiveness compared with the industry average (Fig. 19.5).

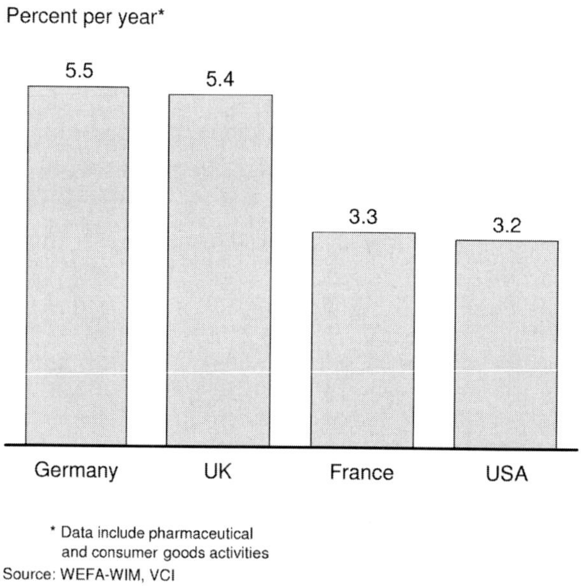

Fig. 19.5 Annual labor productivity increase 1991–2003.

19.2.1 Structural Cost Optimization

As a starting point for structural cost optimization, Clariant first of all has an optimized production network design that leverages scale effects and regional advantages. Second, a review of truly value-creating functions, processes, and process steps is performed on a regular basis.

19.2.1.1 Designing a Global Production Network

As recently as the mid '90s, the production concepts of Hoechst, Sandoz, and other global companies were based on local or regional production sites, supplying the surrounding markets with all the products in a company's portfolio. This paradigm is gone. The market is global, and in order to withstand the increasing price pressure, companies must leverage economies of scale by concentrating their production assets in fewer places. In the future, each plant should have a clear strategic position based on a defined technological profile and clearly defined set of products for which it can play to its strengths. These strengths may include cheap labor, proximity to feedstock or markets, technological expertise, or advantages in the legal system, for instance an efficient planning and building permission process for new plants.

Despite past efforts, most companies still have excessively scattered production networks and too many sites in high-cost countries. This often needs decisive consolidation, but also a shift to low-cost countries. In particular, merger and acquisition (M&A) activities combined with historically grown production structures imposed by the government protectionism mentioned above (i.e., customs, trade restrictions) have led to a proliferating plant network with many inefficient redundancies. Going forward, it is a safe assumption that M&A activities will continue and protectionism will be further reduced – network consolidation therefore cannot be viewed as a one-time effort, but as an ongoing core task in operations management which has to strike a careful balance between the benefits and the costs incurred for writeoffs. This applies even more to a shift of production capacities from high-cost countries in Europe or the USA to China or India.

Does this mean, then, that the ideally structured company has only a handful of plants in China, India, and Russia, or even in an emerging market like Vietnam? Is this the end of a chemical landscape in high-wage countries? Not at all, in the author's opinion: the best network design has to balance savings opportunities against the associated risks, and is likely to include a healthy admixture of developed countries. Currency, trade restrictions, accident rates, and infrastructure problems (such as power cuts, transport bottlenecks, natural disasters, and major diseases) all weigh in the balance. Often, these "extraordinary expenses" are the key driver of financial results. The limited availability of basic raw materials, energy, and utilities seen in China more and more frequently since 2003, as well as growing urbanization and rising living standards (= costs) in urban China and many emerging markets, are only a few examples of unpredictable disruptions over the last five years. It is therefore wise not to rely solely on "cheap" production facilities and product sources, but to maintain a sound production base in developed countries with low risk profiles, despite their high labor costs. The best in class offset this disadvantage as far as possible by pushing lean operations and automation even further, and leveraging new technologies and process efficiency.

At Clariant, for instance, we focus on increasing time yield and efficiency per employee by systematically analyzing throughput time variances and optimizing production processes continuously. Here, a globally operating "rapid process development unit" comprised of experienced senior technologists is key for success.

19.2.1.2 **Focusing on Distinctive Value Creation**

Achieving competitive production costs depends heavily on maintaining a clear focus on those core competences and products that will allow the company to build a distinctive position. This implies that the organizational setup should outsource all non-core functions, including in the portfolio only those production steps or products that show the required margin contribution.

Which areas, then, should be outsourced? Efficiency gains and reduction of capital employed have to be balanced against higher transaction costs and a considerable risk of losing internal knowledge. Services that can be clearly specified and planned would typically be purchased externally if this allows a cost reduction. Detailed engineering is one example, while breakdown maintenance clearly would not fall into this group.

Typical areas for outsourcing are the production support functions, such as logistics, including forwarding and warehousing, or site services, including utility management (electricity, steam, gas), security, and in certain settings maintenance. Tasks that can provide an advantage over competitors, on the other hand, such as highly sophisticated and specialized analytics, product safety and registration, and specific waste water treatment should be performed in-house provided they offer a competitive and transparent cost position based on fixed and sustainable key performance indicators (KPIs).

Outsourcing always bears the risk that service providers present overly optimistic business plans to secure the contract and then raise the prices in the second stage. This is not an issue if the service is commodity-like. But if it is not, reversing the outsourcing decision by an internal skill buildup could pose problems if the required skills are not readily accessible in the region. Therefore, outsourcing will never be an option unless the key unit operations and services are protected from the start from long-term changes or disruptions by reliable partners, precisely defined service levels and service prices, and appropriate exit scenarios.

To set the industry benchmark, however, focusing on core production tasks is not enough. It is equally important to analyze the product-level value chain systematically and continuously and act on the results of the analyses. Substitution of raw materials by cheaper sources achieving the same quality standards is one example; standardized products from which different products can be manufactured via specific unit operations and/or with dedicated technologies is another. A holistic view of the product value chain offers vast opportunities. We identified mixing, filling, and packaging activities that are traditionally seen as core production competences as candidates for cheaper outsourcing, concentration within an effective site network, or subcontracting. The potential threats of such an approach – as discussed above – have to be kept in mind. However, without clear processes to identify the opportunities, companies risk a difficult cost position and loss of market share. Such processes demand methodological experience, solid data, and a cross-functional dialogue between production, product management, and sourcing.

The implications of a "cherry-picking" strategy are twofold: supply chain management becomes even more complex than before, and contracting becomes a key activity whose importance should be reflected appropriately within the organization. Contracts for site services, in particular, have a high upside or downside potential. Service agreements all too often turn out to be the subject of litigation, with considerable costs involved.

Finally, tools need to be in place to track the margin contribution for all products and systematically eliminate those which do not meet the required threshold. Internal transfer price systems are typically detached from market dynamics and distort profitability analysis on the level of a single production plant. Such analysis only produces accurate conclusions when performed on the level of products or franchises.

19.2.2 Operational Cost Optimization

Employment in the chemical industry has fallen worldwide by 35 percent since 1990, to 8.3 million. During the same period, global chemical production increased by 50 percent. In other words, labor productivity has more than doubled. In the USA, for instance, productivity is increasing by about four percent every year, or, to put it another way, each worker produces 50 percent more today than ten years ago. That four percent figure is substantial, and cannot be achieved just by screening the company for cost saving opportunities. Instead, a robust continuous performance improvement (CPI) system must be put in place that drives continuous improvements every day by opening up dialog between the staff involved. In addition to that, plant optimizations will still be necessary in the future to address fundamental changes in the market or production environment.

19.2.2.1 Introducing a Continuous Performance Improvement System

In 1997, Clariant started to focus on the continuous improvement process (CIP) in many areas to optimize effectiveness on a permanent basis. A continuous improvement system consists of several elements: a dedicated organization, a comprehensive KPI system, a target-setting process, a set of tools, and, finally but importantly, the right mindset.

Organization: In the interests of clear targets and efficient work, there has to be a clear split between production staff and those who systematically identify improvement opportunities and refine the standards for the production work.

The "producer" then focuses on supplying the market on time with the right quality. The "improver" works to a target, such as a ten percent productivity increase. They work together as part of the production organization, but with dedicated targets. In our experience, mixing the two tasks at the lowest level of management or even on the shop floor often leads to project delays or failure to deliver the expected benefit.

The cornerstones of an effective CIP organization will combine internal knowledge with a market focus: The "improvers" will get support from process optimization units made up of experienced production specialists who will provide the "outside-in" (production) view. Product managers will keep the production function in touch with the market, and ensure that the "producers" can respond quickly to a changing market environment.

KPI system: To speed up and anchor CIP in the organization, the companies that set the benchmark in this area establish a clear KPI system linking the operational performance metrics consistently and comprehensively to the overall objectives of the company – and these metrics have to be easily understood and reproduced within standard companywide reporting systems. For instance, at Clariant a number of different KPIs have been defined to track maintenance efficiency, but only a few (e.g., repair and maintenance costs/conversion costs) can really be tracked by the production team on the basis of standardized systems (Fig. 19.6).

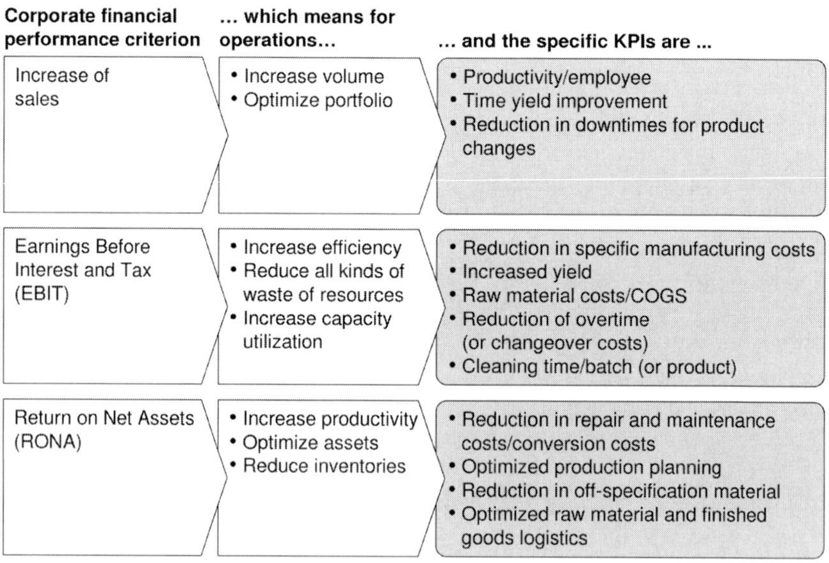

Source: Clariant

Fig. 19.6 Targets are "translated" and broken down to the shop floor level.

An annual target-setting process for operational improvements needs to be established. Very specific targets have to be set for each plant and production unit, that in sum will allow the overall company target to be achieved. The process should start with a top-down target for cost savings, followed by bottom-up action and milestone plans. The improvement opportunities could be reductions in annual plant downtime to below ten percent per year as a sum of numerous measures derived from observations on the shop floor, i.e., reduction of downtime for specific vacuum pumps via preventive maintenance or substitution of a repair-intensive pump by one more suited to the process step.

Mindset: Such a process can only be implemented in an empowered organization. That means that tasks, team organization, and resources are delegated to the lowest managerial level in a plant (e.g., the supervisor or shift leader). It is essential to train staff to recognize the highest-impact cost factors, so that KPIs become transparent and everyone in a plant can have an overview of the progress of project teams and target achievements. "Management by Objectives" is no longer a tool reserved for the top management, it has become a guiding principle for plant organization. In addition, a set of standard tools will be applied uniformly throughout the organization to set off a skill-building process and establish a common language for a company-wide performance and improvement dialogue (Fig. 19.7). Examples of such tools from our operations at Clariant are overall equipment efficiency, total productive maintenance, visual management, and statistical process control. Their application is supported by IT systems.

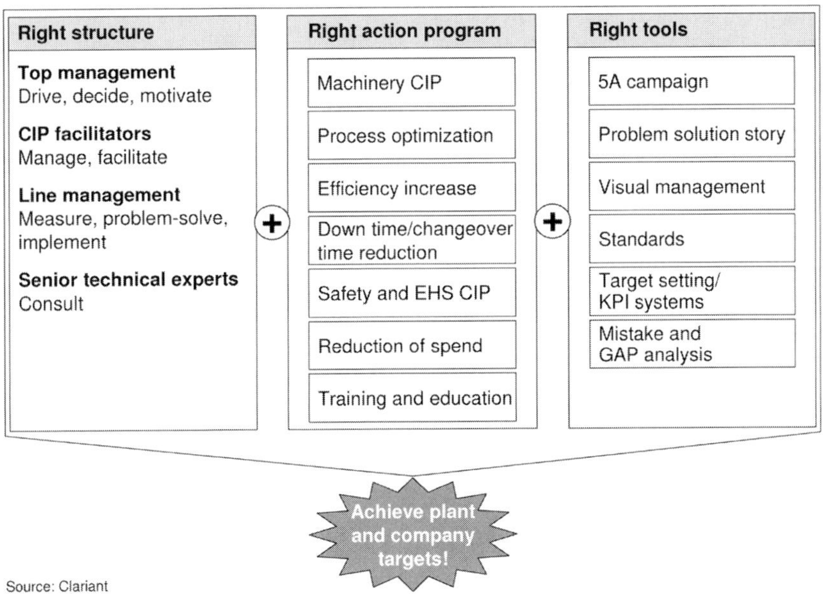

Fig. 19.7 A continuous improvement process (CIP) network and structure – a structured system to make dialogue happen.

In today's business world, nobody likes to spend too much time in meetings. When we started the structured process of CIP implemention, we were countered by objections such as: "We do not have time for meetings or problem-solving stories, our daily tasks are more than sufficient." However, putting structure into meetings and establishing a regular problem-solving organization can put paid to the grumbles. With a working CIP, the effectively available labor time can be increased by ten to 15 percent. This requires regular meetings (e.g., for five minutes at the beginning of a shift) to review performance development and the impact of new initiatives, ideally supported by data from statistical process control (Fig. 19.8). The definition of changes is elaborated in problem-solving sessions of roughly one hour, that take a structured approach to identifying problems and developing solutions (e.g., a 5A campaign). All this needs a higher level of in-plant education, with higher efficiency per worker as the payback.

The biggest mistake a company can make within CIP is for its managers to underestimate the importance of CIP-relevant training and "soft factors" to achieve targets. A CIP system is not a mailbox at the plant manager's office where colleagues can drop suggestions for improvements, and it is not a collection of charts presented by management during a certification audit. It is crucial to have somebody constantly facilitating the process and actions on the shop floor level, someone who acts as a "translator" with a direct reporting line to upper management and a CIP network across businesses and functions in place.

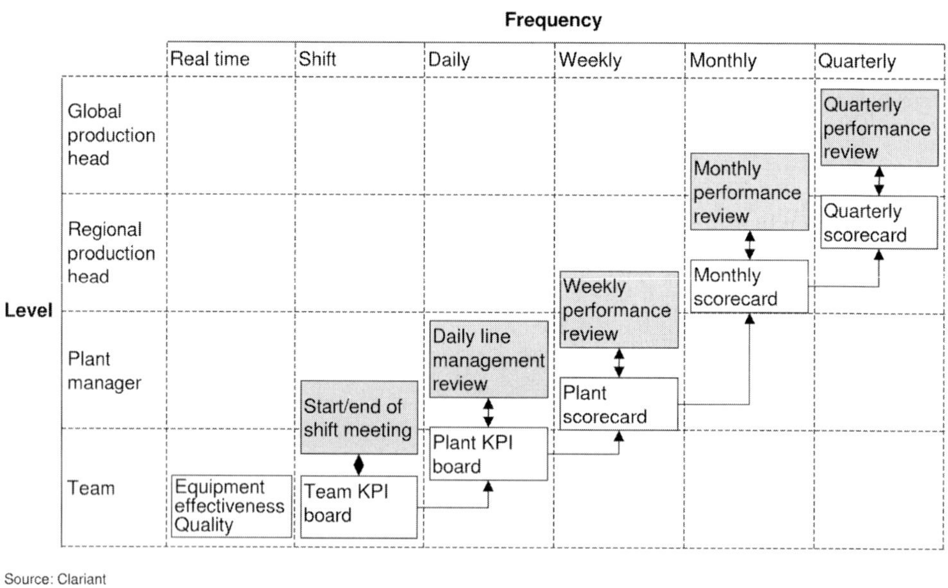

Fig. 19.8 A structured performance review process as one CIP element.

19.2.2.2 Optimizing and Restructuring Plants

Compared with other industries, for example consumer goods or automotive engineering, chemical installations are highly inflexible. The layout, technology, and capacities are largely fixed once they have been built. This does not apply so much to batch-driven operations, but definitely to continuous production systems. While a continuous improvement system can effectively optimize processes within the technical limits of the installation, substantial changes in the market or production environment will also require plant optimization and restructuring in the future, for example a drop in sales for lead products, a shift of products to low-cost countries, or the concentration of product groups in specific plants. These discontinuities always require a top-down adjustment of the plant organization as well as changes in the asset structure. Managing the social matters throughout these transitions is often a challenge.

19.2.3 Managing Complexity

19.2.3.1 Putting Global Supply Chain and Production Network Management in Place

As described above, several effects have led to a supply chain of so far unknown complexity. For example, transportation costs in the USA rose by 50 percent over the last ten years, calculated on stable production volume (according to the Bureau of Census, Bureau of Labor Statistics, and American Chemistry Council). Furthermore, there has been a significant increase in inventories in various parts of the chemical industry. There has been a drop in the chemical industry's operating cash flow, largely due to an increase in finished goods and work-in-progress inventories (Fig. 19.9). Inventory has proven to be a good indicator of the efficiency of supply chain management (SCM) (see also Chapter 22). To bring inventory coverage to an acceptable level, organizational changes are often necessary and the roles and responsibilities of production managers have to be redefined or adapted, depending on how far advanced the company is in its change process. Forecasting and planning processes must become more reliable, and clear accountability and standards for them need to be defined and tightly enforced throughout the whole organization.

For SCM to be relevant in a diversified and global supply chain, it needs a dedicated and globally operating organization. This has to reflect the business environment as well as being able to support a changing production network.

In a global production structure with IT-based information flows that substitute one-to-one relationships, forecasting and planning processes need to be brought to a new level. As mentioned above, this requires clear standards for their execution, support from IT systems, and – most importantly – clear and visible accountability for process quality. More specifically, the forecast accuracy has to be an element of the performance review and incentive system for the product managers and sales force.

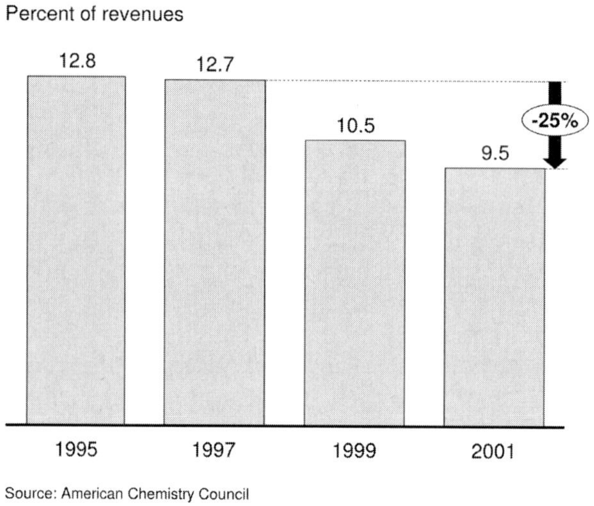

Fig. 19.9 Cash flow in US chemicals.

It is equally important to avoid "psychological safety buffers" in all steps of the supply chain. Since the preceding step is not "next door" any more in a global supply chain setup (e.g., between the raw materials supplier and the manufacturer of the finished goods), processes are a black box and people tend to err on the safe side, driving inventories up.

The only way to avoid this is by strict analysis of the supply chain from the customer order to final product delivery. Definition of the optimized (theoretical) process and sequential work towards a high service level approach allow the identification of gaps, and of opportunities which might not always be the cheapest (ship versus train versus plane) but could be the most effective way to reduce capital costs and shorten planning scope – an important aspect, especially in volatile customer markets with long production processes on the (chemical) supplier side. As in the case of CIP, this needs clear parameters, KPIs, commitment from all players, and regular tracking. The most important parameters are the lead time for all products, optimal lot sizes, replenishment points, and safety inventories.

A state-of-the-art production site and plant network with clear competence profiles for each plant also requires the centralized and structured allocation or reallocation of products to plants, depending on comparative advantages and available capacities. A dedicated organization and systems need to be in place to facilitate efficient and global decision processes.

19.2.3.2 **Establishing Technology and Knowledge Management**

In mature markets, innovation stands alongside mergers and acquisitions as the key to sustainable growth. However, not only new products are innovations, and the definition has to be broadened to include innovations in processes and process chains. The number of new processes and technologies is increasing faster than the number of new molecules. Therefore, knowledge management needs to identify new technologies or innovative production techniques and spread them quickly to the leadership group in production and process development. NIR/VIS spectroscopy in production to reduce sampling and speed up product testing is one example of a quick transfer from R&D into traditional production processes; an intelligent combination of experiments and computer-aided design, and the use of pervaporation to optimize rectification processes are others from the last decade.

All too often, having the knowledge in-house is not so much the issue as applying it everywhere and every day. Doing this effectively requires a mix of expertise and entrepreneurial habits aligned in a straight and target-oriented project management system, run by an efficient project team and supported by an intranet-based knowledge portal that allows for rapid data mining.

A similar approach is effective for the inverse process of transferring questions and problems upstream: from production into pilot plants or development labs. This is the key to optimized plant management, fewer downtimes and a continuous reduction of off-spec material are the rewards.

Manufacturing information systems for real-time process control in the lab and for efficient statistical process control, as well as the right number of lab trials, limits information losses between the plant and the labs. Parallel synthesis, such as units with online analytics in the lab, and the use of new technologies such as Micro Reaction Technology developed by Clariant and a few other companies for application in production mean a step change in reproducibility.

Exchanging best practices, however, is not sufficient. The best companies go beyond that and ensure that best practices are truly applied, by defining clear and mandatory standards which will be redefined as necessary over time and applied throughout the entire organization. Deviation from known best practices for no good reason is well known to be the biggest enemy of an efficient production system. Creativity is still allowed, but not to deviate from a standard, only to improve it.

Training programs must be set up in order to spread knowledge, and tailored to the individual's and organization's development needs. Here, "training by project" is a smart approach. Tools and techniques are provided, a project management architecture is put into place and participants have the task of achieving a defined project target. Examples of tools here are statistical process control, fast changeover, or visual management. When the whole production and product development network is informed about such projects, learnings, and the people involved via an IT-based knowledge portal (e.g., Intranet or Lotus Notes team rooms), momentum is set free and know-how is understood and applied by a much larger number of people – a scale-up factor of the specific kind.

19.3 Outlook

The major trends in the chemical industry are likely to continue over the next years. Globalization will go on, and the huge Asian market will provide further opportunities but also challenges (see Chapters 7, 32, and 33). GDP growth in Asia will lead to a bigger market, but also to higher salaries, at least in the urban centers. Growth in traditional markets will be very limited.

Product offerings will follow distinct business models with either "price" or "service" as the key value proposition. "One size fits all" will be "out", differentiation will be "in". At Clariant we have switched production from efforts to optimize product quality in general towards a two-pronged approach: making "adequate" products at the best price or striving for high-grade specialties that meet specific customer requirements. For the service business model, direct interaction of production and development with customers will require new job profiles that combine production and marketing/sales skills.

On the sourcing side, the slow but steady reduction of fuel and basic raw materials will induce further cost pressure on production. These can only be tackled through flexible and optimized production systems based on concepts that take purchasing power, the supply-demand balance, and strategic orientation into account.

The good old days of production are gone and will never come back. The future can be exciting, though, if a set of different concepts, functions, and professions are orchestrated properly. With music that was written a hundred years ago, it is the conductor and his musicians that make the difference in the interpretation. The same is true in production. It is less about inventing new concepts than about superior execution, in close alignment with internal and external partners.

19.4 Summary

- Decelerating innovation, globalization, and competition from low-cost countries create a challenging environment
- Production management must manage complexity and produce at even lower costs
- Structural cost optimizations are achieved via the design of production networks and value chains
- Operational cost optimizations are achieved via a system of continuous improvements and ongoing plant optimizations
- Complexity will be managed through a global supply chain and production network management as well as technology and knowledge management
- Building institutional skills in production in order to be superior in execution is the key to future success.

20
The Role of Site Services and Infrastructure for Productivity Management

Alejandro Alcalde Rasch

Site services and infrastructure are very often considered a rather unexciting function in chemical corporations. Yet maintenance, logistics, utilities, engineering, analytical services, and SHE (safety, health, and environment) – to name but some of the key tasks – can have a major impact on their competitiveness. Hence, their management needs to be treated by chemical companies as part of their core activities.

Over the last couple of years, site services and infrastructure have undergone a major transition: many chemical companies have established a dedicated service unit, sometimes even as a separate legal entity. As competitive pressure in the chemical industry increases, site services and infrastructure must continue to adapt and thus spare no effort to further increase their own competitiveness and contribute to their customers' success.

20.1
Site Services and Infrastructure: an Important Driver of Manufacturing Productivity

Though often underestimated, site services and infrastructure are an important function in chemical corporations. They account for a large share of total production costs, on average 15 to 20 percent (including energy, raw materials, and the cost of capital). The exact cost share naturally varies from plant to plant (especially for utilities such as steam and electricity), depending on the product and the type of manufacturing process. For example, an electrolysis plant is a clear outlier, where site services can easily make up more than half of total production cost due to the high share of energy cost.

However, site services and infrastructure are not only a major driver of cost competitiveness. Some of the services are also crucial for improving plant effectiveness. Maintenance performance, for instance, can have a huge impact on plant output. Unplanned breakdowns, often caused by failure to implement the most effective maintenance strategy or by poor execution, can lead to as much as a 20 percent loss in overall equipment effectiveness (OEE; see also Chapter 18). On the utilities side, significant cost advantages can be captured, e.g., by providing

Value Creation: Strategies for the Chemical Industry. 2nd Edition. F. Budde, U.-H. Felcht, H. Frankemölle (Eds.)

ISBN: 3-527-31266-8

better opportunities for excess energy recovery as exemplified by integrated "*Verbund*"-type sites such as BASF's Ludwigshafen site. Intelligently managing the advantages of an integrated site as a service and infrastructure provider can become a huge asset for strengthening overall manufacturing productivity and cost position.

Site services and infrastructure also play an important role from a more strategic point of view: over the last couple of years, many chemical businesses have changed ownership. Having a site service and infrastructure organization that is flexible enough to handle these ownership changes – and, nowadays, professional enough to serve external customers – is becoming more and more relevant.

20.2
Transition: Site Services and Infrastructure's Coming of Age

The last couple of years have seen considerable changes in the way chemical companies manage their site services and infrastructure. Until the mid '90s, site services and infrastructure were mainly provided via a functional organization by central departments. There was often no or only little coordination or bundling of the services, with no safeguarding of overall responsibility. Budget adherence was the key imperative. The relationship between the service units and the production plants was characterized by:

- A *monopoly position* of the service units, with production having little or no opportunity to buy services outside
- *Cost allocation* coupled with widespread lack of transparency regarding costs incurred and services rendered
- A sometimes *limited service mentality* in the service units, with production seen rather as a petitioner than a customer.

This led to widespread dissatisfaction with the service departments – and ultimately to a largely untapped performance improvement potential.

Since then, many players in the chemicals industry have initiated huge changes in their site services and infrastructure organization (Fig. 20.1).

These changes were mainly triggered by two developments: first, companies urgently needed to improve their cost position in an increasingly cost-focused environment. Second, the change of ownership of chemicals businesses led to radical rethinking of the best way to organize site services and infrastructure management. This transition has taken several years, typically evolving in sequential stages.

In a *first step*, chemical companies bundled their different services departments in one single service unit. The key challenge at that stage was to strengthen the service mentality. In addition, service agreements with the customers had to be defined and negotiated. Proactively managing service costs together with the customers became the key imperative. Most chemical companies have taken this step by now.

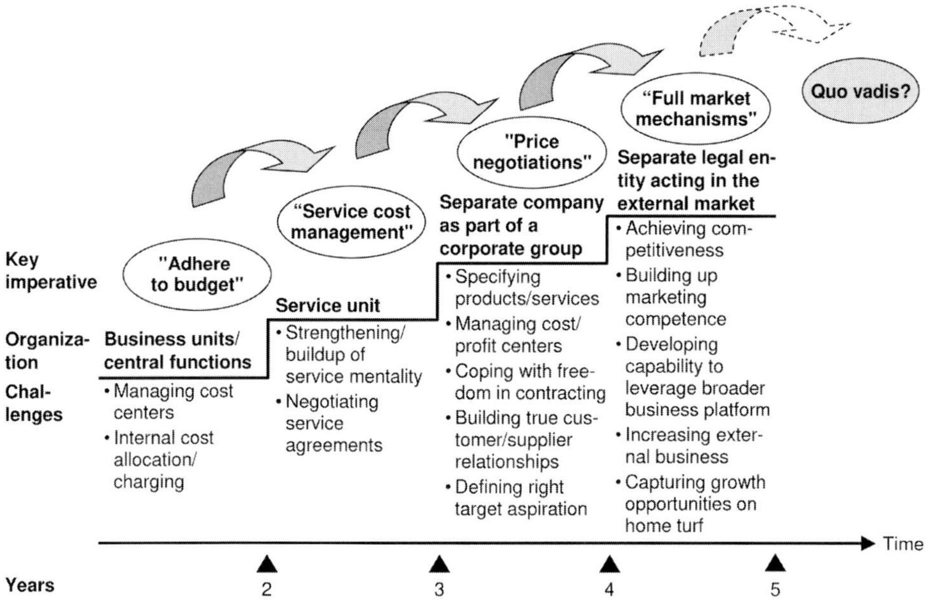

Fig. 20.1 Transition of site services and infrastructure organization.

As a *second step,* some chemical companies, particularly in Europe, established the service unit as a separate company while still retaining ownership. Though this second step might at first glance seem rather unspectacular, it was in fact much more radical than the first one: establishing a separate legal company was not so much required at single user sites, where the businesses and services belonged to the same company. But it was a logical next step at multi-user sites, where issues around the potential change of ownership of assets and fragmentation of value chains had already become prevalent or were expected to do so in the near future. In addition, formerly closed sites opened up for external investors who either took over parts of the existing assets or invested into new plants. Having a separate legal site service entity offered a better way of serving these third parties at multi-user sites. Many services were now offered at price – not at cost, asking all customers to accept a certain margin on many of the services. Consequently, with the set-up of separate legal service companies it became almost inevitable to open up the formerly largely shielded service market and to give production entities freedom in contracting, i.e., in choosing a service provider from the external market if the customers felt external providers had a better, more cost competitive offering.

Having taken that second step, service companies faced new challenges: how could products and services be clearly specified to make them marketable? Which products/services should be run as cost and which ones as profit centers? How should they react to their customers' new freedom in contracting? How should they build and manage true customer/supplier relationships? How should they define the right targets for the service company?

Though taking this second step was often hard, it turned out to be beneficial to all parties involved: the customers, the service providers, and the parent companies.

- Monopoly situations were replaced by *market mechanisms*. The service providers were forced to drastically improve their cost and service positions to be able to keep up with external competitors – often small- to mid-sized entities not only enjoying structural advantages and a lower cost base, but also showing more of an entrepreneurial spirit. Although natural monopolies such as those in utilities have remained, the opening of the site services and infrastructure market has also attracted specialized utilities providers like the Dutch company Nuon who are interested in taking over utility assets – thus further pushing the need for a restructuring process. All in all, the level of maintenance and logistics services outsourced has clearly increased as a consequence of this process.
- Cost allocation schemes were replaced by more flexible *service contracts* that made costs and services rendered much more transparent. These contracts range from more traditional time- and materials-based arrangements to full service agreements, where the service providers and their customers agree on certain output parameters such as uptime/plant performance rather than on input parameters such as maintenance cost. Price negotiations with the customers became the key imperative of this transition phase.
- Merely acknowledging production units as customers was replaced by a determination to strive for a true *service mentality*. Many site service and infrastructure organizations had to undergo massive changes in mindsets and behaviors to catch up with the kind of customer orientation external service providers had developed long before. Massive training programs coupled with revised incentive schemes were needed, yet in many service organizations today there is still a long way to go.

Establishing freedom of contract did have its downsides, though. Often enough, a service company had to realize that some of its offerings were just not competitive: despite all the improvement efforts, aggressive and structurally-advantaged third parties still came out ahead. Many service companies thus had to observe a slow but constant erosion of revenues. For some services such as maintenance or analytical services, however, internal providers often turned out to be the better choice provided they could adapt quickly enough to the changing environment.

Consequently, as a *third step*, service entities increasingly became engaged outside of their own sites. They entered the external market with service offerings they considered competitive. Infraserv and Infracor, which were founded by their parents Hoechst AG and Hüls AG respectively with the mission to professionalize site services, are prominent examples of separate legal entities tackling the external market. The imperative in this step: becoming exposed to full market mechanisms. Key challenges are achieving competitive advantage in the services offered, developing marketing competence, developing the capabilities required to leverage a broader business platform (e.g., finding the right service radius, capturing economies of scale and scope), and ultimately growing the external as well as the internal business. Typical examples for internal growth opportunities are taking over services that the customers have so far performed themselves (like certain decentralized maintenance activities) and/or taking over parts of the customers' value chain (like packaging).

Whether an independent service company acting in the external market is a sustainable business model still remains to be seen, though. Often enough, service companies offer at marginal cost or cover only some of their fixed cost in order to keep excess capacities busy. There are still no compelling success stories favoring the takeover of a complete site by an incumbent service provider. Chemical parent companies are therefore carefully reconsidering the degree of freedom they grant their service affiliates to grow outside of their home turf.

Mastering the transition process requires not only dramatic changes in the service entity's business model, mindsets, and behaviors, it also has huge implications for supporting IT systems. In accounting, for instance, generating itemized invoices for services proved to be a major challenge for the ERP systems. In maintenance, decision support systems needed to be adapted to handle not only time- and materials-based arrangements, but also to be able to offer full service agreements with the associated risks being borne by the service provider alone. Advanced analytical tools such as reliability-centered maintenance or process FMEA (Failure Mode and Effect Analysis) can help in making better-educated maintenance strategy decisions, thus reducing the risk exposure for the service provider. The importance and urgency of such system considerations should not be underestimated. Only with the right systems in place will the service entity be able to fulfill its service obligations successfully.

Not all chemical companies have followed the transition path to the end. Most have taken at least the first step, creating a service unit and professionalizing it. In many cases market mechanisms were established. That alone has increased the pressure to improve service performance and has led to higher levels of outsourcing. Establishing a separate legal entity, though, will be unnecessary in many cases and needs to be considered carefully and with thorough foresight. It is, however, advisable for those chemical companies facing major ownership changes at multi-user sites. And it can inject a lot of entrepreneurial spirit that can lead to a major performance improvement.

From the provider's point of view, site services and infrastructure have in many cases become businesses in their own right. Not only have the incumbent service companies tried to expand their business beyond their existing customer bases, but non-chemical players such as ABB Full Service have been attracted to this market as well and are pushing to capture a larger slice of it.

20.3
Going Forward: Increasing Site Services and Infrastructure's Competitiveness

Competitive pressure in the chemical industry is continuing to rise. For a long time now, the industry has managed to find ways to compensate for the average annual price-cost squeeze of two to three percent associated with the creeping commoditization of its products. Now, however, the competitive dynamics are changing: low-cost suppliers from countries like China and India boast impressive growth rates, increasingly serving Western markets with landed cost 'way below Western manufacturing cost.

To meet the challenge, many chemical companies have launched comprehensive restructuring efforts, often deploying advanced improvement concepts such as lean manufacturing and Six Sigma. On a strategic level, they are rethinking their allocation of production assets and future investments – often enough, investment decisions are no longer between two European sites but rather between Shanghai or the Nanjing area. On top of that, the increased number of open sites in the Western world and the vast availability of infrastructure have increased competition between sites and their service operations.

This development is a massive call for further change: site service and infrastructure organizations will have to bear a substantial part of the restructuring burden. They need to lower costs per single customer served. Moreover, facing the risk of an eroding local customer base, they need to further adjust their resources. This will be particularly challenging, since a large part of the services is very fixed cost-intensive and thus exposed to the risk of a significant share of residual costs. Maintaining the critical mass needed to be able to deliver the services cost-effectively is quickly becoming a relevant issue.

20.3.1
To Divest or Not to Divest Site Services and Infrastructure?

Within that context, some chemical companies are questioning whether site services and infrastructure still form part of their core activities. Willingness to invest into the service company is often very limited, especially when the share of external service customers is very high. In fact, some companies have started to question whether they are the right owners of the services at all. At first sight they have good reasons to consider withdrawing from managing their own site services and infrastructure, such as reducing fixed cost intensity, investing resources into core rather than into supporting activities, reducing complexity, and strengthening the focus on manufacturing. Moreover, increasing the share of external customers is likely to further professionalize the service company and at the same time reduce service costs.

While these chemical companies want to concentrate on their core business "chemicals", the service providers are eager to expand their core business "services". And they do see – at first glance – appealing growth opportunities, ranging from being a partner for purely transactional outsourcing of single services to having complete site service packages handed over to them and to fully managing the infrastructure of additional sites. Furthermore, expanding their external activities would also help the service companies to better utilize existing capacities, and hence achieve economies of scale and fixed cost dilution, as well as to overcome sub-critical mass in know-how-intensive service fields (Fig. 20.2).

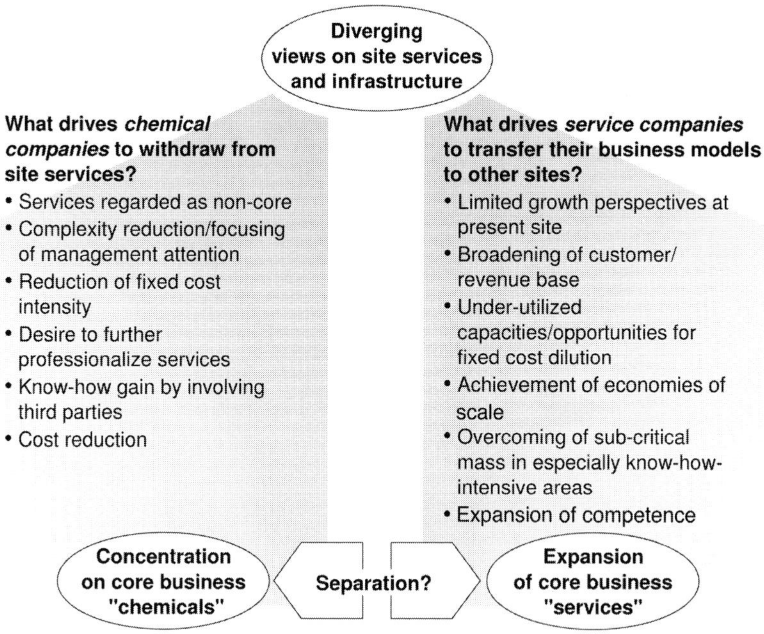

Fig. 20.2 Diverging views on site services and infrastructure.

These diverging views have led to a strong pressure towards separation – predominantly at multi-user sites where service customers are often from different corporations. With service companies increasingly trying to expand into the external market, some chemical companies in turn became convinced that a successful disinvestment would be possible. A number of them have indeed made intense disinvestment efforts. Up to early 2005, however, these efforts have not been successful.

Regardless of the question of site services being core or non-core, a separation is indeed difficult to accomplish, from both the seller's and the buyer's perspective:

The seller's perspective: It is not easy to find a new owner who can help the chemical company seller achieve the cost reductions and performance improvements needed to improve its competitiveness, and who at the same will be able to achieve satisfactory returns. It is very much a matter of trust and confidence, since successful case examples are rare. Often enough, managers in chemical companies already question the right of their internal service providers to generate profit – the idea of switching to an external provider will be even less appealing to them.

Giving up control over services that are natural monopolies (e.g., on-site steam production, waste water treatment, off-gas incineration) is also a concern. Fear of becoming too dependent and being skimmed by the new owner is widespread, although contract provisions could possibly minimize the risks. With marketable services, dependency would be less of a risk, though finding an alternative service provider would still take considerable lead time. In order to safeguard his interests, the seller will be highly interested in securing long-term contracts with the prospective new owner.

The buyer's perspective: One of the challenges will be to identify the short- to long-term value generation opportunities in a very demanding (given the broad range of services offered) due diligence process. Capturing these opportunities might be even more difficult, considering the high labor intensity of site services. In highly unionized environments in particular, transitional arrangements may make it hard to increase personnel productivity in the short term, especially if wage agreements differ considerably between buyer and seller. Though contract provisions once again could provide a way out, this issue might still be a deal breaker, especially in Western European countries. The seller's preference for long-term contracts might also limit the value generation opportunity. Environmental risks are typically another concern. Buyers with no prior experience in chemicals may tend to be overly worried when considering taking over asset-intensive activities. Minimizing possible risks for both the seller and the buyer in a satisfactory manner will certainly require cautious contract negotiations. Selling the property might not always be the best solution and lease arrangements should be considered.

The process of trying to divest or acquire a site services and infrastructure organization is a very tedious exercise, requiring thorough preparation, and with a very uncertain outcome. So far, a successful exit by a major chemical company is still up for realization and the recent failures raise the question of whether this path will be followed in the near future at all.

20.3.2 **Improving Performance and Growing Selectively**

Regardless of the question of ownership, the increased competitive pressure in the chemical industry will be passed on to site services and infrastructure providers. They will need to pull a broad set of levers to improve their own performance in order to help their customers succeed and thus secure their own business (Fig. 20.3).

Redefining the service offering. Based on an evaluation of customer needs and external competitors, site services and infrastructure providers firstly need to reconsider which services they should offer – and which ones they should not. They are clearly not the best owners of activities such as catering, human resources management, information management, or patent and license management – currently often part of their service portfolio. Second, they need to define which services will be provided by their own personnel and which services will be provided by external parties under their management. Many chemical service providers have already outsourced low-end services such as scaffolding or insulation. This trend will continue not only in maintenance, but also for other services such as internal logistics. Site infrastructure providers and specialized logistics companies, for instance, have already formed joint ventures.

1 Redefine service offering

- Define which services to offer
- Decide which of these services to be rendered by own personnel and which services to be sourced externally but managed for the customers

2 Increase level of customer orientation and flexibility

- Create flexibility to accommodate varying business types of the customers and service types requested
- Offer different service agreement types (from work-order-by-work order to pay-for-performance)
- Offer different service levels according to customer needs (e.g., silver, gold, platinum service levels)

3 Improve operational performance

- Operations
  - Adopt best practice examples from companies outside of chemicals (e.g., Total Productive Maintenance)
  - Redesign core processes (e.g., reduce number of interfaces/handovers)
- Structure
  - Reduce number of organizational layers, reconsider allocation of tasks to qualification levels
  - Open up collective labor agreements to eliminate structural disadvantages in wages

Fig. 20.3 Key levers for further improving site services and infrastructure competitiveness.

Increasing the level of customer orientation and flexibility. Although there have been great changes in terms of mindsets and behaviors and customer orientation has over the years become more than just a phrase, service providers still have to make headway. Far too often a "one-size-fits-all" approach is followed that neglects the different development stages and key factors of success of different customer businesses. Commodity and specialty chemicals businesses face different key factors of success and apply different criteria when defining which services they need and where to get them from, and hence will arrive at different make or buy decisions:

- While one business might want to retain a major chunk of decentralized maintenance, another might want to outsource maintenance altogether.
- While one might want to pay on a work-order-by-work-order basis, another might prefer bigger service bundles or even a pay-by-performance service agreement.
- While one production head might feel satisfied with a standard fault clearance time of three hours, another might insist on a service level of thirty minutes.

The site services and infrastructure organization has to be flexible enough to accommodate different types of business and scopes of service, different types of service agreement and different service levels. It also needs solid skills in pricing the service packages according to their value to the customer and according to the distribution of risk between customer and provider. This is not an easy task and requires very detailed handling – plus a key account management to handle the customer relationship holistically.

Improving operational performance. Although many customers prefer to have services bundled and to source as many services as possible from a single provider, they have very little tolerance of cross-subsidies between these services. The service organization thus needs to assess each service's performance from a stand-alone perspective. Customers will certainly do this themselves for those services where markets exist. They will also benchmark natural monopolies, to be able to assert an appropriate level of pressure on the service provider and drive prices down. Service providers, in turn, need to pull operational and structural improvement levers:

- On the *operational* side, they must compare themselves to the leading companies for each service and adopt best practice processes. Often enough, the leaders will be found in other industries. Advanced concepts such as Total Productive Maintenance (TPM) (in which production and maintenance are optimized holistically, often combined with a significant change in the allocation of responsibilities among the parties involved), where the chemical industry is still an adoption laggard, should be transferred more rigorously from industries like automotive manufacturing or aluminum production, which introduced them several years ago. Moreover, in many cases core service processes are still far from being fast and lean, often bloated with too many interfaces and handovers between the organizational units involved. Achieving operational excellence through core process redesign will be the action of choice in many cases.
- On the *structural* side, organizational structures and layers need to be revisited. Comparisons with small- to mid-sized service companies reveal that internal services and infrastructure organizations often suffer from too much overhead and overqualified personnel. Where many smaller companies use craftsmen, internal service organizations use foremen, where smaller companies use technicians, they use engineers. Whether the higher costs involved are always justified by additional value remains doubtful at best – in the end, many services turn out to be just too expensive. A delayering and reallocation of tasks can help improve competitiveness. Another issue is that some services suffer from cost disadvantages due to non-competitive wages that cannot be compensated for by higher productivity levels. Service organizations have the choice of either giving up these services completely or

initiating negotiations with the Works Council. Some providers of services such as logistics and maintenance have already been successful in opening up collective labor agreements.

Improving operational performance must be accompanied by establishing a performance management system that is capable of identifying and cascading the key performance indicators and value drivers down to the front line. Service companies need to identify the gap between the current and target performance for each service, and initiate a dialogue with management and labor on how that gap can be closed.

Finally, the site services and infrastructure organization might selectively pursue external growth opportunities. Carefully determining the price level will be a key factor of success. Are fixed costs sufficiently covered? How will the price affect competition? Offering services at too low a price can be risky: what looks appealing in the short term might lead to eroding margins and eventually to insufficient investments to secure the business. Pursuing growth options will take considerable management attention. The right to grow has to be earned by first pulling the other three improvement levers. Only after having done the "homework" should site service organizations look for external growth opportunities – and only by then will they have developed the skills and the competitiveness required to successfully seize these opportunities.

Overall, given the importance of site services and infrastructure as drivers of manufacturing productivity, the lack of success of attempted divestments in recent years, and the huge opportunities associated with pulling the existing performance improvement levers, chemical corporations should consider treating the management of site services and infrastructure once more as part of their core activities, inescapably linked to operating chemical assets. In this context, outsourcing decisions have to be made on a service-by-service assessment of the economic benefit, whilst keeping the focus on finding the optimal solution from an integrated production and site service/infrastructure perspective which should not leave room for functional egoism.

20.4 Summary

- Site services and infrastructure have undergone a major transition over the last couple of years, gaining increased independence or even being spun off into new entities.
- Site services and infrastructure have a strong impact on total production costs and plant effectiveness, and in times of increased changes of ownership of chemical assets their organizational flexibility is becoming more and more relevant.

- As site services and infrastructure made the transition first to bundled services units and then in some cases even to separate independent service companies, they were faced with the need to significantly professionalize the way in which they rendered services to both internal and external customers.
- Mastering this transition process requires dramatic changes in the service entity's business model, mindsets, and behaviors.
- The rising competitive pressure in the chemical industry, accelerated by the emergence of low-cost suppliers in emerging economies, represents a massive call for further improvement action in the service organizations.
- Key levers to be pulled here are redefining the service offering to be tailored more accurately to the customer needs, increasing the level of customer orientation and flexibility, and, lastly, improving operational performance with regard to both operational cost and the organizational structures and layers.
- After having earned the right to grow by pushing for operational excellence – and only then – the site service and infrastructure organizations might selectively pursue external growth opportunities.

21
Creating a Revenue Advantage through Sales and Marketing Excellence

John Warner, Joël Claret, Ralph Marquardt, and Eric Roegner

In an industry as capital-intensive as chemicals, the focus on operational excellence of the last two decades is understandable. However, a significant additional source of value creation potential exists which is largely untapped: creating a revenue advantage. Excellence in sales and marketing can lead to a tangible incremental improvement in ROS. In addition, a one percent improvement in price leads to significantly more value than a similar reduction in variable costs (Fig. 21.1).

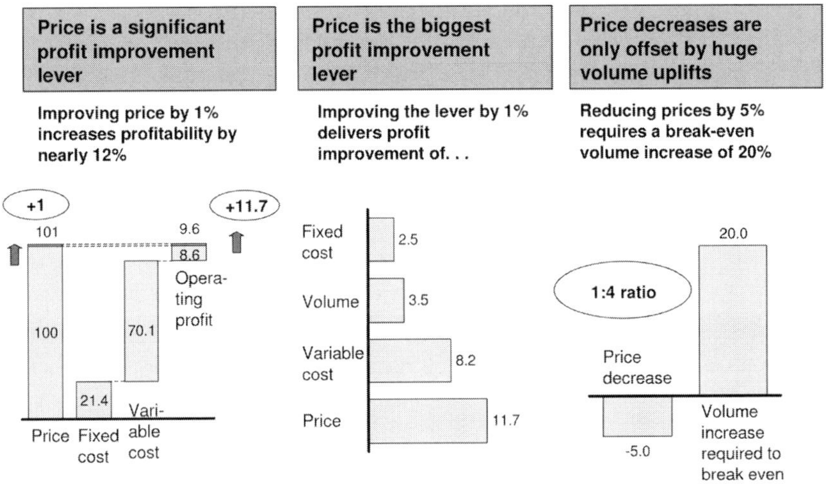

Source: Compustat; composite of 50 public chemical companies; McKinsey

Fig. 21.1 Pricing is a significant improvement lever.

Achieving these benefits requires a paradigm shift and a laser focus on sales and marketing excellence comparable to the operations and capital transformation that took place in the 1980s and 1990s. New tools, processes, and performance management systems are required to successfully implement and institutionalize a revenue advantage – the industry needs to turbocharge the development of the latent revenue gene within its business system.

Value Creation: Strategies for the Chemical Industry. 2nd Edition. F. Budde, U.-H. Felcht, H. Frankemölle (Eds.)

ISBN: 3-527-31266-8

21.1 Defining World Class

Creating a revenue advantage requires the development of a set of interdependent capabilities around six key elements – (1) Go-to-market strategy, (2) Value chain insights/segmentation, (3) Target customer identification/value proposition design, (4) Customer interface model, (5) Pricing, and (6) Performance management (Fig. 21.2).

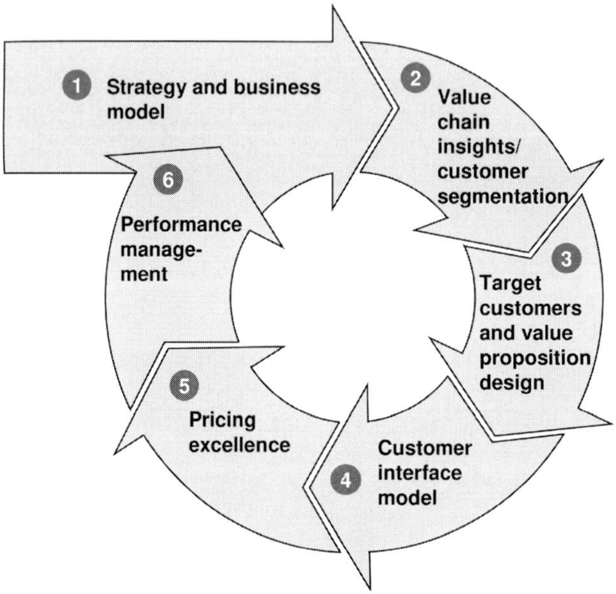

Fig. 21.2 Go-to-market value delivery system.

While it is important to build capabilities in each of these six elements, achieving revenue distinctiveness requires the integration of these capabilities into a differentiated commercial operating system. Experience suggests that 50 percent of the value comes from individual elements, with the remaining 50 percent being achieved through integration across the elements.

21.1.1 Developing a Winning Go-to-market Strategy

The first key challenge is defining a winning go-to-market strategy that will enable a significant and sustainable competitive advantage. A winning model should ideally meet the following criteria:

- It should create significant value for customers and the customers' customers.

- It should create a sustainable position within the value chain that enables the company to capture its fair share of the value.
- It should provide a distinctive competitive advantage from the customer's point of view.
- It should target specific segments and customer needs to enable seamless execution.

In today's economy, companies have to choose between the two fundamental models: a value-added strategy – high benefits at a price premium – or a no-frills strategy – basic benefits at an average industry price. In industry after industry, the squeeze is alive and well for the middle market offering. For example, in retail all of the profitable growth is taking place either at the high end – e.g., Nordstrom – or within the no-frills segment – WalMart, Target. Similarly, in the airline industry, Emirates and Singapore Airlines are prospering at the high end and Southwest Airlines and Ryanair at the low.

- *Value-added provider:* In this model, the goal is to deliver a value proposition that customers and their customers will consider creates higher benefits than competitors' offerings and that they will be willing to pay for.
- *No-frills offering:* In this model, the goal is to seamlessly execute against the customer's basic service and quality requirements at a price that differentiates the offering from that of competitors in the eyes of the customer. As Dell and Southwest have demonstrated in other industries, this does not necessarily mean the lowest price.

The two thematic go-to-market business models can be pursued either on a corporate level or business unit by business unit.

Developing the optimal go-to-market model requires deep customer and value chain insights that are tailored to the unique market intrinsics of different product/market segments. For instance, commodity businesses are likely to have much less sophisticated business models with fewer degrees of freedom for differentiation. The game is likely to revolve around out-executing the competition. On the other hand, more differentiated products are likely to have far greater degrees of freedom and may entail much higher levels of go-to-market sophistication to maximize value capture.

21.1.2
Value Chain Insights/Customer Segmentation

At the heart of creating a revenue advantage is the need for chemical companies to complement their asset-forward view (i.e., maximizing asset utilization) with a market-back view (i.e., maximizing margins based on what customers are willing to pay) of customers' current, latent, and unmet needs. This requires the development of much more sophistication and understanding of the marketplace and customer needs. Revenue-advantaged companies have a much greater understanding of the market, customer needs, and value chain discontinuities than their competitors.

- Where is 100 percent of the global demand for our product by customer and application?
- What is the global market price band and what drives the price bands across regions, segments, and applications?
- What is the value added of the product through the value chain?
- What profit differentials exist across markets and customer segments based on the unique characteristics of the marketplace?

Developing and operationalizing a segmentation approach that changes sales and marketing behavior and leads to bottom-line impact is a significant undertaking. A segmentation approach must be relevant and actionable, leading to changes in behaviors (e.g., different value propositions, alternative sales and service models, different price/benefit positioning of offerings, differential margin expectations).

Different levels of segmentation are required depending on the go-to-market model and the organization's current capabilities. In our experience, the most valuable segmentation approach is based upon profit potential and needs. On the profit potential side, the key issue is whether or not a customer's profit potential warrants a customized, tailored (with a menu of options), or standard offering. On the customer needs side, we see four different behaviors in purchasing: customers will buy on price, service, product attributes, or "win-win" performance partnerships.

21.1.3
Target Customer Identification and Value Proposition Design

Once a differentiated understanding of the marketplace has been obtained, it is important to decide which segments and specific customers the organization aspires to serve and, equally importantly, which ones it will not target. In most chemical companies today, this is a significant mindset shift. In general, most chemical companies are volume-focused, and have never met an unattractive customer.

The dilemma for most chemical companies today is that a 20/90 rule applies to customer performance – 20 percent of customers typically account for 90 percent of profit margin (Fig. 21.3). The challenge is what do with the long tail of customers that add little incremental margin to the overall bottom line. In most instances, companies have decided that the best thing to do is to enhance the profitability of the tail through aggressive pricing actions. In other situations, they have decided to redefine the value proposition and overall offering to enhance the profitability of this customer group or move these customers to distributors.

The challenge is further complicated by the fact that most companies today do not have a good understanding of where to find the untapped business in the marketplace, and what price or margin potential exists for this "white space".

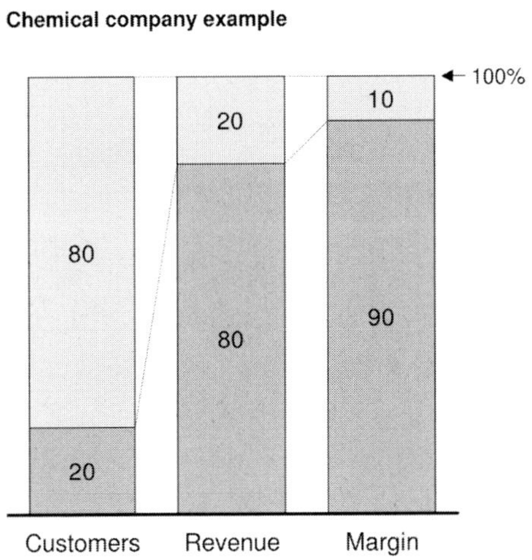

Fig. 21.3 Customer revenue and margin concentration.

In today's environment of sophisticated purchasing techniques and wider-ranging product sources due to entrants from emerging markets, it is critical for chemical companies to create distinctive value propositions for a focused set of target customers by answering three fundamental questions:

1. Where, within the overall market price band, do they wish to compete? In which geographies? Which market segments? Which needs-based segments?
2. Where, within the overall market price band, do they wish NOT to compete?
3. What value proposition will enable enterprises to capture their fair share of the profit pool, and create a differentiated offering versus competitors?

21.1.4
Defining the Customer Interface Model

In light of the targeted customer base and value proposition design, chemical companies must align their delivery models to what their target customers are willing to pay for. In general, the chemical industry has typically given away for free many of its value-added sales and technical services within the current value propositions, for instance high frequency of interactions, high frequency of transactions, visits to multiple plants, and others. These sales and technical services have become an industry norm rather than a way to differentiate offerings.

The majority of chemical companies offer a wide range of channel alternatives – e-channels, direct sales, customer service, third party distribution, technical service. This multi-channel approach has greatly increased customer flexibility and the choice of alternative relationship models, but has done little to reduce the investment in direct customer interface capabilities or to differentiate players' service offerings. Primarily, it has added complexity and costs to the delivery system.

To counteract this phenomenon, chemical companies should consider the following options:

- Differentiating channel offerings to customers based on what they are willing to pay for (e.g., direct, inside sales, e-sales, third party distribution)
- Explicitly charging higher prices or incremental fees for differentiated capabilities that create substantial value for the customers (e.g., rush orders/rapid delivery dates, less than truck load shipments, short lead times)
- Creating a low cost customer interface model for specific segments that provides only the bare necessities to complete a transaction.

For example, Dow Corning has created a specific channel and value proposition in its Xiameter brand that provides a no-frills/low-cost channel to those customers who simply want a basic product/service offering at a low price. Dow Corning reports that this model has been extremely successful in opening up a new customer base, driving volume growth, and generating attractive margins.

21.1.5 Pricing Excellence

At the end of the day, creating a revenue advantage hinges on the ability of the sales force to achieve the right contract terms, at the right price, at the right volume, at the right customers. While the concept is simple in theory, putting it into practice has been an enormous challenge for chemical players. Challenges typically advanced by the sales force include: "We have to cut our price to meet a competitive offering," "We need the volume to base load our facilities," "I know those contract terms are one-sided and expensive, but that is the industry norm."

While these challenges are valid and frequently encountered, a dramatic improvement in their overall pricing capabilities would deliver substantial value to many in the industry. Chemical companies have spent years lowering costs and creating new innovations, involving a great deal of hard work and investment – why give the advantage away by underpricing?

According to Marn, M. et al., excellent pricing requires distinctiveness on three levels:

- *Transaction pricing:* How do you decide the exact price to assign each customer transaction – what base price, discounts, allowances, adjustments, and other incentives?
- *Product/market strategy:* Within each market segment, what price level gives your product the optimal price/benefit position relative to competitors?
- *Industry strategy:* How do industry structure (supply and demand) and competitive dynamics affect overall price levels? What actions

are required to maximize margins in a highly volatile raw material environment?

Achieving pricing excellence requires a significant shift from a volume to a margin mindset, enhanced frontline tools, and significant capability building.

21.1.6
Performance Management

The glue that holds an enhanced revenue capability together is a margin-oriented performance management system. Most chemical companies today have very sophisticated operations and capital allocation performance management systems. On the revenue side, this is often a different story. Capital projects, often worth less than USD 500,000, are debated through rigorous performance dialogues and detailed financial analyses. On the flip side, the sales team makes multimillion dollar decisions on large customer contracts in a relatively ad hoc manner.

Key elements of an enhanced revenue performance management capability include:

- Establishment of a revenue scorecard that includes metrics and targets (both impact- and capability building-oriented)
- KPIs cascaded throughout the organization that clearly define the unique role, metrics, and targets that, when integrated, enable the organization to achieve its margin expectations
- An explicit understanding of historical and expected customer profitability by product and by transaction, with a forward-looking view on raw material costs.

21.2
Building World Class Revenue Capability

Developing a world class revenue capability is a complex process and requires a paradigm shift in most organizations. To achieve this change, companies need to develop a commercial system, the set of core capabilities and organizational enablers that work together to create significant value and sustainable competitive advantage. Five key components are required to create a world class commercial system (Fig. 21.4):

- *Performance expectations:* Significantly change the aspirations on the expected performance trajectory
- *Program design:* Focus on what matters
- *Commercial toolkit:* Put in place the right tools to solve the right problems
- *Capability-building agenda:* Build the right skills and provide coaching

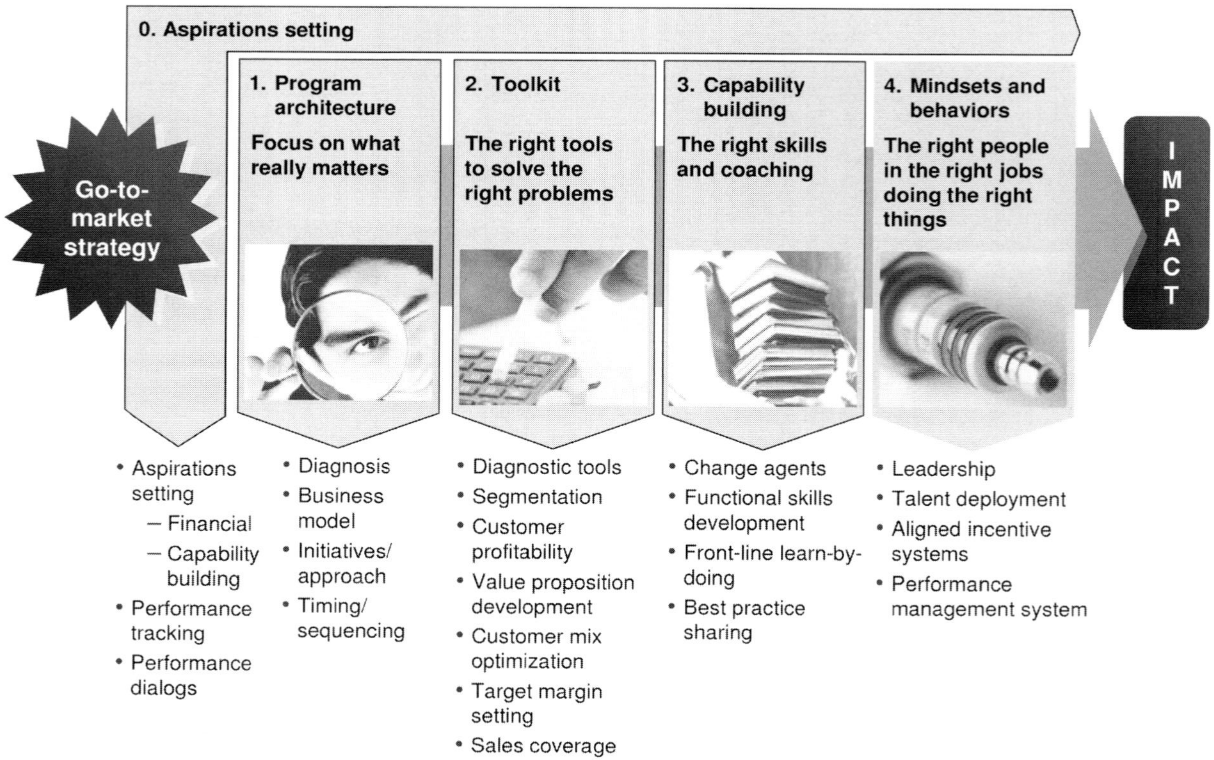

Fig. 21.4 Commercial operating system.

- *Mindsets and behaviors:* Make sure you have the right people in the right jobs doing the right activities.

Our experience suggests that all five of these elements are critical to success. Any weak link in the commercial system will lead to suboptimal results.

21.2.1 Performance Expectations

The basic principles of successful change management also apply to creating a commercial system – set stretch aspirations (i.e., performance expectations that require the organization to think radically and NOT incrementally about becoming excellent in sales and marketing), make it one of the top three strategic priorities, ensure senior management leadership, sustain focus to ensure the journey is completed, and invest sufficiently in skill building. The challenge lies in living by these guidelines. This must be viewed as a way of changing the way the company does things rather than as a new initiative.

21.2.2 Program Design

It is important up front to spend time on developing a game plan to embed the new commercial system into the organization. A critical aspect of this design phase is developing an explicit articulation of what the organization is trying to accomplish through an enhanced go-to-market value delivery system. There are two thematic options:

- *Integrated front-line sales and marketing performance improvement:* Step-change, immediate performance improvement through deliberate action and process change
- *New commercial business model:* Radical change/innovation in the value proposition and delivery system.

21.2.3 Commercial Toolkit

Each element of the go-to-market value delivery system is likely to require the development of new tools and techniques that enable enhanced transparency and faster and improved fact-based decision making. This starts with enhancing the transparency of customer profitability, to gain an understanding that goes deeper than annual overall profitability by customer and provides transparency by customer, by product, and by transaction. Averages quite often disguise improvement opportunities – 50 percent of transactions are below the annual customer profitability.

In addition to customer profitability tools at the transaction level, additional tools will need to be implemented and delivered to the accountable individuals in areas such as customer segmentation, target margin setting, contract terms policies, raw material cost movement expectations, and customer/product mix optimization.

21.2.4
Capability Building Agenda

The commercial system must address the central question of how the organization is going to build new and sometimes radically different capabilities within the entire sales and marketing organization. In many instances, these capabilities need to be instilled in individuals who have many years of experience in the old way of thinking and are resistant to change. In addition, successful implementation may require an influx of new talent.

Embedding these new capabilities in the organization requires a learn-by-doing approach, with constant reinforcement of the desired behaviors and consequence management for non-compliance. Allowing people to opt out because of seniority or good past performance is a potentially fatal sin. Elements of a successful capability building program include:

- Leadership by credible change agents
- A learn-by-doing training approach that requires the application of new tools and techniques to current customer issues and opportunities
- Reinforcement through an ongoing capability building agenda that includes getting repeated practice in using the new tools and processes.

21.2.5
Mindsets and Behaviors

Quite often, the greatest challenge companies have to overcome in creating a revenue advantage is the organization's existing mental model of the right way to conduct commercial activities. This is often reinforced by the conversations process and performance dialogues that take place within the organization. It is critical to develop a tailored influence model to make the required change happen and stick. This requires an explicit understanding of the fundamental changes in the mindsets and behaviors of marketing and sales decision makers that will be needed to enable the successful execution of the new commercial system.

There are four key elements to reprogramming the mindsets and behaviors of the organization (see Chapter 24, section 3.2 for more detail):

- *Fostering understanding and conviction:* Ensure each individual clearly understands the performance requirements and believes that they are meaningful and appropriate

- *Providing effective role models:* Ensure respected people at all levels of the organization behave in the new way
- *Reinforcing with formal mechanisms:* Align the structures, processes, and systems to reinforce the desired changes in behavior
- *Developing skills and talent:* Ensure each individual has the skills and competences to perform with the new commercial system.

While these elements may seem basic and simple, it is extremely difficult to implement them within large, complex organizations. Quite often, it will require significant mindset shifts and symbolic actions to overcome institutional barriers such as: "No one has ever been fired for agreeing to a lower price...but many have been fired for losing volume"; "Our customers and competitors set the price... we just have to be competitive"; Our customer contract terms are industry standard ...they have existed for years... there is no way we can change"; "I realize that we don't make much money on this large customer.... but they are strategically important to us as we have been their preferred supplier for 20 years".

21.3 Summary

- Executives of chemical companies today must ask themselves the central question – is it time to undertake a second paradigm shift to create a revenue advantage to complement an operational one?
- A revenue advantage is a significant opportunity with large potential financial rewards. However, it is not easy to achieve.
- It requires the development of an integrated go-to-market value delivery system based on six elements.
- Creating a revenue advantage through sales and marketing also requires a multi-year transformation that turbocharges the development of the latent marketing and sales gene within the organization.

References

Marn, M., Roegner, E., and Zawada, C.: The Price Advantage, John Wiley & Sons, 2004

22
Achieving Top Performance in Supply Chain Management

Andrea Cappello, Martin Lösch, and Christoph Schmitz

Over the last few years, chemical companies have focused more and more on supply chain management (SCM). Some have already achieved significant improvements and are continuously striving towards excellence. However, the chemical industry in general has not been an early adopter of SCM techniques, whereas others have captured huge potential. The key elements and success factors of top performance in this area are clear today, making it the right time for a further evolution of SCM practices across the entire industry. An SCM improvement program is complex and protracted, as it addresses the entire company business system and integrates and trades off different and sometimes conflicting functional goals, trying to orient them towards the common objectives of higher service level and lower operating costs. Therefore, in order to be successful, it needs to set ambitious targets, systematically address issues and organizational units by subprojects, and carefully manage the required cultural and skill change.

Companies able to win such a challenge can release the pressure on their financials and secure sustainable success in the long term.

22.1
Supply Chain Management as a Strategic Lever for the Chemical Industry

Return on invested capital (ROIC) in the chemical industry is under pressure from both falling prices and rising costs, and some of the trends responsible, already mentioned in previous chapters, have simultaneously made SCM vital. Chemical companies, traditionally focused on manufacturing productivity and very busy with mergers and acquisitions (M&A), have recognized in the past few years that SCM can play an important role in fulfilling customer expectations cost-effectively, through complex global operations linking all parts of their business system and including suppliers and customers.

SCM can be a compelling force driving all factors of the ROIC equation: cost, revenues, and capital.

Value Creation: Strategies for the Chemical Industry. 2nd Edition. F. Budde, U.-H. Felcht, H. Frankemölle (Eds.)

ISBN: 3-527-31266-8

Directly or indirectly, it can influence as much as approximately 20 percent of total costs[1], and a 20 to 30 percent reduction of total supply chain costs can be tremendous for companies that, in some cases, are operating on profit margins of around five percent.

In addition, because of improved service levels, SCM can make a significant contribution towards preventing lost sales and become a strategic factor for differentiation, enhanced customer value propositions, and market share gains.

Finally, SCM has a direct impact on production and logistics assets utilization and on the entire inventory level, which represents on average around ten to 20 percent of sales. An improved supply chain performance, such as a potential ten to 20 percent increase in production asset utilization and/or a ten to 15 percent reduction in inventory levels, enables companies to postpone or even cancel planned capital expenditure.

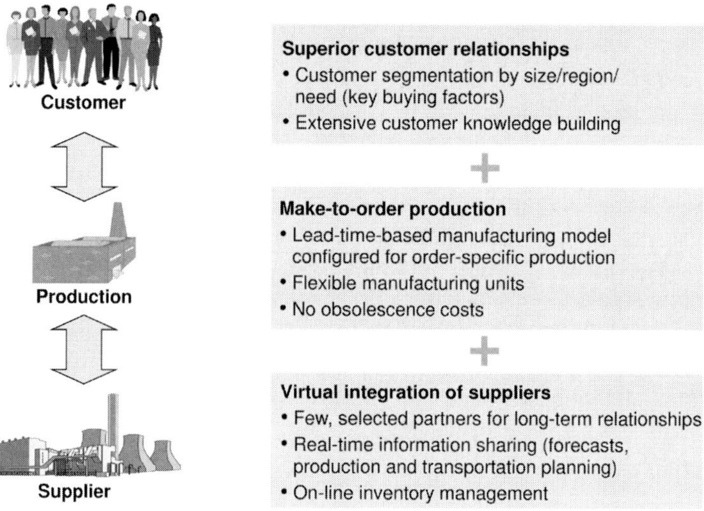

Source: Dell press releases, analyst reports, McKinsey Supply Chain Management Practice

Fig. 22.1 SCM as a strategic factor for differentiation, value creation, and market share gain – the Dell case.

1) A typical breakdown would show five to eight percent for logistics, one to two percent for the financial cost of inventory, five to seven percent for SCM-related material and conversion costs, and one to three percent for customer service, planning, and related IT costs.

Well known examples in other, even mature, industries such as Procter & Gamble in packaged consumer goods, Dell in consumer electronics, Wal-Mart in retailing, and Toyota in the automotive sector prove that outstanding performance is achievable and can contribute to the company's overall success through tailored service levels, fast and reliable deliveries, and lean and cost-effective operations, even in a complex environment. Though not directly comparable, these industries have developed solutions that, appropriately selected and tailored, can be of tremendous value and serve as a reference for all branches of the chemical industry. Dell, for example (Fig. 22.1), became one of the biggest and the most valuable PC producers during the '90s; while there are obvious differences between the PC and chemical businesses, approaches like product-customer segmentation or on-line inventory management can definitely be successfully borrowed.

Some steps have already been taken in the chemical industry: during the deep restructuring of the '90s, many players changed their organization from a functional to a customer and market orientation and, at that time, some also abandoned the traditional focus on simple functional excellence in manufacturing or logistics, and developed a more customer-centered and integrated approach to their operations. These companies have discovered the potential benefits for all company stakeholders inherent in SCM.

In comparison with progress in other industries, this transition is not complete yet. Companies that fully understand the key elements of SCM (see section 22.2) and successfully orchestrate the technical and cultural change required to achieve top SCM performance (see section 22.3) can further improve their ROIC and unlock further value, thus creating the platform for their long-term and sustainable success.

22.2 Key Supply Chain Management Elements and Opportunities for the Chemical Industry

In general, SCM requires two key components: first, functional excellence in the materials and information flows along the entire value chain from the supplier to the customer, and second, overarching capabilities in cross-functional planning and execution so that all parties involved take consistent and coordinated actions.

This section describes each of the six elements illustrated in Figure 22.2 and highlights the opportunities that have been, and in some cases still can be, successfully tapped.

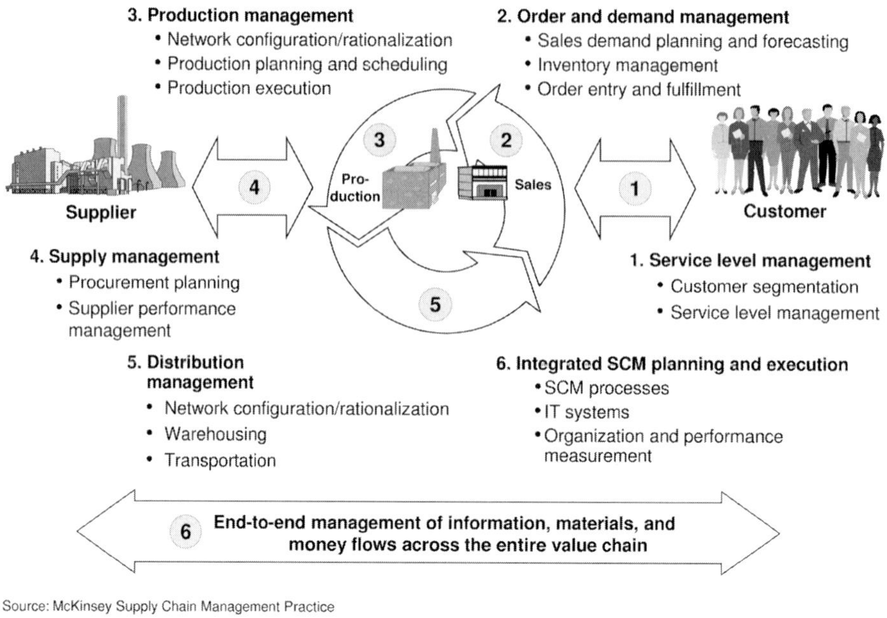

Fig. 22.2 Key SCM elements.

22.2.1
Service Level Management

Service level management has to ensure that customer requirements are understood and fulfilled, if necessary, through differentiated supply chain models and service level propositions (see also Chapter 21).

It is crucial to know the requirements and service breakpoints of each customer segment (Fig. 22.3) and the related economic opportunities. Most successful players go as far as designing and tailoring differentiated supply chains and service level offers (Fig. 22.4), and tracking service levels and profitability to address service issues and extra costs.

Some chemical companies have yet to abandon their focus on manufacturing – understandably enough, in view of the high capital tied in. However, they should be wary of a mentality that assumes the supply chain ends at the factory gate and can lead to service levels sometimes in the range of 65 to 70 percent of order lines delivered on-time-in-full. Best practice reaches up to 85 to 95 percent. In addition, some companies have yet to explore and understand their customers' real requirements – they may be pursuing higher product quality standards when what the customer really wants is higher delivery compliance and shorter delivery times.

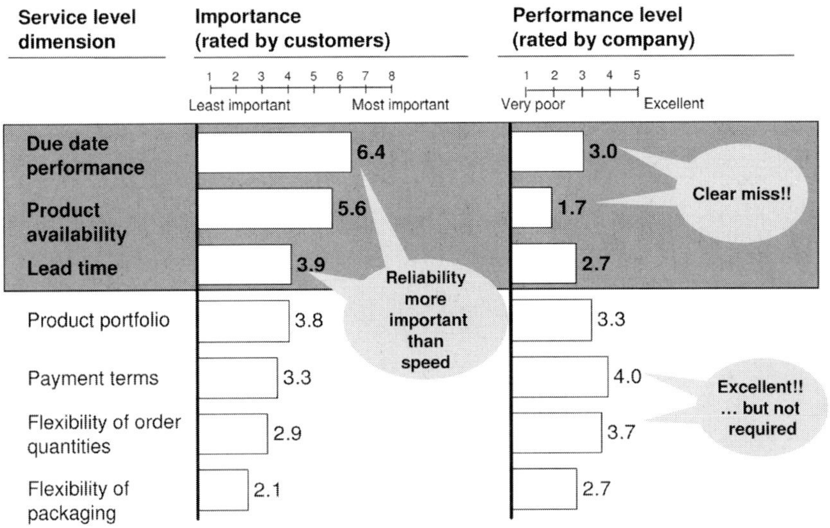

Source: McKinsey Supply Chain Management Practice

Fig. 22.3 Service level breakpoints determining supply chain design.

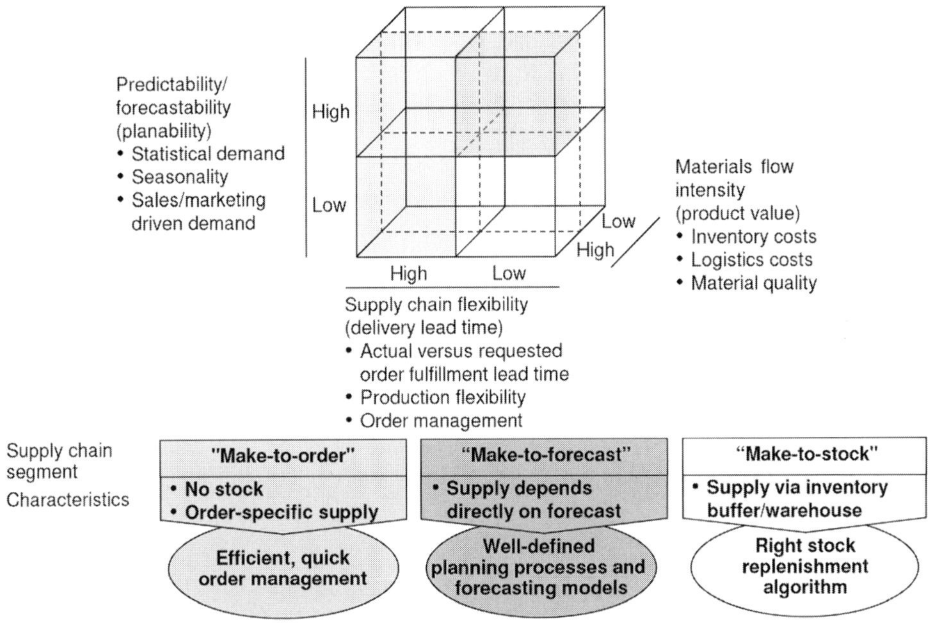

Source: McKinsey Supply Chain Management Practice

Fig. 22.4 Differentiated supply chains and service level offers.

One supply chain model clearly emerging in the industry is vendor managed inventory (VMI). Although this is often used by suppliers to lock-in customers and by customers to relieve themselves of the responsibility and of the financial and management costs of their inventory, as time goes on business partners are finding that more can be gained from seeking mutual advantages.

22.2.2
Order and Demand Management

The objective of order and demand management is to forecast future sales as precisely as possible, manage customer orders from entry to delivery, and consolidate forecast and orders with supply chain inventory requirements in order to generate a holistic demand plan for purchasing, production, and distribution planning.

Forecasting, inventory management, and order management each have specific key success factors.

Forecasting. This should be a real priority for chemical companies, as at times forecast accuracy is lower than 50 percent and deeply influences customer order and inventory management. If the replenishment lead time of a product is longer than the lead time available to fulfill customer orders, forecasting is key to planning inventory and guaranteeing the expected service level. As a rule of thumb, an acceptable forecast accuracy, defined as the ratio between sales and forecast measured at the stock keeping unit (SKU) and monthly levels, is higher than 70 to 80 percent: this allows accurate inventory planning with reduced safety and buffer stocks, a higher service level due to more stable and reliable supply, and a higher utilization of production assets because of more stable production plans, and leads to fewer consequential issues in logistics, production, and procurement.

Whenever customers cannot directly provide an accurate forecast, for example using Collaborative Planning, Forecasting, and Replenishment (CPFR®), and lead times cannot be reduced to a level where no forecasting is needed (e.g., through make-to-order), forecasting is based on IT-based statistical forecasting and/or manual market intelligence, depending on the product's "forecastability" and importance as defined, for example, by revenues or margins.

The rigorous management of the number of SKUs reinforces this approach: structured programs often result in a 20 to 30 percent SKU reduction potential and improved focus, and reduce the complexity of the entire supply chain. A further key to success is to hold sales and marketing accountable and incentivized for forecast accuracy, and to keep regular track of it.

Inventory management. High-performing inventory management has four fundamentals: trying to avoid inventory ownership through supplier VMI or just-in-time (JIT) delivery; using statistical tools to optimize safety stocks and order lot sizes; assigning clear and individual responsibilities for each item of inventory held, ideally for each SKU; and carefully managing slow-rotating inventory.

Customer order management and fulfillment. Effective and efficient order management relies on transparent policies and decision rules reflecting customer needs and supply chain constraints, for example, for prioritization in case of shortages, and on real time IT tools, including enabling on-line confirmations to customers and continuous order tracking.

22.2.3 Production Management

Production management is responsible in the short term for planning production in order to cover the overall supply chain demand, and for delivering timely, complete, and quality-checked production volumes accordingly. In the long term, it optimizes the production network configuration and available capacity.

In the short term, more advanced companies meet the overall supply chain demand while simultaneously minimizing the overall costs of inventory, production, and missed sales opportunities. Further, with frozen production periods they avoid unnecessary changeovers, smoothen the whole production, purchasing, and maintenance process, and achieve superior adherence-to-plan and productivity performance. Nevertheless, if required, they can react promptly to unforeseen changes thanks to reduced changeover times, flexible suppliers, and powerful IT tools that adjust the plans quickly.

Over the long term, operational excellence comes from balancing long-term capacity availability with demand by reducing or increasing production operations, and smoothening medium-term fluctuations by insourcing and outsourcing.

The main opportunities for chemical players lie in the following four areas:

1. *Production scheduling.* By implementing frozen production periods while achieving a good forecast accuracy, chemical companies could reach an adherence-to-plan of over 90 percent and a three to four percent increase in productivity through a reduction of last-minute changes to the production schedule.
2. *Complexity management.* A typical batch-oriented chemical plant has hundreds of finished products, five- to ten-level bills of materials, configurable plant layouts, multi-plant production routings, days-long setup times – production scheduling can be very complex, executed in some cases only on a monthly basis. Some companies keep buffer stocks to decouple such inflexible and slow-reacting production from fluctuating customer demand; naturally, reducing complexity and increasing flexibility would shorten reaction times and reduce inventory.
3. *Total supply chain cost optimization perspective.* Forever striving to fully utilize production capacity and minimize manufacturing cost, some chemical companies tend to run very long campaigns, usually resulting in excess inventory and increased cost. The break-even of the tradeoff decision between campaign length and inventory level can be determined if the cost of excess inventory is considered as well when minimizing total supply chain costs.
4. *Production network rationalization.* The M&A deals that took place in the '90s, not always followed by full production network consolidation, the

opportunity to move to cheaper areas (e.g., the Far East, particularly China), and sometimes even excess production capacity make production network rationalization an opportunity that should be periodically addressed, including all aspects of available production capacity and technologies, specific market requirements, and a full understanding of the possible scenarios and related economics (e.g., reclamation and social costs).

22.2.4 Supply Management

Supply management has to ensure a smooth and uninterrupted supply of raw materials by defining clear service level agreements with suppliers, planning raw materials, issuing procurement orders, and tracking and managing supplier performance.

To ensure best performance, suppliers should be well integrated into operational processes in order to guarantee a smooth two-way exchange of information and of goods, as in most advanced CPFR® or VMI programs. A further key to success is the Total Cost of Ownership (TCO) approach in purchasing: the mutual benefits of integrating operations with suppliers are then actively sought, rather than focusing solely on minimum purchase cost.

Many players are going through purchasing spend reduction programs and leveraging sourcing opportunities from low-cost regions. However, with a real TCO minimization approach, they could keep in mind inventory-holding costs and the consequences of longer procurement lead times as well.

22.2.5 Distribution Management

Distribution management is responsible for planning and executing distribution to local subsidiaries and customers and for regularly reviewing and optimizing the distribution network.

In order to minimize the total distribution costs represented by logistics and inventory expenses, many companies are adapting their networks to the shifts of the geographical distribution of sales volumes or to changing tariffs, and optimizing the use of full or less than truck (or container) loads.

Moreover, warehousing and transportation operations are being outsourced more and more, as well as an increasing number of bundled services including administration, quality assurance, finance, legal affairs, and IT.

In the chemical industry, increasing service level expectations and Environment, Safety, Health, & Assurance regulations have helped to make logistics one of the biggest SCM cost components. Most chemical companies have already reduced distribution costs, usually by outsourcing logistics activities. Nevertheless, room still exists for developing more value-added services with logistics service providers and distributors (see also Chapter 12).

Furthermore, chemical companies should go on reducing their large number of distribution centers and warehouses, the legacy of an historically grown network based on M&A history or reflecting old national boundaries, like country-specific warehouses in Europe.

22.2.6
Integrated SCM Planning and Execution

As mentioned above, in addition to functional excellence, the second dimension of SCM is integrated planning and execution. A number of overarching elements enable the integration of our five SCM functional process elements:
- *SCM processes:* Clear, standardized and optimized SCM processes along the five dimensions described in the previous sections are the first prerequisite for integrated SCM planning and execution, as they align the multiple parties involved, their data and actions, and their perspectives and objectives.
- *SCM organization and performance management systems:* In addition, a well-designed organization for the end-to-end integration of supply chain operations encompasses clear responsibilities, transparent performance measures, and a performance management system with non-conflicting incentives, fostering discipline and a continuous improvement culture. It supports trade-off decision making and provides mechanisms to reconcile conflicts: at the core of SCM are a series of delicate balances that need to be managed, like those between inventory costs and immediate availability, inventory and production costs, or warehousing and transportation costs.
- *Supporting IT systems:* Finally, the accuracy and speed of an integrated process is heavily influenced by the availability of IT systems, which provide a common database, automated functions like Materials Requirements Planning (MRP), and opportunities for simulation and optimization (for instance, real time finite capacity production scheduling, or on-line customer order confirmation based on existing overall supply chain inventory or capacity).

Some chemical companies may still be facing the challenge of removing thinking in functional silos. They should establish fully cross-functional, collaborative planning processes, set up or strengthen central SCM coordination, and ensure the consistency and completeness of the goals and incentives set by the performance management system.

22.3 Delivering on the Opportunities – the Key Success Factors for Achieving Top Performance

There is clearly no standard recipe for achieving SCM excellence, but experience has shown that the full SCM potential can be captured only with a thorough and consistent investigation of all improvement levers, followed by their implementation. This requires a learning process which is individual to each company, and whose key success factors will be discussed in this section.

Whereas section 22.2 described "what" top-class SCM should be, this section focuses on "how" to set up such a system, in terms of both the design principles and the management of a possible improvement program.

22.3.1 Designing Top-performing SCM

Top-performing SCM can be an all-embracing theme that effectively and efficiently supports a company in reaching its performance goals. This requires a full understanding of and alignment with the business strategy and product-market-customer characteristics, a set of structural elements that make all functions involved work together in unison – transparency of performance, clearly defined processes, accountabilities, and organization – and, just as importantly, the appropriate approaches and tools.

22.3.1.1 Alignment of SCM with Business Strategy and Product/Market/Customer Characteristics

SCM needs to be aligned with and support the overall business strategy, and therefore to be in line with product, market, and customer characteristics. As different products, customers, and markets might need to be supported with different strategies, SCM needs to be tailored accordingly.

Figure 22.5 shows how product, customer, and market characteristics can determine different SCM strategies and design implications. For example, the strategic choice to serve niche customers with high price specialty chemicals is only tenable if the high-value inventory is centralized to reduce costs, and if a fast, responsive supply chain is set up to fill customer orders within the requested lead time from central inventory.

22.3.1.2 Transparency of Performance – "Inventory as the SCM Thermometer"

Transparency in supply chain performance is a prerequisite for analyzing performance hitches, prioritizing any improvement efforts, monitoring the impact of the measures taken, and focusing the organization's attention – which by itself can lead to huge improvements.

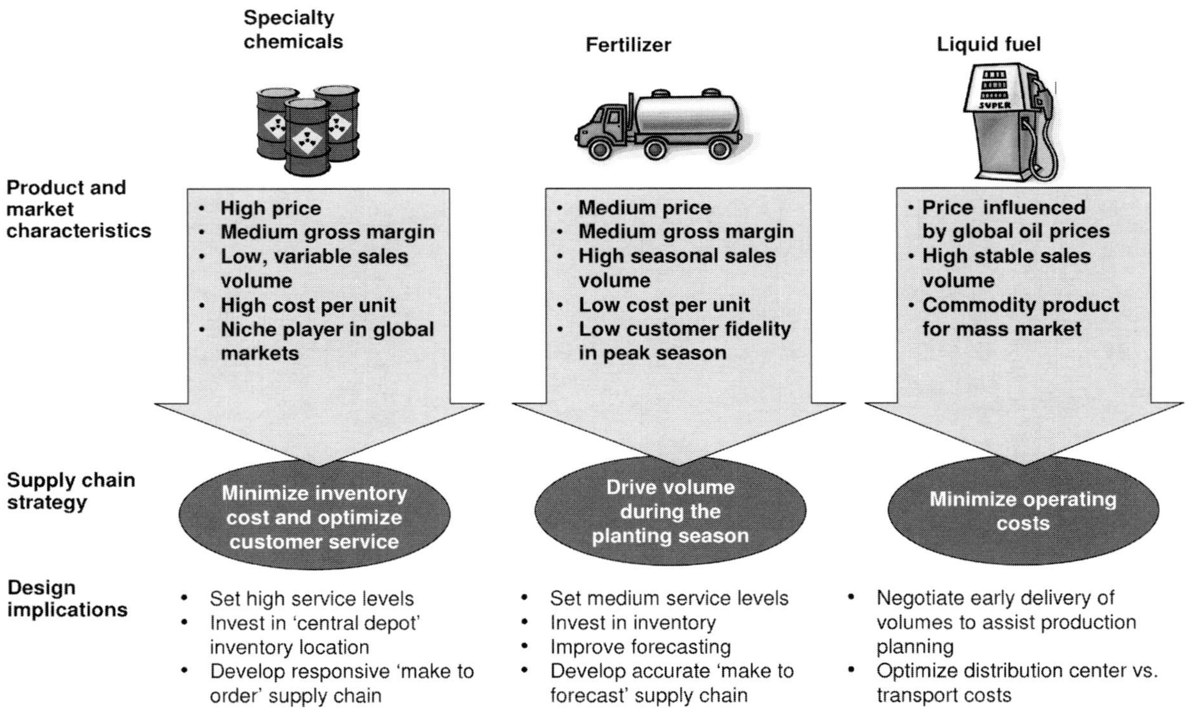

Source: McKinsey Supply Chain Management Practice

Fig. 22.5 Alignment of SCM with business strategy and product/market/customer characteristics.

Ideally, the most critical financial and operational key performance indicators (KPIs) centering around service level, quality, and cost should be monitored for each function, but it is important for *all* SCM-influenced costs to be reported to avoid distortion: for example, more frequent procurement orders would reduce the cost of raw materials inventory at the expense of inbound logistics costs, and omission of the latter would give a misleading picture.

A final observation here concerns inventory, usually one of the main cost drivers of chemical supply chains. Excess inventory of raw materials, intermediates, and finished goods is often present in plants and local warehouses and aged or dead stock can be up to five percent of total inventory. As the root causes can be widely spread, from inaccurate forecasting to long production lead times, from lack of integrated SCM processes to insufficient IT coverage, from lack of clear accountability to lack of performance management, inventory can be taken as a good thermometer in SCM health checks.

22.3.1.3 **Clear Cross-functional Interfaces and Coordinated Decision Making**

Whereas it is obvious that SCM processes must fully leverage functional core competences, cross-functional interfaces are less unequivocal and need to be clearly defined, with synchronized timing, agreed rules for interaction, and standard templates. In addition, seamless access to up-to-date and consistent information is required to allow effective decision making. Moreover, to achieve coordinated decision-making, the different functional perspectives and objectives have to be reconciled and rules set which will prevent optimization in individual functions, which is counter-productive in terms of the overall supply chain optimum. Regular cross-functional sales and operations planning meetings, held to share current end-to-end supply chain plans, can jointly address issues and decide on the trade-offs that represent the best opportunity for coordination and integrated decision taking.

22.3.1.4 **Careful Design of the Performance Management System**

A number of prerequisites for a performance management system can make all SCM actors work together effectively and achieve excellent performance: a complete set of KPIs, non-conflicting and clear accountabilities, and rigorous performance management. All SCM-relevant KPIs need to be unambiguously defined and monitored to ensure a holistic picture: disregarding one component might endanger the performance of the entire supply chain ("a supply chain is as strong as its weakest link").

Non-conflicting and unambiguous accountabilities are key to success. Supply chain managers are often held responsible for inventory levels but are unable to influence production lot sizes, which are normally under the control of the production managers responsible for production costs. As lot size adversely affects inventory levels and production costs, this is a common example of conflicting and ambiguous accountabilities, where responsibilities are not backed up by control of all the required levers. Carrying on with this example, the issue could be resolved with formal agreements between SCM and production, e.g., about the maximum number of changeovers, allowed changes to the schedule, and number of products simultaneously produced per month. This way, the scope of action of each manager would be clear and accountabilities would cease to be in conflict. This specific issue could also be addressed in a more fundamental, organizational way, by forming a joint "Supply Organization" that includes both production and supply chain functions.

Last but not least, a management system intended to steer people's performance needs to rigorously address achievements and failures, for example by actually linking KPI performance with incentive schemes and career opportunities.

22.3.1.5 Tailored SCM Organization Structure

The "organizational boxes" should only be drawn after the critical processes and accountabilities have been defined, and during this process a very great deal of care should be given to getting buy-in from all affected stakeholders along the entire value chain – such as heads of functions and of major production, distribution, and sales units.

When integrating multiple legal entities into an SCM organization, ownership of inventory should be strongly consolidated to simplify geographical mobility and avoid local optimization moves. An SCM organization with full responsibility for operational performance should also develop, standardize, and implement company-wide practices that go beyond the responsibility of single functions.

22.3.1.6 Detailed Approaches and Sophisticated Mathematics

SCM requires complex and highly detailed modeling as well as sophisticated optimization techniques. For example, planning parameters such as safety stocks and order lot sizes need to be regularly optimized using statistical tools that require sound analytical skills; slow-moving and dead inventory – often real "leftovers" – needs ad-hoc solutions for each material, sometimes demanding major efforts (e.g., for obsolete finished goods, suitable customers have to be identified and specific deals negotiated); distribution and production network consolidation, make-or-buy decisions, and allocation of products to plants are problems that go beyond spreadsheet-based modeling and require sophisticated operational research techniques for which specific skills are necessary.

22.3.1.7 Priority of Process over IT

IT should be adapted to process, not process to IT. In some cases, a kind of supremacy of IT over processes and change management can be observed: vague business cases and plans, forced introduction of standardized IT tools, quick training programs, poor results. A correct re-prioritization would make it possible to focus (and spend) on what is likely to provide real benefits: IT should only be an enabler.

22.3.2 Managing an Improvement Program

An SCM improvement program should create momentum, break down the scope into manageable pieces, and explicitly manage the cultural and skill change necessary to ensure sustainability and continuous improvement. In addition, top management conviction and support are unshirkable prerequisites for success.

22.3.2.1 Ambitious Targets and Systematic Approach

In order to be successful, SCM redesign or improvement programs should be inspired by some setup principles.

Ambitious and measurable targets should be defined and transparency on performance should be regularly provided, to mobilize the organization and track progress objectively.

Furthermore, sub-projects addressing individual organizational units should be put in place, to cover the whole company in subsequent steps. Focus and continuous prioritization are also key in order to achieve higher impact: due to their cross-functional nature, SCM projects tend to raise countless issues, and companies should avoid "biting off more than they can chew".

The resources responsible for implementing SCM techniques and processes should be chosen carefully and developed using a train-the-trainer concept, and should regularly be assigned to subsequent sub-projects within the program, to achieve the highest impact in the shortest possible time.

22.3.2.2 Strong Basis for Long-term Sustainability and Continuous Improvement

An SCM redesign involves many people at multiple levels and usually requires a change in culture and mindsets. This must be addressed explicitly, with strong leadership, a compelling and fact-based vision and value proposition, and extensive involvement of stakeholders in designing the solutions.

In addition, the necessary understanding, motivation, and capabilities have to be created to ensure sustainability and continuous improvement by means of formal training as well as personnel development planning supported by regular coaching and feedback sessions.

Finally, the chemical industry, traditionally rather closed, should possibly consider the introduction of SCM experts from other more advanced industries to facilitate change and ensure long term sustainability.

22.4 Conclusion

Reaching world class SCM is no easy job: a program to achieve it could even take a couple of years, involving a lot of stakeholders from many different functions and incorporating many simultaneous and interactive initiatives.

Nevertheless, improving or even innovating SCM design and operations is a reliable way to boost ROIC and therefore market value, as supply chains link nearly all parts of a company's business system. When Dell developed its radically new supply chain model, it was really "thinking out of the box", and probably no one would have believed it would soon become the reference in (and outside) its industry, and one of the biggest and most valuable players in its sector. Similarly, other industries have indicated ways that the chemical industry can take in the near future.

Chemical industry players have only recently started transforming their supply chains, and enormous potential still remains to be captured. Being an early adopter, moving first, and taking a creative approach can really ensure distinctiveness, competitive advantage, and long-term, sustainable value creation.

22.5 Summary

- Improving or even innovating SCM design and operations is a reliable way to boost ROIC and therefore market value, as SCM links nearly all parts of a company's business system and can be a compelling force driving all factors of the ROIC equation: cost, revenues, and capital.
- Examples in other, even mature, industries prove that outstanding performance is achievable and show the way and the potential left to capture for the chemical industry, where the transition to world-class SCM is not complete yet.
- From an operational point of view, world-class SCM requires two key components: first, functional excellence in service level management, order and demand management, production management, supply management, and distribution management; and second, overarching capability in cross-functional planning and execution so that all parties involved take consistent and coordinated actions.
- From a structural point of view, top-performing SCM requires a full understanding of and alignment with the business strategy and product-market-customer characteristics, a set of structural elements that make all functions involved work together in unison – transparency of performance, clearly defined processes,

accountabilities, and organization – and, just as importantly, the appropriate approaches and tools.

- An SCM transformation program can be complex and protracted, as it addresses the entire company business system and integrates and trades off different and sometimes conflicting functional goals, trying to orient them towards the common objectives of higher service levels and lower operating costs.
- In order to be successful, such a program needs to set aggressive and ambitious targets, systematically address issues and organizational units by sub-projects, employ rigorous project management, and carefully manage the required cultural and skill change. In addition, top management conviction and support are unshirkable prerequisites for success.

23
Right Second Time – Unlocking Value with IT

Peter Peters and Detlev Ruland

The hallways of chemical companies' IT departments have been echoing to a similar refrain of late. Talk of atoning for the sins of the recent past mingles with hand-wringing over shrinking business margins, and there is an all-pervasive feeling that IT has to play catchup after failing to deliver on the promises of the Internet hype. There is no doubt that there has been a fundamental shift in mindset since the bubble burst. The digitization of business processes that would seamlessly integrate providers, marketplaces, and customers around the world has not happened. Business returned to normal after the spending spree of the late 1990s, but the IT investments remained and consolidation is now the name of the game.

Within the chemical industry, two percent of revenue is an accepted IT cost benchmark. Breaking through this cost barrier – sometimes from the lofty heights of three or even four percent – became the target for almost all multinationals. Some companies, such as Dow Chemical, BASF, Eastman, Celanese, and Degussa claim to have achieved this, or are well on the way to doing so.

23.1
Breaking the Barrier

Although two percent is a healthy spending rate for IT in the chemical industry, achieving this benchmark rapidly after a period of high spending does bring some risks. Removing the IT legacy is harder than it initially seems as the systems created for globalization, process automation or supply chain integration have become – by their very nature – deeply entrenched within the company's business processes and IT system landscape. Therefore, spending has been cut where it is easiest to do so: investments in new systems or functionalities.

The trouble is that after three years of consolidation and massive cuts in investment there is no more money to be squeezed out. Continuous cost management needs to take over. In fact, the major tasks facing the industry, such as reducing the overall business cost, protecting and enhancing revenues, and complying with regulatory requirements, actually require significant new IT investments (see Fig. 23.1).

Value Creation: Strategies for the Chemical Industry. 2nd Edition. F. Budde, U.-H. Felcht, H. Frankemölle (Eds.)

ISBN: 3-527-31266-8

Drivers of IT-enabled improvements

- Rising overall business cost, partly due to increasing energy/raw material cost
  - Complex business setup
  - Shift from regional to global business
  - Decreasing return on capital/ underutilized assets
  - High order errors/inventories
- Protection of revenue as margins are squeezed
  - Customer-specific servicing needed
  - Too high transaction cost for customer
  - Limited information exchange with business partners
  - Demand/supply chain improvement in multi-tier environments
- Management of regulation to stay in the game

IT priorities for the chemical industry

- Standardization of IT support
  - Support processes like accounting, HR
  - Compliance with regulation
  - Operation processes across legacy systems
- Global integration
  - Connecting plant management, MES, ERP
  - Business integration (especially Far East)
  - Supply chain integration with customers and suppliers
- Demand/supply chain excellence
  - Multi-channel access
  - Collaborative planning and forecasting
  - Customer relationship management

Source: IDC, McKinsey

Fig. 23.1 Business focus requires new IT investments.

Does this mean that the two percent target is in jeopardy before it has even been achieved? Not necessarily. The solution, we believe, lies in the balance between the relatively high spend on maintaining and running IT systems (operational IT), which can account for up to 90 percent of IT costs, versus the IT investments that actually create new business opportunities or significantly improve the performance of the current business (expansion IT) (Fig. 23.2).[1] Continuously managing down operational IT can release enough money to allow for investments while still keeping the two percent IT cost target in sight.

To maintain this balance, CEOs and CIOs should focus their attention on three major IT topics in the next two to three years:

- Manage down IT infrastructure costs. These are the biggest component of operational IT costs, accounting for up to 60 percent of the IT budget;
- Establish global solutions to integrate business processes within the global company, as well as towards global suppliers and customers;
- Establish CRM solutions to manage customer-specific product packages and services.

In the rest of this chapter we will elaborate on each of these topics and discuss their viability within the chemical industry. We will also address the general problems of IT management and organization especially at the IT customer/supplier

1) Expansion IT does not equal investments, as companies invest heavily in enhancing and maintaining the *existing* system landscape.

interface, which we still consider a major roadblock for achieving the value outlined here.

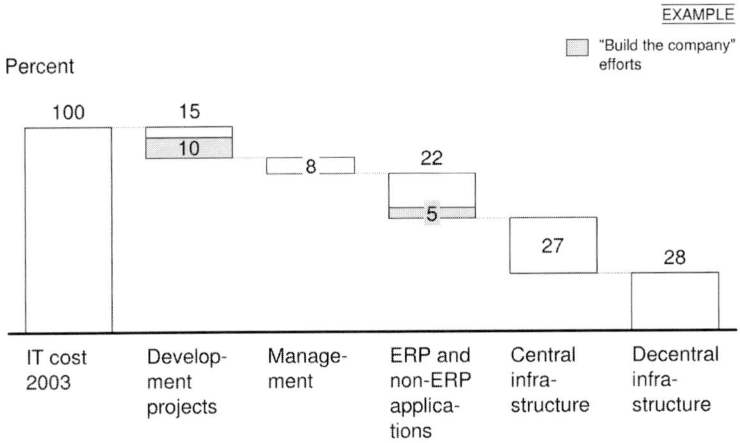

Source: McKinsey

Fig. 23.2 Only 15 percent of overall IT cost is spent on building the company.

23.2 World Class IT Infrastructure Management

Top managers often think IT infrastructure is the machines and cables in the basement. They are wrong. IT infrastructure has become a complex system that drives IT performance and accounts for over half of the IT budget (Fig. 23.3). On top of that is the complexity of assimilating innovations such as voice-over-IP, thin clients, IT security standards, blade servers, or Linux at a near-breakneck pace.

Removing this complexity and buying IT infrastructure 'on demand' is the next big promise of the IT outsourcing industry, but a lot of companies are not yet ready to buy these commoditized – and often only partially implemented – services because of the very complexity outlined above. Five strategic elements of infrastructure management need to be in place to simplify the overall infrastructure landscape and to establish clear accountability for infrastructure cost on the IT customer side.

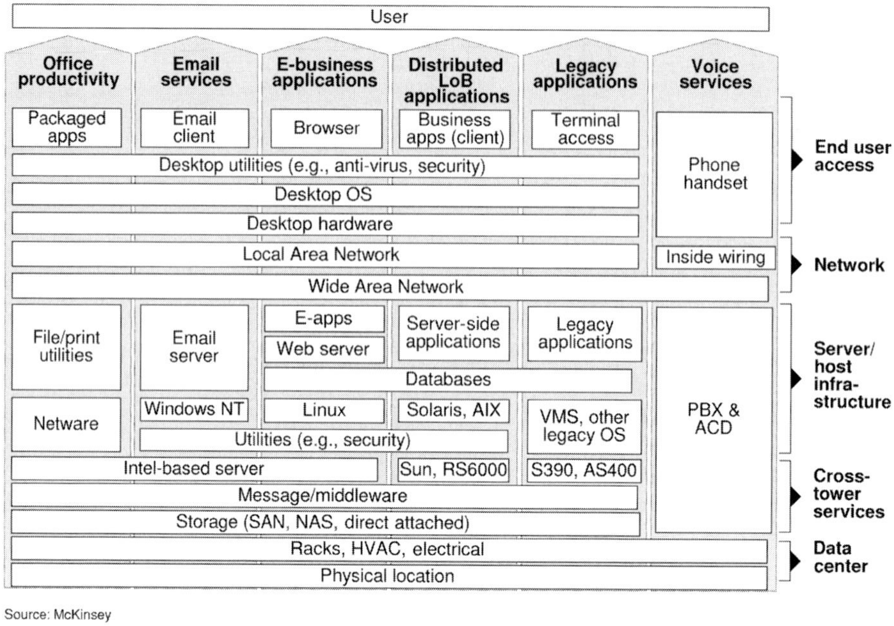

Fig. 23.3 Vast array of technologies in IT infrastructure creates management complexity.

Implementing the five infrastructure management elements described below can create significant cost savings potential of above 20 percent for most companies (Figure 23.4). This translates to twelve percent of the overall IT budget that can be moved to the 'expansion IT' share without increasing the overall budget (assuming that IT infrastructure accounts for a 60 percent share of the budget). Furthermore, the standardization of service structure and delivery that goes with the cost reduction means IT infrastructure services can be seen as commodities – a necessary step for buying IT infrastructure on demand.

The first element is to *set clear performance imperatives* by stratifying service levels by business lines, applications, user groups, and even time of day/week. By establishing such transparency and stability regarding services, IT can create an operating environment in which it can work continuously on the operating cost and focus on the highest-priority applications and users. Of course, the services profile must reflect the company's financial imperatives as well as the quality priorities of the business operating model. Therefore, in the chemical industry, the service profile will typically distinguish between the administrative functions and the production sites in terms of service and technology standardization and service quality levels.

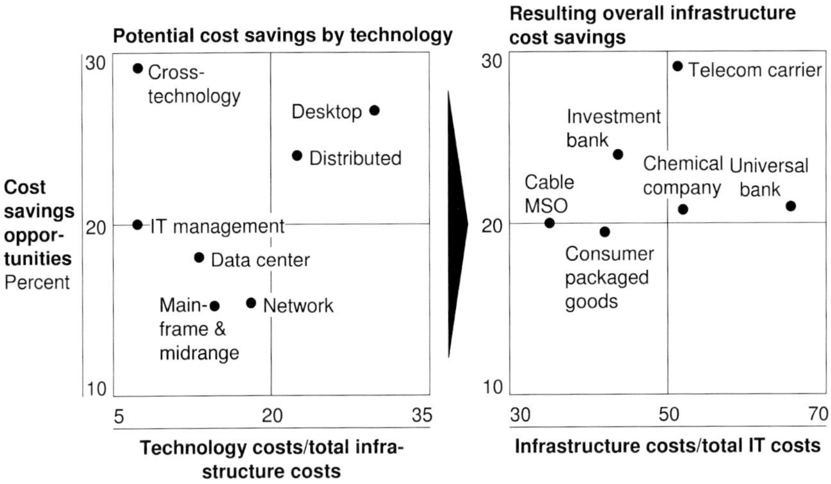

Fig. 23.4 Five strategic elements create savings opportunities of 20–30 percent.

The second element is to *consolidate (almost) all IT infrastructure activities in one organization*. IT infrastructure is increasingly becoming a scale game as competences become more commoditized. Consolidating data centers is cheaper and less risky as network costs drop and network security improves. Remote management tools and global helpdesk management allow more and more field service tasks to be consolidated. Leverage with vendors also increases as purchasing volumes are bundled. This type of consolidation typically implies a shared services governance model, and therefore the IT infrastructure tasks linked to production are not normally consolidated. Maintaining direct access and organizational integration with the production organization ensures service quality, which is more critical than cost-driven consolidation. One company, for example, decided to keep the whole site IT infrastructure (administration and production) under local management to avoid interfaces and quality gaps within the site, even though the cost reduction potential in alternatives was substantial.

The third element is to *consolidate service offerings into highly standardized infrastructure products*. Once service offerings are stable and the organization is bundled, internal operations can be standardized. The objective of service standardization is to improve stability by reducing the number of technical platforms, and to reduce engineering and deployment costs as the products are built only once. The effects of this can be seen with the introduction of limited desktop profiles for all employees or the standardization of server platforms/services (one Unix, one Linux, one Windows). Drastic standardization is necessary as cost does not scale down in line with complexity reduction. As described in the previous two elements, environments that require high flexibility are not suited to this level of service standardization.

Element number four is to *establish an outsourcing strategy.* Once points one to three have been implemented, the debate over outsourcing IT infrastructure can begin. The critical issue is partner portfolio management and the trade-off is between creating a large deal that maximizes accountability and leverage for the outsourcer or managing multiple deals to foster competition and maintain management control and flexibility. The decisive factor is very often the amount of IT management talent available at the customer end. A survey of 70 companies in 2003 showed that, given the greater availability of IT talent following the Internet hype, companies tend to lean towards a portfolio of deals and that the overall volume of outsourced activities is growing.

The final component of our five strategic elements is to *create cost accountability* with IT users by producing a catalogue that restricts them to buying a defined service at a defined price for any given volume. This final step creates maximum transparency and separates accountability for demand management from unit cost efficiency. However, it will work only if the first four points have been implemented and have led to a stable IT infrastructure operating model that allows IT managers to adapt to changes in demand patterns without jeopardizing quality and cost KPIs. Companies that require 'market prices' from their IT shared services too early (it usually takes one to two years depending on the starting point) have often been frustrated with the outcome and have restructured again before the expected results had sufficient time to be realized.

One word of caution regarding these five steps: company dynamics can influence the scope of simplification, as can the separation between administrative and production IT infrastructure. M&A has been an important lever in restructuring and growth in the chemical industry in the last 20 years. Therefore, keeping units independent to support divestment should be considered before maximizing the standardization and simplification of IT infrastructure.

23.3
ERP Harmonization as the Basis for Global Process Architectures

Globalization has been a key trend in the chemical industry for many years, with regard not only to business with suppliers and customers on all continents, but also to operations. The shift of production to the Far East and business process outsourcing to India or Eastern Europe are only the latest examples of this trend.

The digitization of business processes in the 1990s has been the driving force behind the recent boost in globalizing operations. For the chemical industry, this has mostly meant process automation and integration by ERP and SCM software, which is dominated by SAP. According to IT research firm AMR, 60 percent of the application spending in 2004 was on ERP and SCM systems (Fig. 23.5) and 85 percent of this spend in chemical multinationals was on SAP systems (55 percent for the mid-tier market).

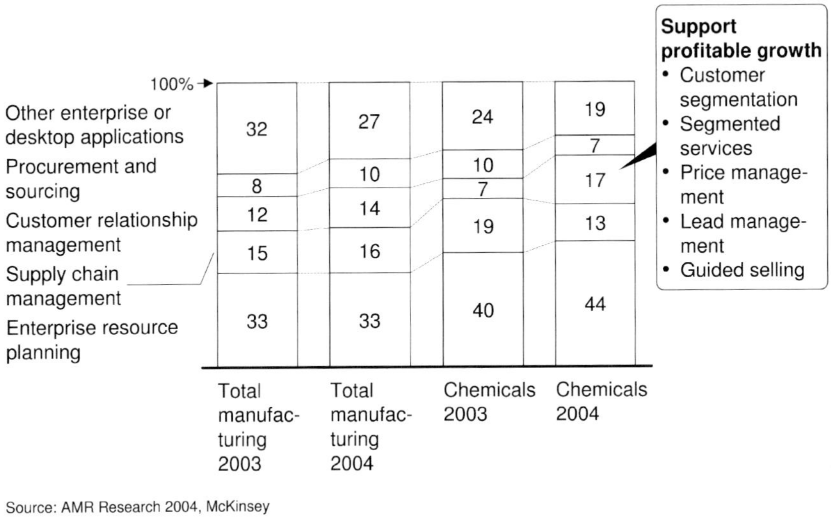

Fig. 23.5 Application spending in chemicals.

It is not surprising, therefore, that overhauling and extending ERP systems is very high on the CIO's agenda, and a new trend coined Single Instance ERP is emerging. The underlying assumption is that consolidating all the company's transactional business processes onto one ERP solution will dramatically drive down process cost, improve the quality of management information, and ensure compliance with regulations such as the Sarbanes-Oxley Act.

Harmonizing ERP systems in line with a global business set-up is definitely a good thing. Many chemical companies maintain a fragmented, patchy landscape of up to 60 ERP systems that are not all aligned with the business process or organizational setup. This creates huge barriers when responding to regulatory requirements and both business improvements and IT performance can hit brick walls due to the high IT investment needed and unacceptable timelines.

On the other hand, implementing a single ERP system is tough. It can be done – DuPont, Sun Microsystems, and Texas Instruments have all achieved it – but the investment is in line with the SAP mega-projects of the late 1990s. Companies need to be thinking about costs of up to a billion US dollars when talking about ERP consolidation, and the memories of big project failures are still at the back of their minds, not to mention the huge business disruption such a consolidation would create.

So, facing such gargantuan tasks, the question has to be whether the project can be sliced up.

The general answer, of course, is 'yes'. There are three dimensions to a global ERP landscape: the business processes, the organizations, and the geographies (Fig. 23.6).

In order to determine the right solution for a harmonized ERP landscape, a company has to find the right tradeoff for three decision elements: value creation

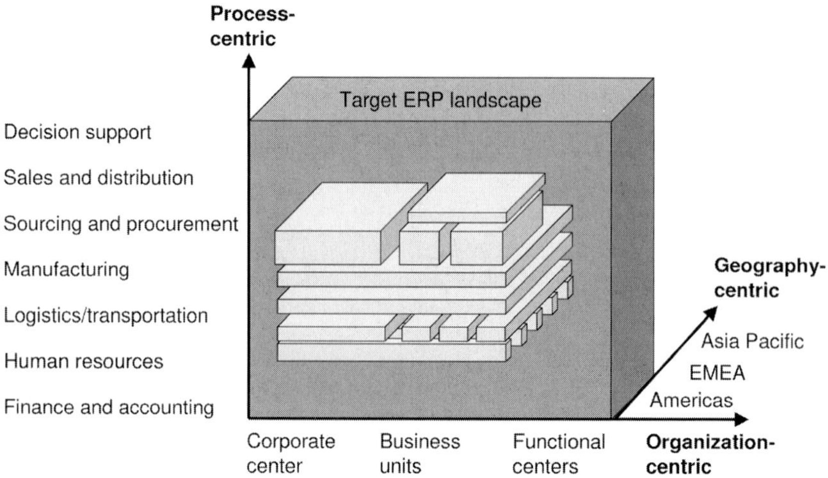

Source: McKinsey

Fig. 23.6 "Slicing the elephant" from the business perspective is critical.

potential, organizational ownership of the process change, and implementation constraints and enablers.

The *value creation potential* should be the starting point for any harmonization discussion. From our experience with other ERP value creation projects, there are a few key considerations when looking for value:

- Operational urgency and strategic considerations should never be considered as sole reasons for ERP harmonization – it is too risky and too expensive;
- Only 20 percent of the cost savings can be related directly to the automation of processes and interfaces by the system – and most of that has already been captured;
- IT cost savings have never justified an ERP implementation, and there is no reason to believe this time will be any different;
- Rarely has a correlation been found between revenue enhancement and ERP implementations;
- There are typically a few major contributors to the cost savings from the business side, such as consolidated purchasing, inventory management, supply chain integration, offshoring etc. Look for those to justify the harmonization decision – the rest is 'windfall profit'.

However, even if there is great value creation potential in principle, companies must ensure *single organizational ownership* of that potential. The following example illustrates our thinking. The main value driver of an ERP harmonization in finance was the setup of an offshore shared service center for accounting. All the divisional CFOs agreed and a joint ERP project was set up. First, during the course of the project, the process design standardization fell apart due to a multiplicity of specific divisional requirements, and further changes resulted from the localization of the system in the different countries. Second, the CFOs could not agree on the scope and the processes of the shared service center and finally voted against the offshore shared service in favor of an onshore location. The resulting savings potential was reduced to a quarter of the original assumption, but even this target was not met because of the high process and systems complexity.

Thus, even if the business case promises great savings, the scope of those savings must be aligned with an ownership that is empowered to implement the necessary measures to achieve them.

Finally, *implementation must be ensured.* The scenarios resulting from the first two criteria can range from one mega project, to a set of multiple projects to consolidate smaller instances, to setting up a single instance.

In the first case the implementation approach is clear. The critical issue is to limit cost and risks. Starting from an efficient existing solution is, when possible, definitely a good start from both a technical and an implementation skills perspective. The second step is to ensure clear standardization and harmonization empowerment and capabilities. Finally, companies should cut the solution into stable releases that can support operations for a fairly long period of time in case of financial pressure or changes brought about by M&A.

Under the single instance scenario, the company has to decide whether it wants to set up a dedicated program to implement the harmonization projects, or whether it wants to include the projects in an existing IT application development roadmap. The advantage of the first option is the bundling of capabilities, a common approach, faster completion, and higher management visibility to ensure the necessary business changes. The second option can be financed from the IT budget and gives more flexibility regarding implementation timing and prioritization.

Irrespective of the implementation approach, consolidating ERP systems will be the single most important topic for companies wanting to create a viable platform for IT investments that will finally start to fulfill the promises offered by tools such as CRM (Customer Relationship Management).

23.4
Making CRM Work to Create Profitable Growth

Many of the first wave of CRM implementations failed (Fig. 23.7). Large packages with a wide spectrum of functionalities were implemented without a target being defined or the necessary culture and process change being supported. Halfway through these projects it became clear that the business cases were fragile and the amount of change needed was simply too much for the organization. As a result, most of the systems created limited value at best.

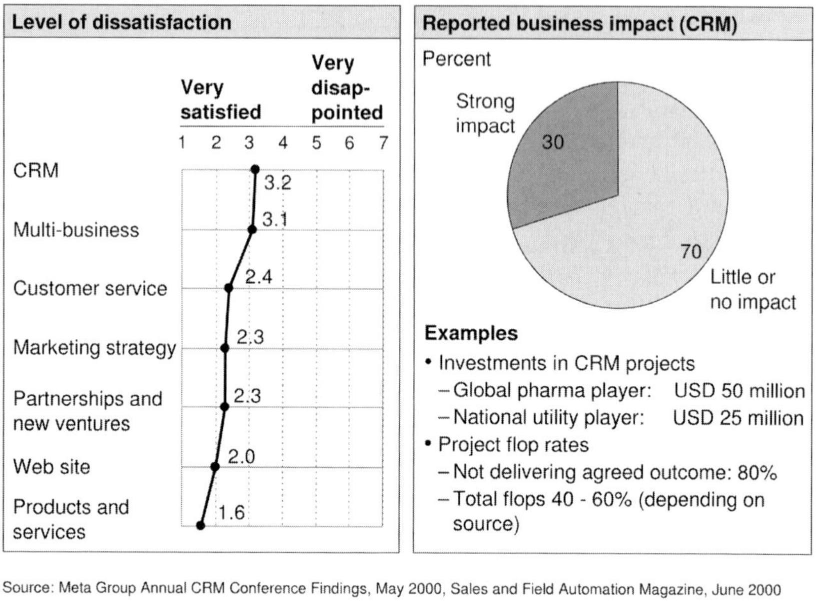

Fig. 23.7 CRM satisfaction level versus business impact.

After recent consolidation and globalization efforts, many chemical companies are now geared towards creating better value for customers. The main way of achieving this is to offer customer-specific products and services. Bundling commoditized products with customer-specific value-added services allows customers to outsource manufacturing tasks and services.

According to AMR, this trend means that chemical company spending on CRM systems that enable the definition and management of services increased from seven to 17 percent of the overall IT budget from 2003 to 2004, and will grow even further (Fig. 23.5). Implementation focuses on applications that support customer segmentation, customer-specific service design, price management, lead management, and guided selling. These leaner, focused solutions should increase the value from CRM while still keeping implementation risks low.

To avoid the pitfalls of previous CRM projects, implementation should take three success factors into account: a clear focus regarding CRM functions, customer channels, and analytic needs; careful platform and provider selection; and a pay-as-you-go strategy.

In the past, CRM has often been both perceived and tackled 'ERP style', i.e., large rollouts of complete integrated packages that led to expensive, monolithic projects. But CRM, by its very nature, is much more diverse in its functionality, more dynamic in its process evolution, and less clear in its organizational ownership (Fig. 23.8). Clear focus on the most beneficial functionalities and an evolutionary roadmap allow better learning and more flexibility when implementing CRM. This suggests three key questions to be answered up front:

- *Which functions are needed?* For example, customer service support, campaign management, key account management, or loyalty management. A solid understanding of the key drivers of sales success will help focus on the right topics for each region and product portfolio.
- *Which channels should be addressed or integrated?* From stores to mailing to web access to call centers there is a variety of options, and the prioritization should be clear. Most companies address channels that are new or underperforming in order to create early success stories before addressing the key channels.
- *What kind of analytics and reports are needed to create optimal decision support?* It is not necessary to set up sophisticated data mining capabilities if a small management dashboard to manage the sales pipeline is sufficient to improve sales performance.

While SAP dominates the ERP vendor landscape, the CRM provider and solution landscape is much more diverse. The big players such as Siebel and SAP cover more than 30 percent of the market, but there is a wide range of vendors serving niche and mid-tier markets or simply providing specialized solutions for indus-

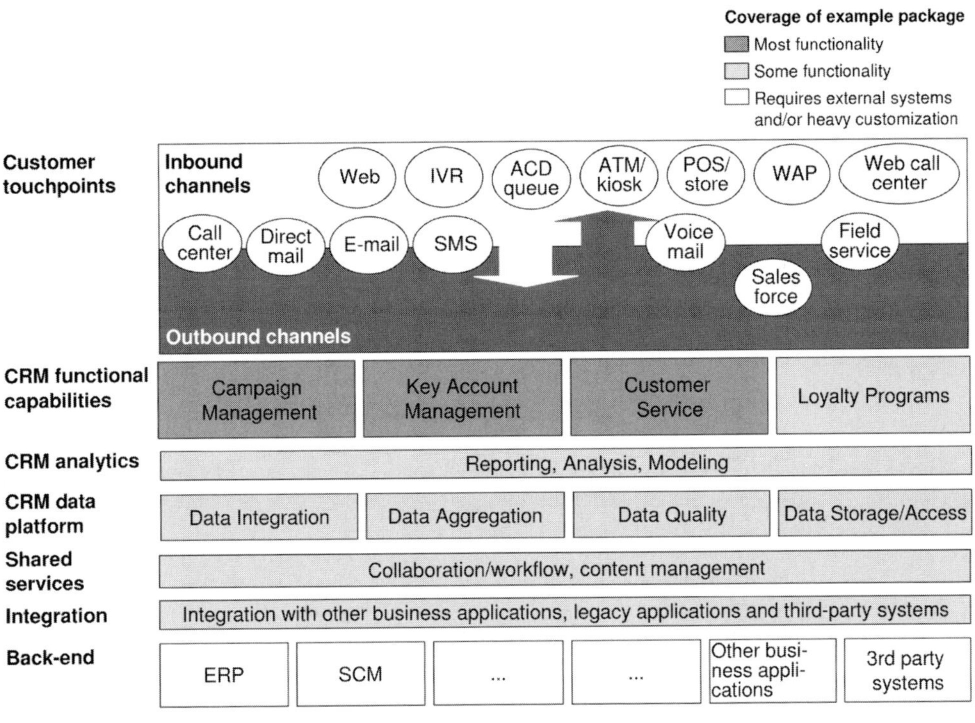

Source: AMR, Doculabs, McKinsey team analysis

Fig. 23.8 No single solution for all CRM requirements.

tries or functionalities. This is also a sign of the market's relative immaturity compared to ERP, and companies should choose a vendor/system integrator carefully. On the one hand the specific needs of the project must be met, but the solution must also be scalable in terms of functional, technical, and business coverage.

Finally, the business cases that drive those projects are becoming more conservative. Pay-as-you-go approaches can keep the risks low; investment in new projects occurs only if the previous projects cover the costs. Even though CRM projects can become quite lengthy (three to five years), pay-as-you-go ensures budget stability and enhances organizational acceptance and ownership of the implemented solutions.

Alongside the technical aspects of solution scoping and roadmap planning, the organizational change needed to establish a more customer-specific go-to-market approach is still in development. The sales force that sells production volume generally needs to be restructured and equipped to design an overall product and service package together with the customer. Yet whatever the situation, it is clear that CRM can fundamentally alter the way chemical companies do business. IT really can change the company, and even the industry.

23.5 Key Success Factors for the Journey

We have discussed how companies can achieve the two percent IT cost barrier, while still investing in real business improvements. However, to improve effectiveness while maintaining efficiency, it is clear that the rigidity and the effort of management processes needs improvement. We consider that three processes clearly sit at the heart of that improvement:

- *IT demand management:* transparency of incoming demand, clear evaluation and decision processes that align all stakeholders (IT customers, IT suppliers, and IT governance), and tracking of actual spending are prerequisites for staying in control;
- *Project portfolio management:* besides the obvious prioritization by value, it is necessary to have a clear understanding of foreseeable risks and available capabilities as well as the fit with the application roadmap;
- *Improved value management:* this process is not about writing business cases but about ensuring that the promised value is actually delivered. Business ownership and business change management also need to be improved.

Some of these processes may already exist in the organization, some may not. But in almost all companies, the interface between business and IT is treated like a poor relation. Improving competences and clarifying accountability on both sides, for example establishing business-driven IT demand organizations and internal IT service providers can dramatically improve IT performance. Most chemical companies are well aware of these interface issues from previous experience, especially with large-scale ERP projects. This should mean they are better prepared for the second time around, when it comes to breaking through the two percent barrier.

23.6 Summary

- After a spell of consolidation and investment cuts, the way forward in IT lies in continuous cost management, and three topics in particular: managing down infrastructure cost, establishing global solutions, and introducing CRM solutions.
- There are five strategic elements to managing infrastructure cost: setting clear performance imperatives, consolidating all infrastructure activities into one organization, consolidating service offerings into highly standardized products, establishing an outsourcing strategy, and creating cost accountability.
- Establishing global solutions requires companies to find the trade-offs between value creation potential, organizational ownership of process change, and implementation constraints or enablers.
- To implement a CRM solution, companies need a clear focus on its functions, careful selection of both platform and provider, and a pay-as-you-go strategy.
- To succeed in the process of unlocking value with IT, three processes in particular need special attention: IT demand management, project portfolio management, and improved value management.

24
Managing Organizational Performance

Karsten Hofmann and Heiner Frankemölle

CEOs and their top management teams are increasingly aware that many of the big strategic and operational questions on their plate involve organizational issues: what they do to get – and keep – the right people moving in the right direction. Studies have shown not only that more than 70 percent of corporate change initiatives fail to reach their full potential, but also that over 70 percent of those failures occur because organizational issues have not been managed successfully.

McKinsey research has also found a significant correlation between a strong performance culture and an above-average industry TRS and ROIC (Fig. 24.1). The rub, of course, is that getting organizational performance right is not easy. When managers think about organization they tend to focus on tangible components such as structure (as manifested in organizational charts) and processes. The intangible side – thoughts, feelings, values, and mindsets that underlie organizational behavior – is often neglected. This limited ability to see, interpret, and influence mental models and social patterns, however, often turns out to be the root cause of failure to establish a true performance ethic.

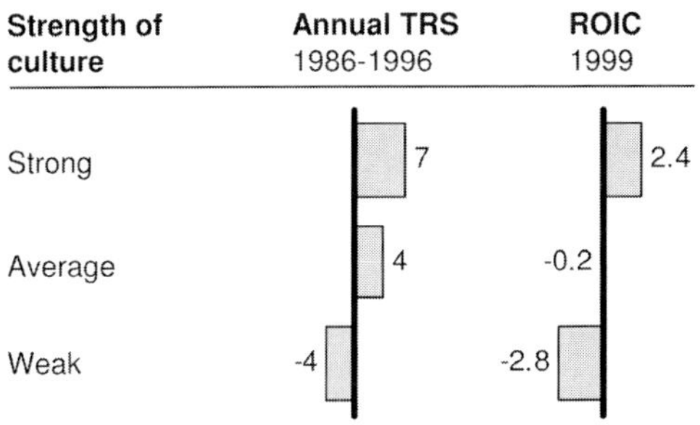

Source: McKinsey Performance Ethic Research

Fig. 24.1 Correlation between performance culture and economic performance.

In this chapter we will look at the essential ingredients in the makeup of any international chemical company that aspires to be truly excellent. We will first consider the "hard-wiring" of the company, to examine whether certain structures

Value Creation: Strategies for the Chemical Industry. 2nd Edition. F. Budde, U.-H. Felcht, H. Frankemölle (Eds.)

ISBN: 3-527-31266-8

work better than others in supporting a chemical company's strategic purpose. We will then focus on the essential elements of effective performance and organizational health management. Finally, we will describe how companies can make lasting organizational change happen.

24.1
Supporting Strategy by Structure

A well-designed formal organization continues to be a vitally important part of successful companies. For decades, the challenge for multi-product chemical conglomerates has been to strike the right balance between leveraging scale in order to gain competitive advantage, whilst at the same time keeping entrepreneurship intact and remaining nimble and responsive to the markets and customers they serve.

Organizational research has not yet clearly resolved the dilemma of global reach and local responsiveness. To many managers in the chemical industry, concepts such as triple matrix organization, competence-based organization, or virtual organization appear too far removed from the very practical challenge of optimizing effectiveness and efficiency. They are looking for a structure that is both simple and robust, and that delivers their global and regional strategies.

Our own research into best practice in this area does not surface one right answer, and given the vast differences between chemical companies, we would be very surprised if it did. Nevertheless, one organizational model seems to fit the needs of the chemical industry particularly well: in our experience, this is an organization with global business divisions (i.e., product-based business units with worldwide control of all critically important business functions) as the main axle, guided by the corporate center and supported by shared services. About three quarters of the global players in chemicals have adopted this structure.

Compared to more monolithic organization models, global business divisions have some clear advantages:

- All the industry's major players have a global presence, and more than 40 percent of chemical sales are typically outside the companies' respective home markets. In such a context, global business divisions allow for coordinated product strategies and customer service across the world's markets.
- The real profitability of different products becomes more transparent as cross-subsidies among product lines are exposed. This is a problem intrinsic to the highly integrated production processes of many chemical companies.
- Clear points of P&L accountability are established, thus promoting personal ownership and responsibility for results along a clearly defined range of products – a prerequisite for a "no excuses" policy (with reference to implications for the design of site services, the reader is referred to Chapter 20).

- Opportunities for building global businesses arise, and these help to attract and retain entrepreneurial talent. This is especially important in an industry which is in an uphill battle against other industries in the reignited "war for talent".
- Internal service providers have to serve a number of different customers with quite diverse demands. Business divisions allow the tradeoffs between the demands to be optimized, often resulting in substantial savings and more tailored services.

However, the challenges that need to be addressed to make this model work should not be underestimated: it takes special mechanisms to translate corporate strategic intent to the business unit level and to roll out key strategic initiatives across the corporation. For example, if head office decides that understanding the customer is to become a corporate spike, it will take a major effort to make sure that this message is understood, adapted, implemented, and sustained across a wide range of businesses with differing cultures. It is crucial to create a meaningful, content-based, and challenging dialog between the corporate center and the business units – many chemical companies are still struggling to make this work.

Knowledge transfer, capability building, and mutual learning are even less likely than usual to happen automatically and naturally if colleagues rarely see each other; here, appropriate incentives need to be defined to make sure that people communicate with and support one another. Finally, there is a danger that overhead will be duplicated in each business division. This should be kept to a minimum by finding ways to tap into synergies across business units, in local sales operations, for instance.

One crucial factor in establishing winning global business divisions is to scope them appropriately, making sure that special circumstances are taken into account where appropriate. Four main criteria should be considered here:

1. *Size:* In commodities, most businesses reach the critical size required for an independent product division almost by definition. The strong drive for economies of scale would otherwise quickly put them out of the game. In specialties, however, the situation is different. Small businesses may well meet the market conditions required for independence, yet still be too small to achieve any significance in a corporate portfolio (while still bearing the burden of overhead costs tailored more to larger units). These should be candidates for sale if they cannot be brought to critical size.
2. *Customers:* Companies have to design their structures according to their customer needs. Marketing and sales units, for instance, may be organized in very different ways even in a global business unit setup to mirror the customers' preferred business model. At a given fine chemicals producer, for instance, the following may be found in parallel:
   - Local account managers for selected local customers with very country-specific needs (e.g., language requirements, understanding of local regulation)

- Global key account managers for customers with global organization structures, such as pharmaceutical originators, which typically want to have one single point of contact with a layer of local/regional contacts behind it
- Global lead key account managers across some or all business units for selected global customers whose buying needs span several chemical businesses. This helps to avoid pitfalls such as refusing business unit X's most important customer a product from business unit Y in times of product shortage because unit Y rates the customer as a second-tier customer.

3. *Maturity of market and organization:* Chemical companies may want to make an exception to the principle of global business divisions in the case of emerging markets. In the startup phase of a business, it makes a lot of sense to focus the attention and responsibility of managers on local or regional business development. Appointing a "Vice President China" or a "Vice President Middle East" – typically without profit and loss responsibility – or an "Asia Pacific Council" consisting of business unit and country managers can help companies to tap into synergies of shared infrastructures in the target markets. Careful observation is required to decide when this operating model should make the transition back to the global business logic.
4. *Competences and product portfolio*: Finally, careful consideration should be given to the combination of different smaller businesses that make up a business division. There are numerous examples of cases where this selection has been driven not so much by matching product portfolios and shared competence platforms to serve customer needs, but rather by vague beliefs about what should fit together "in principle".

With the global business unit set up, what, then, is the appropriate role and design for the corporate center and the shared services?

Ideally, the corporate center should focus on those key governance functions that allow it to create real value for the corporation, i.e., tasks with high value added or, if performed badly, high potential to destroy value. The definition of which tasks should be included will depend on the overall strategy and the center's self-conception, i.e., whether it defines its role more as that of a strategic architect or an operator (Fig. 24.2).

In most cases, the basic corporate center setup includes corporate finance (financial control, risk management, taxation, and treasury), external affairs and investor relations, legal and corporate affairs, (group) portfolio management, and corporate human resources. With regard to size, it is advisable to determine the minimum required for the parent role, which could be as low as 30 for a mid-sized global chemical company, and then add the headcount required for tasks that add distinctive value to the entire group.

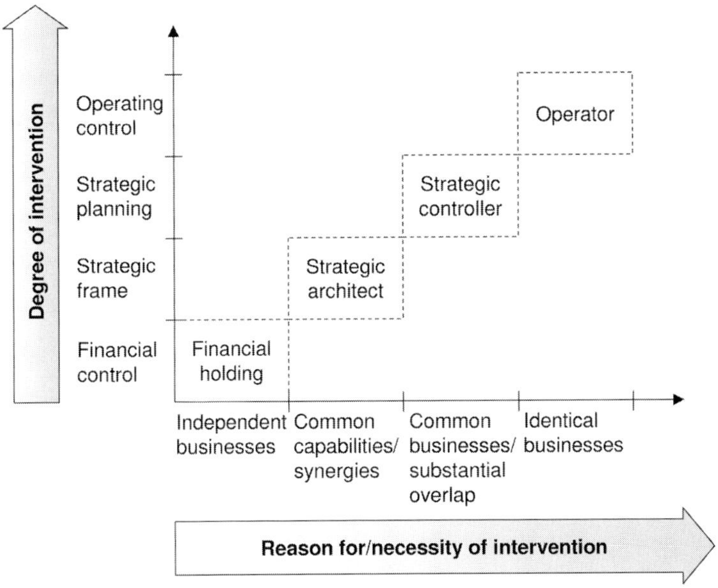

Source: McKinsey

Fig. 24.2 Potential corporate center roles.

There is a clear trend towards including those units at a corporate level that are essential to help the CEO deliver against the company's strategic priorities. The center may thus also comprise specialized competence centers or project offices for Marketing & Sales Excellence, Innovation (mainly for new technologies and processes, e.g., project houses), Lean Manufacturing/Six Sigma Academy, Global Sourcing, or new business entities.

24.2 Understanding the Performance Challenge

Even when the company has the right strategy and the right structure, it may still not be one of the small band of truly excellent companies – the ones that outperform the competition year in, year out. To find out more about how to create an organization that can excel on a sustained basis, McKinsey interviewed over 600 senior executives in 50 high-performing and 100 average-performing companies, and analyzed numerous survey responses from executives in a representative sample of companies.

We found that an aspirant to top performance, whether in chemicals or other industries, must follow a balanced approach to the management of *performance*, i.e., what the organization needs to accomplish today, and *health*, or how it can improve or adjust its performance over time. Integrated performance and health management enables an organization to deliver value today while simultaneously building the capacity and capability to sustain and evolve performance over time. We also learned that leading companies tightly integrate business-performance with people-performance management: they diligently track performance and capability building at the levels of both the organizational unit and the individual.

Our research shows that organizations should adhere to six principles in order to build an effective performance and health management system, centered around metrics, targets, reviews, links, rhythm, and culture.

1. *Cascade a focused set of balanced metrics that are linked to the key value drivers.* A focused set of metrics improves alignment and direction within the organization, helps avoid "initiative overload", and keeps the system practical. In the early 1990s, General Motors followed several hundred individual metrics on a monthly basis. Staff referred to these as DRIP – "data rich, information poor". Not surprisingly, the metrics failed to lead to the desired performance. Balancing metrics ensures that the organization focuses on delivering impact today while at the same time not neglecting its preparations for the future. Linking metrics to the real value drivers is another key component. A major North American wireless telecommunications firm focused on what it saw as one of its key performance metrics – customer churn. In an effort to reduce its monthly churn rate of nearly three percent, the company was spending lots of money on customer retention. However, further analysis revealed that 75 percent of its customers appeared to be either unprofitable or only marginally profitable. Cascading metrics from the top of the company to the shop floor helps ensure that the whole organization is focused on and aligned with what needs to be done. It also creates a common performance language across the organization.

   BP Chemicals can be quoted as a good example in the chemical industry for having implemented a set of cascading scorecards with leading indicators and lagging measures, taking both performance and health as well as different stakeholder interests into consideration.
2. *Create opportunity-based stretch targets owned by those who will deliver them.* Target-setting is a crucial component of the performance management process. Without an opportunity basis, targets can feel arbitrary and unrealistic. For instance, the Chrysler Group's monthly sales targets appeared, according to dealers, to be "plucked from the sky". Fifty percent of the brand dealers were not able to hit their targets in 2003. Ownership of targets by those who have to deliver them significantly increases the probability of delivery and improves organizational alignment. However, stretch targets, "reflecting the dreams" of a company, stimulate teams to find fundamentally better ways to perform work. Obviously, there is no single right way to set opportunity-based stretch targets, as organizations' target-setting processes differ according to their distinct management philosophies and cultures. In any case, the key objective for an organization is to stretch targets enough to raise performance while making sure managers still own them.

Some petrochemical companies – faced with a very cyclical market environment – have made great efforts to set "real" targets by differentiating between nominal improvement (i.e., performance compared with last year, often driven by external factors) and real improvement (i.e., performance vs. competition, neutralizing overall market development).

3. *Orchestrate fact-based performance reviews to help leaders and managers "face reality together".* Fact-based reviews keep leaders and managers honest and help make corrective action effective. When Electronic Data Systems introduced a new ranking system in 1999, some employees suddenly found themselves closer to the bottom of the roster than the top where they expected to be. Immediate superiors had given them a less-than-honest review the year before rather than tackle the unpleasant task of telling them where they were falling short. Reviewing performance together in a constructive environment with peers encourages managers to be open and to take responsibility in their own realm. Orchestrating the review with care helps ensure it motivates managers. Using reviews as problem-solving sessions helps establish the root causes of good and not so good performance, and provides an opportunity to coach managers or take other corrective action.

   GE Advanced Materials has been for several years – and still is – a benchmark in the chemical industry when it comes to making individual evaluation the key driving element of corporate performance management.

4. *Make the critical links between people management and the performance and health processes of the business so that they are complementary and mutually reinforcing.* Aligning business goals and personal objectives helps people focus on the right activities. In the early 1980s, Philips' informal rules and personal relationships dominated formal systems for performance evaluations and career advancement. Position and perceived power in the company network determined who got what. Consequently, people were reluctant to tackle the problems the company was facing so as not to block their personal growth opportunities. Linking compensation and reward mechanisms to the performance and health of the organization is essential and far from trivial, offering significant potential.

   Account managers at high-tech companies such as Oracle or Sun Microsystems can almost double their base salary depending on their performance – while many chemical companies, especially in Europe, are still struggling to give their managers meaningful and differentiating incentives for superior performance.

5. *Drive performance and health processes with the right rhythm to allow the organization to be directed and operated effectively and efficiently.* A chemical company can have in place the appropriate review processes, value-drivers, and targets, but not the proper sequence, timing, and pace of performance management events for its particular businesses.

   For slow-moving segments such as chlorine-alkali or cellulose acetate it may be sufficient to hold in-depth strategy reviews as infrequently as every second year, while more dynamic areas in specialty chemicals may require sessions every couple of months. Deliberately setting a pace for the performance management process drives the organization "to get things done" at a speed appropriate for its industry environment.

6. *Ensure the performance management system acts in concert with existing culture and other management interventions to create a powerful, positive mechanism for cultural evolution.* When Unilever HPC-NA introduced its new performance management system to its employees in 2003, it learned that just offering the scorecard online did not assure adoption. Employees were not tapping into the intranet anything like as often as expected, but preferred to read about performance in newsletters and acquire information verbally. Integrating the performance management system with the company culture amplifies its effect.

24.3 Making Organizational Change Happen

Like individuals, organizations change continuously, reacting to developments in their markets and to the arrival and departure of key people. In a large company, most of these changes go on more or less unnoticed. But sometimes a company has to change more quickly or more fundamentally than this gradual evolution allows; it needs a break with the past, an accelerated pace of change – a real transformation.

Successful corporate transformations and their leaders, such as Lou Gerstner at IBM, George Cain at Abbott Laboratories, or Darwin E. Smith at Kimberly-Clark, become the stuff of business legend. Rather oddly, it is the leaders of companies in crisis who may be best placed to achieve a true transformation. David Simon and John Browne were able to transform British Petroleum from one of Britain's weakest industrials into one of its strongest because the company faced imminent ruin. By contrast, most transformations undertaken in non-crisis conditions end up failing: employees' attitudes and behavior remain unchanged, ambitious targets slip downward, and the program is finally inconspicuously abandoned.

Can a company be transformed without first experiencing a crisis? We believe that the answer is yes, but it is a veritable challenge. Leaders need to have a detailed understanding of what makes individuals and groups transform their view of reality. The initiation of organizational change requires a clear answer to three key questions:

- Where are you today and where do you want to go? This means understanding current organizational performance and gaps to the aspired performance, and identifying the main changes required.
- What do you need to do to get there? This includes developing a set of initiatives to drive outcomes through behavior and mindset shifts and creating a compelling transformation story.

- How do you lead implementation? This entails architecting a program to maximize impact and minimize risk during implementation, and navigating the organization through the change process.

24.3.1 Diagnosing the Status Quo

Organizations often assume they know themselves inside-out and fail to acknowledge white spots. In particular, mindsets and attitudes important for change programs are often not fully understood. A diagnostic makes the performance culture transparent and identifies possible performance gaps and road blocks.

Diagnostic approaches typically involve collecting data from a variety of sources. In addition to fact-based analyses, such as competitive benchmarking or value driver analyses, surveys and interviews can help in gaining a better understanding of the company's workings and uncovering the complex roots of performance problems:

Surveys: There are numerous surveys in existence that give insights into the current performance culture of an organization and the management practices/behaviors that sustain it. We use the organizational performance profile that encompasses nine elements that are the outcomes of an effective organization. Every organization should understand its current effectiveness in each of these reinforcing elements and set aspirations for them (Fig. 24.3):

- Three elements deal with *alignment*. Leadership addresses how well all leaders, not just the CEO or members of the senior management team, approach their responsibility to inspire and shape the actions of others, set the appropriate tone for execution, and ensure that the company pays sufficient attention to development and renewal. Direction is management's articulation of where the company is heading and how it will get there. Environment and values help foster a shared understanding of the company's core values and shape the quality of employee interactions.
- Four elements focus on *execution*. Accountability deals with how the organization designs its structure and reporting relationships and manages and evaluates individual performance. Coordination and control spells out the company's perspective on measuring and evaluating business performance and risk and on coordinating the activities of various parts of the organization, e.g., its business units. The capability element focuses on ensuring that the company has the skills and talent it needs to execute its strategy. Motivation deals with how the company inspires, encourages, and rewards employees.

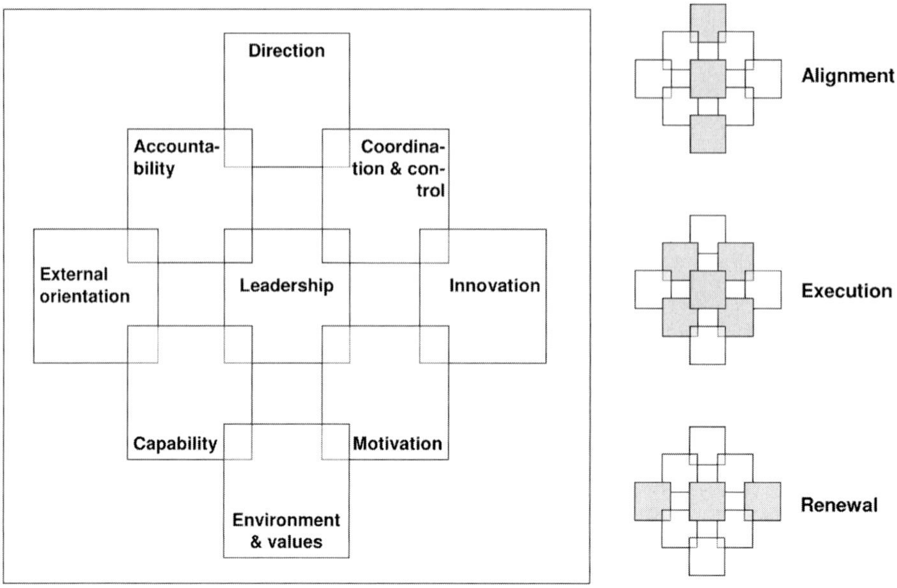

Source: McKinsey

Fig. 24.3 Fundamental elements of organizational performance.

- The remaining two elements address *renewal*. External orientation refers to how the organization interacts with customers, suppliers, government regulators, and other external groups. Innovation devotes specific attention to anticipating opportunities and challenges and generating ideas for dealing with them.

While all companies need to have at least some level of effectiveness in each element, management does need to make decisions about *where* it places its focus – those elements where it must be distinctive. Some elements will be more important to some strategies and operational objectives than to others. For instance, a commodity chemicals player in a segment where operational excellence is the name of the game may choose to push hard on coordination and control, but place less emphasis on renewal. In addition, it is difficult, and indeed probably impossible, to work on all or even many of the elements at the same time. A company also needs to make choices within each element. Will its strategy be better served by hands-off or hands-on leadership? By top-down or more broadly-based direction setting? Is it better to build the capability that lies at the heart of a critical operational effort internally than find it externally? Is the rate of innovation best encouraged via a bottom-up, continuous improvement process or by a cross-functional, targeted team approach? And so forth.

Interviews: Deep structure interviews are an effective tool for surfacing the more hidden drivers of behaviors. These highly personal interviews explore key opinion shapers' firmly held mindsets, assumptions, and basic values – their "deep structure" – and engage them in a dialogue around critical issues. The interviews are built around open questions using non-guiding interviewing techniques. Other tools include focus groups that engage a small number of managers in an interactive real-time assessment of critical organizational issues or intensive dialogue workshops that use digital voting systems to achieve full transparency of the individual beliefs and stimulate discussion within the group.

The results of these surveys and interviews need to be carefully synthesized to identify the core issues for the different stakeholders. They should then be discussed in cascading workshops to uncover complex, systematic connections between the problems and their causes and build alignment within the teams. The performance profile discussion and analyses will yield insights into gaps in the company's organizational elements and offer a perspective on their relative prioritization. These insights can be further synthesized and syndicated to arrive at a manageable number of performance improvement themes, e.g., "increase collaboration across the organization" or "clarify responsibilities and increase personal accountability for results", and very concrete actions behind each of these.

24.3.2
Designing the Program Architecture

A company's success at achieving its target organizational performance profile depends on its success in translating each performance theme into specific mindset shifts (e.g., from "it's not my fault" to "I'm accountable for results" and behavior shifts (e.g., from managers responding reactively to cost issues to managers proactively monitoring costs and seeking improvement opportunities) for appropriate sets of employees. Change programs that do not address mindset and behavior shifts may work for a while, but they generally do not achieve lasting change. Therefore, changing mindsets and behaviors should be one of the major aims of any change program. The *Influence Model* describes how this change can be driven (Fig. 24.4).

The model consists of four influencers, for each of which there are specific levers that can be used to define the set of actions that will lead to the desired change. Effective change programs will incorporate complementary and consistent actions from all four.

1. *Fostering understanding and conviction.* This influencer is based on the premise that people need to understand what they are expected to do and why they should be doing it – otherwise they will not act. Management needs to develop a compelling change story, articulate it in language that brings it alive, personalize it for employees, and then discuss it widely, repeatedly, in a variety of forums.
2. *Role-modeling.* Seeing leads to believing – and to doing. This influencer puts a premium on ensuring that respected people, at all levels of the orga-

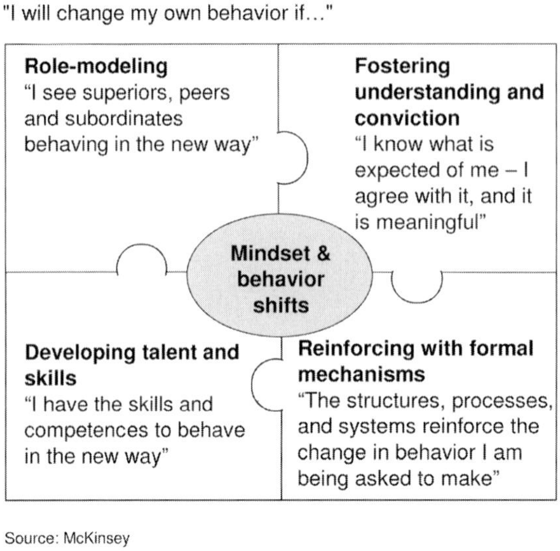

Fig. 24.4 The influence model: change levers.

nization, demonstrate the desired mindsets and behaviors in their day-to-day activities. Quickly identifying and winning a critical mass of visible role models helps "infect" the rest of the organization with the desired changes.

3. *Reinforcing with formal mechanisms.* Management needs to ensure that roles and relationships, basic management and business processes, reward and recognition systems, and information systems all support and reinforce the requisite mindsets and behaviors. It is close to impossible to convince employees that more risk taking will be a desired behavior in the future, if just one person who then takes a risk and fails is demoted.
4. *Developing talent and skills.* Obviously, people need to be able to do what they are actually being asked to do. And they must believe that they can do it. Depending on the specific situation, initiatives for this influencer take the form of programs to build capabilities (on-the-job development, training programs, support tools) and/or upgrade talent (hiring, replacing). For the chemical industry, talent management is a particular critical issue. Surveys show over and over again that most chemical companies are considered relatively unattractive employers by many high-potential individuals. To attract and retain more of the best people, chemical corporations must make talent management a top corporate priority and create and perpetually refine an "employee value proposition" that sets the company apart from its competitors even beyond its own industry.

The first influencer, fostering understanding and conviction by developing a change story, often does not get enough management attention. Many like to think that careful detailing and sequencing of the required actions is enough to make a great improvement program. However, a program that cannot be told in terms that are compelling and exciting for employees is unlikely to gain support in its implementation. A great corporate story creates and sustains commitment to the company and its course. One has the patience to see the program through because it is not just a sequence of events: an exciting plot is unfolding. A great corporate story has a natural rightness about it that makes it feel compelling. It ought to come true, people believe in it, people want to help realize it.

A transformation born of crisis writes its own story. Before a crisis hits, it is much harder to create an authentic story explaining why a company should transform itself. Nonetheless, even if a transformation is not driven by a crisis, it is important to clarify and write down the story. Although each story's specific form will vary, we envision a transformation story in three chapters, corresponding to the why, where, and how of the transformation.

1. *Why does the company need this?* Almost always, the story of a transformation acknowledges the events that triggered it: the company must take action because its balance sheet is weak, say, or because competitors are offering better prices to customers, or because technology has revolutionized the segment. But these are often only symptoms of deeper problems. These, too, must be included in the story which must explain, for instance, why and how the financials became weak. Putting hard truths on the table might make some people uncomfortable, but avoiding such truths puts success at risk. A shared understanding of the actual cause of the current state of affairs is essential to a transformation.
2. *Where is the company heading?* The second chapter outlines the company's future and makes it so compelling that is seems destined to happen. Much of the power of a great corporate story is drawn from a central exciting idea. It does not need to be a novel insight or a startling innovation. A strongly held vision can energize a story even if that vision appears unexciting at first sight. There must be a clear sense that the story leads the company to a position that is a radical improvement compared with its current one, and not just in terms of profitability.
3. *How can the company reach its goal?* There are many technical details to spell out concerning tasks, phases, timing, and responsibility. But while detail is important, it does not adequately answer the question of how the company will achieve its goals. Transformation learning comes, ultimately, from personal experience. The leaders' experience, which should be embedded in the story, must be internalized within every participant in the process. Each participant must undergo an "identity transition" to make the vision come to life.

Whatever the individual components, it is crucial that the CEO "owns" the story. If the CEO does not fully believe it, it is unlikely that anyone else will. When the CEO tells the story with genuine force, in his or her own natural language, the story often develops unpredicted power.

Once a story is developed, it is crucial to "stress test" it. This can be achieved by looking at it from the viewpoint of the different audiences, such as customers, investors, or employees: how would this particular audience respond to the telling of the current story? What does it take to turn skeptics into believers and passive onlookers into active participants? The next test would be to develop a full story variant for each audience. The variants for employees may be extended further in a "story cascade". What form does the story take for key groups, such as business units, functional units, and front-line teams? Is this a good story for the R&D personnel, one that will increase the company's momentum in innovation? Will the sales force be sufficiently excited and motivated by the story to reach the required sales volumes? Of course, it is not really possible to define precisely what the story will mean to an audience, group, or particular individual. In the end, everyone connected with the corporation will tell their own version, but ideally in close alignment with the general theme.

The transformation story forms the core of the program architecture. The highest level is the corporate vision, the key transformation theme. Beneath it are the core performance themes (typically three to six). Closest to the ground are the individual initiatives that should deliver the happy ending – about five to ten for each theme. These concrete initiatives turn the change agenda into reality at the front lines and make it tangible.

24.3.3 Ensuring Effective Implementation

With a compelling transformation story, a complete set of individual initiatives, and actions to influence mindset and behavior change in place and integrated into a rigorous change architecture, the following practical guidelines have been found useful by many senior executives in navigating the difficult waters of implementation management:

- Design the pace, sequence, and mix of initiatives according to the organization's urgency for change and capacity for change.
- Learn from what has worked and what has not and make adjustments to the plan along the way.
- Execute with discipline. Implementation management requires rigor – at least as much as managing the core businesses. Appoint strong steering committees to oversee the program.
- Align the team. Alignment and commitment happen when top executives work together; team building efforts should emphasize work planning and execution over symbolic action and ceremony.
- Build change champions. Identify line managers who are potential change agents and provide training on leadership and on specific change management tools and techniques.
- Build and sustain momentum. Leaders can and should take active steps to generate and sustain momentum. These steps include

identifying and explicitly addressing resistance to the initiatives, and publicizing quick wins and successes along the way.
- Maintain an appropriate focus on the day-to-day demands of running the core business.
- Over-communicate. Communication is the single most underrated component of implementation. Organizations need to explicitly define a communication strategy that will last throughout the entire change program.

Creating a consistently high-performing organization is the central, and perhaps the most daunting, challenge facing today's CEOs and top management teams. But it can be done. We are convinced that success in corporate transformations is more than a matter of luck, and the art of leading them can be learned. In a turbulent competitive environment, this art may be the most important "core competence" of all.

24.4 Summary

The achievement of true excellence in any international chemical company depends heavily on organizational aspects. This chapter examines the best structures for this purpose, the importance of addressing performance both now and in the future, and ways to make lasting organizational change happen:
- Although there are challenges in making it work, an organization with global business divisions as the main axle, guided by the corporate center and supported by shared services, seems to fit the needs of the chemical industry particularly well. About three quarters of the global players in chemicals have adopted this structure.
- Aspirants to top performance must follow a balanced approach to the management of *performance*, i.e., what the organization needs to accomplish today, and *health*, or how it can improve or adjust its performance over time. Leading companies tightly integrate business-performance with people-performance management.
- Most organizational transformation programs that are not crisis-driven lose impetus and fail. We believe that organizations can change without being in crisis by developing a clear picture of their current situation and a compelling vision of what they desire; shifting mindsets and behaviors to accept, welcome, and implement change; and setting up a project-based implementation program which is strong and disciplined, but nevertheless flexible enough to make adjustments along the way.

References

Allen, P., Ghislanzoni, G., Alegri, J., Price, C., Shearn, J., Vinson, M., Wendler, J., *Performance Transformations: Client-Shareable Thought Piece,* **2004**

Benson-Armer, R., Birch, D., Howse, R., Leslie, K., Todd, P., Turnbull, D., Van Wijck, J., Integrated Performance and Health Management *Internal working paper,* **2004**

Day, J., Jung, M., Corporate Transformation without a Crisis, *McKinsey Quarterly* **2000**, *4*, 117–127

Getting Organization Right: The Ultimate Performance Challenge, *McKinsey Global Organization and Leadership Practice Client Brochure,* **2003**

25
Post-Merger Management: it's All in the Design

Eric Bartels, Tomas Koch, and Philip Eykerman

Many companies generate value by growing through mergers or acquisitions. When, often after months of negotiating, the parties concerned announce that they have reached an agreement and state the purchase price, the benchmark is set against which the success of the transaction will be judged. The price, including the strategic premium, reflects the negotiating power of both parties and their expectations of the potential synergies. Post-merger integration cannot turn a bad deal into a winner; however, if properly implemented, it can realize the estimated synergies and often identify additional potential. Successful post-merger integration also stabilizes the organization and secures the business, which prevents the merger from causing short-term falls in profit that can end up destroying more value than the synergies create.

In the first edition of this book, we estimated the capital market's expectations on the value creation of a merger by the development of the total return to shareholders in both the short term (five days) and the long term (two to three years). In that edition, our chart showed that only half of all mergers in the chemical industry succeeded in creating value. We have updated this analysis, and worryingly the success rate has fallen to just 40 percent (Fig. 25.1). We believe that much of the blame for this can be placed on poor post-merger management rather than on poor deals being struck in the first place.

There is, of course, no instant recipe for success in post-merger integration. Companies must find the path that is right for them, even if they have enjoyed successful mergers in the past. Each situation is different, with industry dynamics, the legal context, and the starting position of the companies all playing a part in determining the best approach. Perhaps the only universal truth is that careful planning and staying power are certain to be indispensable.

Value Creation: Strategies for the Chemical Industry. 2nd Edition. F. Budde, U.-H. Felcht, H. Frankemölle (Eds.)

ISBN: 3-527-31266-8

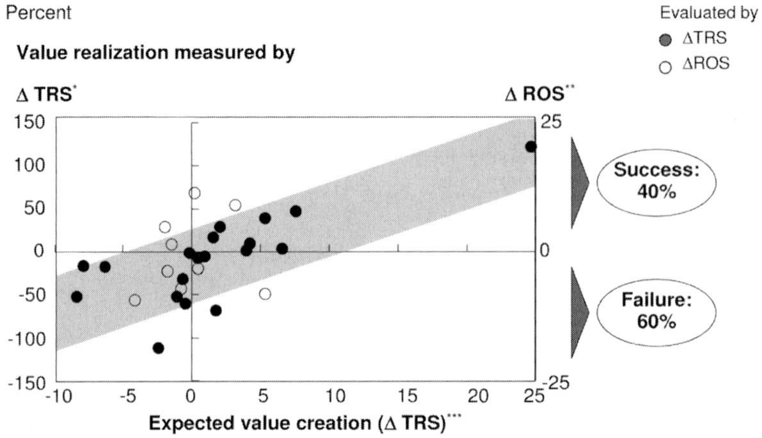

* Change in total return to shareholders relative to index two to three years before and after announcement of the merger
** Change in return on sales relative to peer group three years before and after announcement of the merger
*** Change in TRS performance five days before and after announcement of the merger
Source: PMI team

Fig. 25.1 Classification of major mergers in the chemical industry (1993–2004).

25.1
Phases of an Integration

The integration process is typically split into two phases. During the first, the design phase, management makes all the fundamental decisions on the orientation of the new company and plans the implementation. This can take two to three months depending on the project's scope and complexity. The second phase, the execution phase, is when the company works out the individual integration actions in concrete terms and implements them. This tends to take one to two years.

The design phase is critical – in most cases the first hundred days are decisive in determining whether the merger will be a success. During this period the company needs to clarify the form that the integration should take. There are four possibilities:

1. In a *bolt-on*, the acquired firm is added to the acquiring firm but largely retains its own structures. Value is generated by cross-selling the acquired company's products through the acquirer's sales channels. Internet networking company Cisco Systems is one of the great exponents of this model;
2. In a *takeover*, the acquiring firm simply integrates the target;
3. A *best-of-both* approach combines the best structures, approaches, and processes of the two organizations;
4. The final option is a *transformation*, where the existing structures, approaches, and processes of both companies are, at least partially, replaced

by new and better ones. The merger of Ciba-Geigy and Sandoz, for example, was a transformation that eventually allowed the companies to achieve a completely new level of performance as Novartis.

The first two or three months are also crucial for winning the support of employees affected by the merger. The companies should provide convincing reasons for why the deal is taking place and seek to create an atmosphere of starting anew.

Many companies fail to implement this first phase with enough rigor. Some even overlook it altogether. This is often because top management is still involved in the final throes of the transaction itself, or has started immediately implementing integration actions. Many of the problems encountered during mergers can be traced back to mistakes made in this initial design phase, in particular failure to plan the integration in a systematic and focused manner.

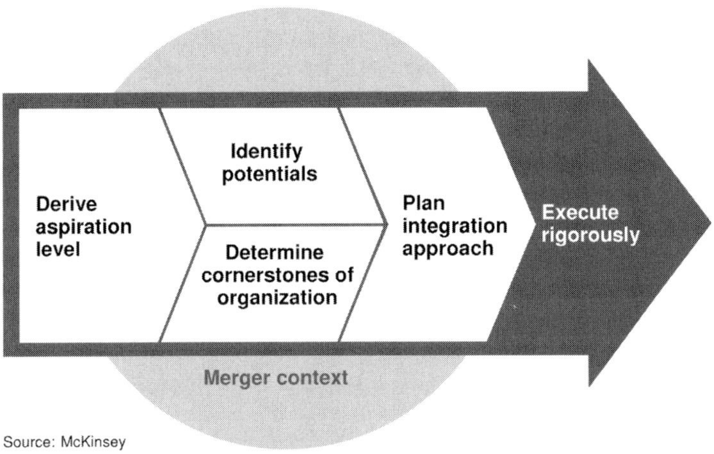

Source: McKinsey

Fig. 25.2 Main tasks in the design phase.

In the first few weeks, top management should concentrate on quickly completing four key tasks (Figure 25.2):

1. Deriving the new company's aspirations and translating these into binding targets and values;
2. Identifying all the value creation potential that can be tapped over the short, medium, and long term;
3. Determining the cornerstones of the organization;
4. Drawing up the integration plan.

In the second phase, management must ensure that the planned integration is implemented rigorously. The success factors during this stage are essentially the same as those associated with other major change processes (such as turnarounds or restructurings), so the rest of this chapter focuses specifically on the tasks carried out in the design phase.

25.2
Defining and Communicating the Aspirations

The new company's aspirations should be announced quickly because a clear vision prevents uncertainty among staff, sets the priorities that lie ahead, and helps guide any decisions that need to be made. A clear vision also allows management to explain to the capital markets that it firmly intends to follow a different course. Veba and Viag formulated such aspirations. The day the merger was announced, management presented a clear perspective on two core areas of business – energy and chemicals. The energy business would be expanded over the long term to create a company operating at the European level. The chemicals business would focus even more on specialty chemicals through intensive streamlining of the portfolio, and would also be expanded. The merger eventually led to the formation of Degussa, now the world's largest multi-specialty chemicals company.

Once the company's aspirations have been set, they have to be broken down into measurable targets. These can be cascaded down the organization by management. At the front line, these targets create a direct link between the company's aspirations and day-to-day operations; and also serve to make the expectations comprehensible and tangible for all employees.

Management should also lay down and communicate the values that reflect the new company's orientation and help the organization to behave consistently. When Ford bought Volvo's passenger car division, CEO Jacques Nasser made clear from the outset that the 'Swedish' aspects of the Volvo brand would be retained, and explained what this meant for the product and service range and the corporate culture.

Forming the aspirations, targets, and values and communicating them throughout the organization is generally an iterative process requiring repeated correction and adjustment (see section 25.5.3). However, it is well worth the effort and helps direct employees' attention towards opportunities rather than potential difficulties. As a result, the stability of the entire integration process increases from day one.

25.3
Fully Identifying Value Creation Potential

A merger's success is ultimately measured by the value it generates. Before announcing the transaction, senior management will have estimated the short-term synergies between the two organizations, justifying the transaction premium and the integration risks. During the merger, the emphasis is on realizing these synergies. However, it is often overlooked that an integration usually offers other, much more extensive opportunities for value creation.

There are three different types of value creation potential:

1. *Short-term cost or revenue synergies,* resulting directly from the combination of both businesses;

2. *Standalone potential* in the individual businesses that could be tapped irrespective of the merger, but that is easier to address because of the discontinuity of the merger;
3. Potential from *new strategic opportunities* created by combining the strengths of the two companies.

It is essential to acquire an in-depth and exhaustive understanding of these potential value creation areas, even in the design phase. Only then can the integration process be geared towards tapping them.

An iterative top-down and bottom-up method has proven successful in quantifying the potential. Management sets demanding but realistic goals derived from the new company's aspiration and comparisons with top-performing companies in the industry (or other industries). These goals send out a clear signal to the organization that the integration is not just 'business as usual'; extraordinary things are both possible and expected. These top-down goals then need to be underpinned by the organization at an actionable level of detail. Individual analyses must identify concrete value drivers and projects that need to be implemented. This bottom-up process serves to make the abstract top-down goals tangible for line management. As a result, the readiness to change increases. Figure 25.3 shows a typical process used to identify value drivers.

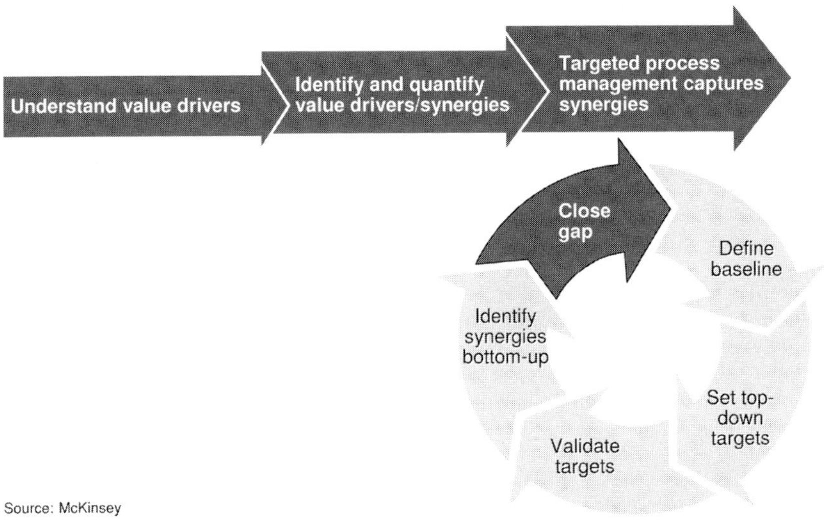

Fig. 25.3 Identification and activation of value drivers.

This process can begin in the design phase – unless there are restrictions on the exchange of information prior to the official closing of the transaction. In cases where such restrictions exist, calling in a 'clean team' (with no employees from the companies concerned) to put the company-specific data into an aggregated, comparable form and then evaluate them can save considerable time.

25.3.1
Determining the Synergy Potentials that are Directly Business-related

Merging two companies generally leads to substantial synergies – redundant costs can be reduced and revenues increased due to the expanded customer base. Additional value also results from the targeted exchange of know-how: the two companies, previously separate entities and in many cases competitors, exchange best-practice experiences, which allows them to boost efficiency even further. If two agrochemicals players merge, for example, the efficiency and effectiveness of their research and development often increases simply through the merging of the two R&D teams with their own experiences, approaches, and data records.

Figure 25.4 shows typical savings for commodity and specialty chemicals companies. It demonstrates that specialty players can extract greater savings from their core areas of expertise – such as marketing and sales – while their commodity counterparts can create value in purchasing and administration.

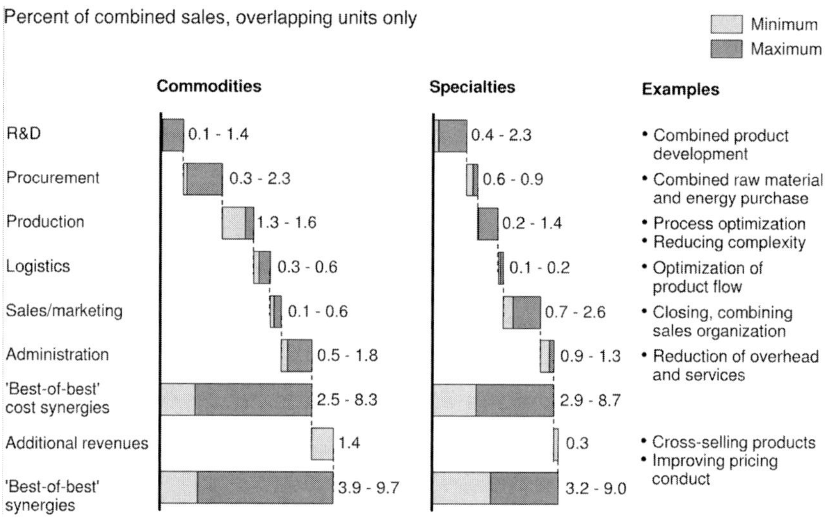

Fig. 25.4 Direct merger synergies.

The extent to which synergies may be available depends largely on the starting point of the companies. Where two large companies have little overlap beyond centralized functions – as was the case with Degussa – synergies will inevitably be lower than where the overlap is substantial, as may occur when a small company is acquired.

The market context is also very important. If the company has especially high aspirations then the CEO must set aggressive targets, which means extracting maximum synergies in all areas. Companies that find themselves in an 'acquire or exit' position or that merge during a downturn must also work that bit harder to capture all the potential.

As a rule, the business-related synergy potential can be identified using a simplified business system focusing on four areas.

Reinforcing purchasing strengths. Substantial purchasing savings can usually be achieved soon after a merger. The main levers are:

- Selecting the most cost-efficient supplier, while consolidating the supplier portfolio;
- Realizing discounts on the basis of increased volumes;
- Redesigning purchasing processes and transferring best practice to the new company.

The potential largely depends on the class of material, the volumes purchased, and the purchasing skills already present. If two companies of equal size and with average skills are merged, the total savings potential is usually in the region of ten percent of the purchase volume. If the companies are of different sizes and already have excellent purchasing skills, the potential is generally limited to around four to six percent.

Optimizing production and technology. Savings in operational areas result from three approaches:

1. Developing a long-term site strategy: when similar-sized companies merge, the number of production sites and thus the complexity usually doubles. Management must therefore check the extent to which savings are possible by closing certain sites. Likely candidates are similar production plants with surplus capacities, high administration costs per site, and redundant research facilities;
2. Optimizing production plants through knowledge transfer: if both companies have comparable production plants and/or production processes, benchmarking on the basis of a 'virtual plant' is suitable. This comes from combining the cost leaders in the individual process steps;
3. Introducing new service concepts: restructuring product ranges and production sites usually raises the issue of outsourcing. Many chemical firms bundle tasks such as power supply, maintenance, and logistics in independent companies with a pronounced service mentality. This allows the business units to concentrate on their core business (see also Chapter 20).

Reinforcing marketing and sales. Combining these two activities is the most delicate aspect of the entire merger, as they directly affect the customer. Value can be created from the global consolidation of the customer portfolio and the sales organizations. Companies should not only adopt the best of the existing systems but should also consider new sales structures, perhaps reorganizing by market and/or customer segments. The expanded customer base also makes it possible to exploit cross-selling opportunities.

Trimming down the management team. Probably the most obvious area for value creation is the corporate center. First, define what value the center should add, for example should it be confined to the role of a financial holding company or, at the other extreme, should it also have operational control over the business units? Once this has been decided, the tasks, competences, and scale of the corporate center can be redefined.

25.3.2
Securing the Business and Uncovering Additional Potential

Mergers are associated not only with opportunities, but also with considerable risks. In the first few months following a merger, many companies often notice a decline in the organization's efficiency. Operational targets set before the merger are missed and improvement initiatives fizzle out. This is due to day-to-day operations being neglected: management becomes absorbed in integration planning, the decision-making processes employed thus far begin to falter, and employee motivation suffers due to fears of possible rationalization. McKinsey studies have shown that a decline in productivity of between five and ten percent is not unusual during the integration phase.

During the design phase, therefore, management should take steps to counter this trend. This includes communicating that original budgets and targets are still valid and that management will be held accountable for them, and strictly controlling target achievement. The merger partners should also keep a close watch on the market and competitors. The introspection of a merger can lead to customers being neglected and lost – especially as competitors often try to lure attractive customers away. Targeted customer-retention actions can reduce this risk.

Along with securing the current business during the design phase, companies also need to identify standalone sources of value creation in individual businesses. Employees know that a merger usually brings fundamental change and management should capitalize on this feeling of starting anew by tackling older problems not directly connected with the integration. This can be a unique opportunity to push through changes previously blocked by organizational inertia. It may involve redesigning well-worn processes, centralizing tasks, or refilling key positions. It is much easier to push through radical measures when other changes are already on the agenda than during a normal everyday situation.

Top management should also attempt to quantify this additional potential in the top-down and bottom-up process, even though it is usually much less obvious than the short-term synergy potential. Cross-functional teams, external benchmarking comparisons, and the systematic revisiting of the strategic plans of both companies are typical tools for this task.

25.3.3
Determining the Potential from New Strategic Opportunities

Merging two companies can open up new strategic opportunities that the companies could never have captured independently. When, for example, Linde's industrial gases division acquired AGA of Sweden, the expected cost savings were quickly determined. Another factor, however, was very important for the sustained success of the organization: the merger gave Linde access to AGA's gaseous therapeutics activities, and the combined companies were able to build a strong healthcare business. Today, this business is one of the strongest growth options for Linde's industrial gases.

Other companies have found that mergers have enabled them to reach the critical mass required to become leading and profitable players in a particular market or region. Mergers can also create a basis for further acquisitions or the development of new technologies. For example, Bayer's acquisition of Aventis CropScience strengthened its herbicide research and gave it access to biotechnology know-how.

Estimating the potential of these strategic options is difficult, but nevertheless essential. It can have a decisive impact on the new company's set-up and on where the emphasis of the integration program should lie. A separate team should be assigned to this task at the earliest opportunity.

25.4 Determining the Cornerstones of an Effective Organization

The cornerstones of the organization need to be determined in the design phase to make the new entity functional as fast as possible. Management must tackle four closely related issues: aligning the management team behind a common goal, quickly defining the organizational structures, building a performance culture, and managing top talent. If any of these 'soft' topics are neglected, the success of the merger can be placed in jeopardy (Fig. 25.5). Sir Richard Evans, former chairman of BAE Systems – a very acquisitive company – captured this idea when he said, "Managing a merger is a case of complex change management. If you get the people part wrong, you will never recover."

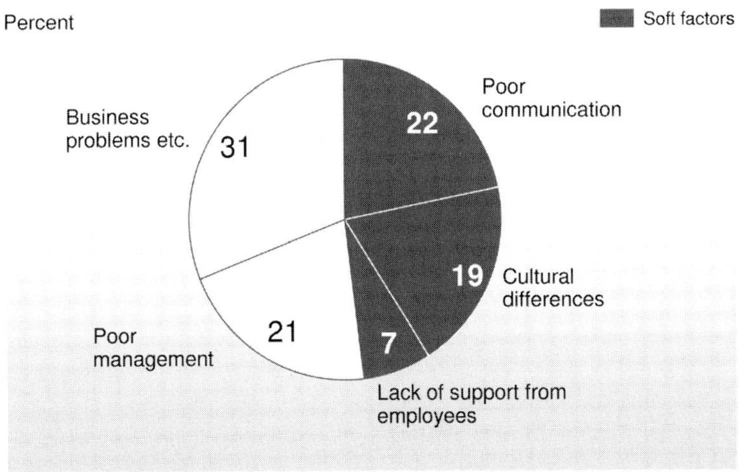

Fig. 25.5 Why mergers fail.

25.4.1
Aligning Management Behind a Common Goal

There is often a lack of unity at many management levels, even in day-to-day operations. Rarely are all top managers completely aligned; perhaps because personal targets differ, or because of a lack of clarity around business priorities. The result, however, is always the same: a lack of coordination, prolonged decision-making processes, and a poor level of performance.

Unity is particularly important for post-merger integration. The new management team usually lacks common roots or experiences, which means its interests and ideas may be divergent. But various decisions must be made during an integration, often quickly and decentrally. This can happen only if top managers all pull in the same direction and are speaking the same language.

Interviews and consensus workshops are good ways of ensuring that top management is focused on common goals. The exact method, scale, and time required will depend on how divergent the managers' ideas are at the outset. The end result, though, should be a company able to rely on a committed and aligned management team. All members must know their role within the team and be aware of the team's contribution to achieving the targets.

25.4.2
Rapidly Determining New Organizational Structures

Determining the new company's organizational structures is not easy. Even a normal reorganization can be a political minefield – let alone a full-blown merger. There is also a danger of getting bogged down in detailed discussions. In the design phase management should concentrate on a few key decisions relating to the structure:

- *Creating a shared understanding of the organizational principles* of the new company, dividing up responsibilities along the core processes, and establishing the main performance indicators;
- *Marking out the tasks and responsibilities* of the corporate center, (shared) service units, and business units;
- *Defining the top level of the organization* to allow the board to take control of all critical functions;
- Deriving the *headcount of the individual units,* ideally on the basis of a business plan.

With these cornerstones in place, the company can refine the structures and detail the key processes in the second phase. Using a staggered procedure allows all the appointed managers to determine their relevant reporting structures in cascades from top to bottom.

25.4.3
Building a Shared Performance Culture

The integration of the corporate cultures does not happen by itself. Even if all the other success factors are met, a merger can still fail due to the lack of a shared corporate culture. In the short term, management needs to focus on identifying and addressing potential cultural flashpoints. These may include the form and scope of communication, or relate to the location of the new corporate center. When two chemical companies merged, the relocation of the headquarters to a new, third location had a significant impact on the corporate culture. Employees used to working in the nerve center suddenly felt they had been demoted to incidental roles, and even employees not directly affected by the relocation left the company after the decision. If the merger is cross-border then integrating the corporate cultures is even harder. However, there are success stories. When pharmaceutical companies SmithKline and Beecham merged, they tackled cultural integration systematically from the outset and successfully avoided major conflicts.

Even in the design phase, top management must actively control the process of building the future, shared corporate culture, and take the following points into account:

- *Fact-based decisions:* differences between the companies must be made transparent. Possible approaches range from in-depth interviews with managers through to broadly-structured employee surveys;
- *Consideration for internal differences:* there are often cultural differences within a company. Sales, for example, will typically have quite different working methods and back-office requirements. These differences also need to be highlighted;
- *Role-model function of the management team:* management must take up the cause of cultural integration and ensure that it stays on the agenda. The team should not only actively drive forward the necessary changes, it should serve as a role model for the entire organization;
- *Communication:* communicating the new culture within the organization is also important. This may involve symbolic actions, such as introducing new department names and titles (see section 25.5.3).

The long-term goal is to build a performance culture that will enable targets to be met. This may mean completely redesigning management and evaluation processes.

25.4.4
Managing Top Talent

Employees can understandably become insecure in the face of a looming merger. Their professional future is uncertain and many in the acquired company will feel let down. Competitors will happily take advantage of this and lure away top talent, so companies need to take immediate action. The following steps are particularly important:

- *Identify key talent early:* both in top management and at other levels of the organization, especially in critical areas such as R&D, IT, or the field service;
- *Build employee loyalty:* perhaps through financial incentives, involvement in integration activities, or special development programs;
- *Fill key positions quickly:* this can counteract uncertainty. When BP and Amoco merged, several hundred senior management positions were filled within a few weeks of the announcement.

It is also important to check how many staff will need to be recruited. Boards tend to hesitate in this area. A merger creates a company that is usually substantially larger and has different areas of focus – and therefore may require different skills. Reviewing the available pool of talent with the help of an external personnel consultant can help create an objective basis for a decision.

25.5
Tailoring the Integration Approach

Once all the starting parameters – aspirations, value creation potential, and organizational cornerstones – have been determined for the new company, the rest of the integration approach needs to be planned in detail. This involves defining the project organization, selecting the project management tools, and steering the communication.

25.5.1
Setting up a Powerful Project Organization

A clearly structured project organization is required to deal with the numerous complex individual integration tasks. This typically has three levels: (1) a steering committee, responsible for laying down guidelines, making human and financial resources available, and deciding on the integration actions; (2) an integration office that provides general support to and manages the project teams; and (3) the project teams, which work out the individual integration actions and have responsibility for implementing them. Special steering committees are often used for large mergers with numerous project teams, in order to decide on the actions to be taken by the various sub-projects.

The individual elements and the division of tasks will depend on the particular context of the integration. The following points, among others, need to be clarified.

Role of the CEO and the integration manager: management can carry out the integration itself or use a separate team with an integration manager. In a best-of-both integration or a transformation the CEO usually plays a very active, visible role, setting the direction and driving the change – as in the case of Daniel Vasella at Novartis. In these situations, the integration manager concentrates on process control and is supported by the integration office. In a bolt-on or a takeover, the integration manager usually assumes the managerial role, while the CEO and management team concentrate on running the business. In either case the integration manager is important. He or she should be someone with a good knowledge of both organizations, able to bring extensive experience in project management to the role and with excellent soft skills.

There is no single answer as to which form of decision-making is most appropriate for the integration. Those that create a large proportion of their value through operational improvements need the support of on-site employees, suggesting largely bottom-up driven decision-making. Structural synergies, on the other hand, can be realized more quickly using top-down decisions.

Role of the integration office: the integration office can actively drive the integration or simply coordinate activities. In most cases, it will push forward the integration activities. One of the most important tasks in this context is exercising overall control over the merger. This means developing and managing a master plan that contains all the projects and their targets, the responsibilities, the status of the workout and implementation of actions, and the deadlines. Using the master plan, the integration office can monitor the progress and quality of the integration activities, and prepare decision drafts for the steering committee.

The integration office may also assume an arbitration function and moderate discussions in difficult situations, for example when an optimum organizational design is being derived for individual units. Additional tasks may include developing the communication strategy or clarifying personnel matters.

Such an active role is particularly appropriate if the integration activities are very extensive and/or the companies concerned have little M&A experience. In the example of an automotive industry merger, however, the integration office was deliberately given only a coordinating role and integration activities were largely carried out within the context of day-to-day operations. This meant it was possible to maintain a general overview of the scope of the changes (both companies had already experienced mergers). The members of the integration office should always be managers with a good reputation in both organizations.

Role of the project team: individual project teams are the cells of the merger. They devise the actions that ensure the new company can function and capture value. The number of project teams can vary substantially. In most cases there are initially around ten, but in major global mergers the number can rise very rapidly. The project teams are geared toward specific issues such as sources of value creation, business units, functions, regions, locations, or special tasks, e.g., communication. Each team usually has one project leader; having two team leaders makes sense only in exceptional cases, for example to secure access to information or in a merger of equals. The teams themselves should be structured such that both companies are represented. Attention should be paid so that only the best people are selected as team members – not just those who happen to be available. Nomination for a project is also a good way of binding top performers to the organization.

Every team should have a charter that describes its targets, tasks, and resources, lists the end products, and contains an initial rough schedule.

25.5.2
Selecting Project Management Tools

Controlling an integration program can be a Herculean task. With projects (and sub-projects) running in parallel and countless individual actions documented with concrete end products, responsibilities and deadlines, a high level of complexity is guaranteed. In addition, individual project teams often work at different locations in different countries.

It is therefore all the more important to monitor systematically the progress of integration activities and the achievement of the targets. Two instruments in particular help here. Implementation controlling makes it possible to track whether all the actions are being implemented as planned, whether they are being driven forward with the necessary rigor, whether individual actions, projects, or the overall program are on schedule, and when each action will be realized. Effect controlling, on the other hand, should be set up so that the monetary impact of the actions can be traced in the company's cost centers. This allows management to check whether synergy effects have kicked in as planned, and when an individual action will have an impact on net income.

Specialist software can help with these tasks. Standardized reporting structures, for example, allow each team to report the project status regularly to the integration office. An automated comparison with the master plan highlights possible deviations and makes transparent interfaces and dependencies between actions.

To help ensure acceptance, existing instruments should be used or modified appropriately wherever possible. Web-based tools are particularly suitable. If new instruments are required, then pilots and training programs should be planned in good time. In principle, preference should be given to quick, pragmatic, individual solutions.

25.5.3
Communicating Permanently

As already mentioned, the importance of continuous communication in ensuring the success of a merger cannot be overestimated. Attention should be devoted to the issue right from the start. When initial negotiations get under way, for example, it must be clear how the companies will react if any rumors of the merger leak. When the merger is announced, the messages conveyed to shareholders, customers, and employees must be clear and aligned. When integration begins, it must be clearly communicated how the workforce can be motivated to cooperate. Furthermore, there are legal provisions in the area of communication that must be taken into account, such as disclosure of information by listed companies.

It is important to decide as early as possible who will be responsible for developing the communication strategy and who will release the specific content. Agreeing on uniform wording and a few clear core messages early on has proven worthwhile. This would cover topics such as why the merger makes sense, what it offers the companies concerned, and what will change for stakeholders.

When developing the communication strategy, the company needs to consider all the relevant internal and external target groups and prioritize them according to their communication requirements. A detailed communication plan should be drawn up for each target group that describes the aim of the communication, the specific messages, the new or existing media to be used, the timing of individual communication actions, and the people responsible. Feedback mechanisms, such as regular employee surveys, can be used to keep a constant check on whether the messages are reaching the target groups. If necessary, the communication plans can then be revised and adapted accordingly. Communication plans for the individual target groups need to be coordinated with one another in terms of their content and timing to ensure that no contradictory messages are given out.

25.6 Summary

- Mergers and acquisitions have changed the chemical industry landscape over the last few years. However, the transactions do not always bring about the desired increase in value. Success becomes predicated on making the integration work.
- The first hundred days of the process are vital, as this is when all the important decisions on the company structure are decided, and the post-merger integration program is set up.
- The following six success factors need to be considered: (1) value creation should be seen as the driving force behind the integration, (2) ambitious goals should be set without losing sight of what is feasible, (3) the existing business should not be neglected, (4) cultural differences should be taken into account at an early stage, (5) the fears and resistance of employees should be taken seriously, and (6) all stakeholders should receive comprehensive, continuous communication.
- The second phase of the integration process is then concerned with ensuring that the integration program is implemented rigorously, with sustained momentum.

26 M&A – the UCB Case

Georges Jacobs

Created in 1928 by the merger of thirteen chemical intermediate factories, Union Chimique Belge (UCB) has continuously used mergers, acquisitions, and divestments to strengthen its position in those markets that were consistent with its strategy of "innovation and globalization", first in chemicals and films and later also in pharmaceuticals.

Over the last two years, however, the pace accelerated dramatically and the company went through a major transition from a hybrid company operating in specialty chemicals, chemical films, and pharmaceuticals to a pure bio-pharmaceutical player. The unique feature of the situation is that the company was not just broken apart to give each of the parts more focus: UCB took effective action, including the acquisition of another player, to significantly strengthen the specialty chemicals and films sectors before divesting them. The proceeds were then used to repay an important part of the debt incurred by a major acquisition in the pharmaceutical sector.

Indeed, in December 2002, UCB surprised all analysts – who had expected an investment in the pharmaceuticals sector – by more than doubling the turnover of its specialty chemicals business, about EUR 465 million at that time, with the acquisition for EUR 500 million (EBITDA multiple of 6.8) of Solutia's resins, additives, and adhesives business. After the new purchase had been integrated and whilst the synergies were being realized according to plan, the whole chemicals entity, renamed Surface Specialties, was sold to Cytec at the beginning of 2005 for EUR 1.4 billion (EBITDA multiple of 9.6), creating a EUR two billion focused resin player. This sale immediately followed the divestment of UCB's films activity for EUR 0.3 billion to a UK principal investor (EBITDA multiple of 7). Most of the EUR 1.7 billion proceeds from the two divestments was used to repay the debt taken on to finance the acquisition in May 2004 of Celltech, the leading UK biotech company, for EUR 2.3 billion.

First and foremost, these moves created three stronger, more focused entities and removed any doubts about their individual futures. In addition to that, however, UCB – through the transactions and effective post-merger management in chemicals – created approximately EUR 600 million of capital gains on its books. Finally, the company leveraged the post-merger integration in pharmaceuticals to

Value Creation: Strategies for the Chemical Industry. 2nd Edition. F. Budde, U.-H. Felcht, H. Frankemölle (Eds.)

ISBN: 3-527-31266-8

fundamentally rethink the organization and its main processes, and better prepare itself for the challenges ahead in terms of product transitions and increased external pressure on the industry.

This chapter will:

1. Briefly cover the history of UCB, and how it developed from a hybrid to a pure bio-pharmaceutical player;
2. Describe the mechanistic, almost military, but very people-centric process followed for the post-merger integration of UCB Chemical and UCB Films with Solutia's resins, additives, and adhesives business;
3. Explain the main differences between the above post-merger integration and the integration of UCB Pharma and Celltech, executed against even more ambitious targets;
4. Close with an overview of the key learnings from both post-merger integrations.

26.1 A Brief History of UCB – Eight Decades of Mergers, Acquisitions, and Divestments

Union Chimique Belge was founded in 1928 by, among others, Baron Emmanuel Janssen, who was married to a granddaughter of Ernest Solvay, the founder of Solvay, the other Belgian hybrid chemical/pharmaceutical company. At that time, Union Chimique Belge was producing several heavy chemical intermediates such as ammonia, sulfuric acid, sodium sulfide, and coal derivatives.

During the Second World War, Union Chimique Belge started producing basic pharmaceuticals such as calcium, phosphor, vitamins, insulin, and sulfamides. However, it was only in the late fifties that it launched, with limited market impact, its first real pharmaceutical products – Postafene (an anti-histamine), Atarax (a tranquilizer) and Tuclase (a cough and cold drug) – and concluded a first license agreement with Pfizer for the distribution of Atarax in the U.S.

In 1961, the current UCB was formed through the merger of Union Chimique Belge (briefly described above), Fabelta (an artificial textile fiber producer) and Sidac (a cellulose film manufacturer).

Eight years later, in 1969, and after the sale of Fabelta to AKU, UCB was re-organized into four divisions: Films, Chemical Products, Chemical Specialties and Pharmaceuticals. The first two divisions accounted for eighty percent of the turnover, while Chemical Specialties and Pharmaceuticals each represented approximately ten percent. From that moment onwards, UCB started to focus increasingly on specialty products and international expansion, first in Europe and later also outside Europe. It initiated a long-term strategy aimed at reinforcing those activities and fulfilling two basic criteria – innovation and globalization – while consistently divesting activities that did not fulfill them.

- In chemicals, UCB sold, among others, its fertilizer business to BASF, its agrochemicals to Rhône-Poulenc, and its phthalates to Sisas. It also divested its methylamines interests to NIB Capital in

2003. Having sold all these base chemicals – most of them at peak value – UCB focused on innovative specialty resins such as UV cure resins, powder coating resins, and high-solid resins for automotive, graphic arts, and industrial applications. At the end of 2004, just before the closure of the sale to Cytec and two years after the acquisition of the Solutia activities, UCB's chemicals activity had a turnover of EUR 1.1 billion, around 40 percent of which was outside Europe, an EBITDA of EUR 147 million, and a leading position in most of its markets.

- In films, UCB sold its conversion activities to Amcor in 1996 after having reinforced them by the acquisition of Pabeltec's conversion business a few years earlier. The strategy was to concentrate on the production of specialized films – cellulose as well as bi-oriented polypropylene – on a global basis. UCB, however, anticipated the decline of cellulose in favor of polypropylene film well ahead of time. As a defense, it consolidated the cellulose market until it was the only supplier, acquired ICI's BOPP business, and focused on higher value-added segments such as labels, overwrap and innovative security products such as OPP banknotes. At the time of the sale in 2004 – the divestment and corresponding disentanglement from the group are not further covered in this chapter, but went just as swiftly as the post-merger integrations – UCB's films activity had a turnover of EUR 0.36 billion and an EBITDA of EUR 45 million.
- UCB systematically re-invested part of the cash flows from its chemicals and films activities into R&D for the pharmaceutical sector. This yielded its first big success in 1987 when the company launched the anti-allergy drug Zyrtec, which propelled UCB onto the global pharmaceuticals stage through partnerships with Pfizer in the USA and Daiichi and Sumitomo in Japan. Today, it is the world's most widely used antiallergic drug, with global sales of over EUR 1.2 billion in 2004. In 1999, UCB confirmed its capabilities in pharmaceuticals once more with the launch of an anti-epileptic drug called Keppra. Keppra became the most successful anti-epileptic launch ever, and is the current market leader in the USA ahead of products from GSK, J&J, Novartis, and Pfizer. UCB's strong and very profitable growth in pharmaceuticals, however, has also been its biggest challenge over recent years. Since 2002, the Zyrtec patent has been gradually expiring and by the end of 2007, Zyrtec will no longer be protected in the major countries. The Keppra patents will expire in 2009–10. In addition, UCB had little in its pipeline – beyond specific successors to Zyrtec and Keppra – to fuel continued double-digit future growth. The company therefore embarked on an extensive target scan,

and in May 2004 announced its bid on Celltech. UCB and Celltech complemented each other well for the following two reasons:

- Celltech had a product against Crohn's disease and against rheumatoid arthritis in its pipeline (Cimzia) that has blockbuster potential and should fill the growth gap left by Zyrtec and Keppra. However, it lacked UCB's drug development and commercial expertise, needed to launch this product successfully;
- Celltech brought unique technology leadership in molecular antibody (MAB) research, which nicely complemented UCB's chemical synthesis know-how. MABs are expected to be the main growth driver in pharmaceuticals over the next decade.

At the end of 2004 (including five months of Celltech's added revenues), UCB's pharmaceutical activities had a turnover of EUR 1.7 billion, and an EBITA of EUR 402 million.

Throughout the history of UCB, the Janssen family has retained an important say in the strategy of the group, and to date it still controls just over 40 percent of the company. Strong corporate governance, however, has made the reference shareholder[1] structure a supporting rather than a constraining element for the transition.

26.2
The Formation of Surface Specialties – A Three-way Integration

When it announced the acquisition of Solutia's resins, additives, and adhesives business in December 2002, UCB surprised analysts as well as employees, who were all expecting an acquisition in pharmaceuticals rather than in chemicals. The acquisition, however, was not completely opportunistic. UCB Chemical's business development department was continuously on the lookout for value-creating opportunities to give the business more critical mass in an industry where suppliers and customers were rapidly consolidating, which made a concentration wave in the resins industry inevitable over time. In this context, the acquisition was a clear no-regret move: the acquisition price was attractive and could easily be financed by UCB, and there were many complementarities as well as synergies. In addition – and as clearly stated at the time – the increased scale created more options for the future, including a possible spin-off or divestment.

The post-merger integration of the activities acquired from Solutia, however, was somewhat complicated by the fact that it became a three-way rather than just a two-way merger. By the middle of 2002, it was becoming clear that the films sector was not performing as well as expected. UCB therefore decided to leverage some of the available management capacity in chemicals by integrating the films

1) Reference shareholder: in Belgium, an important shareholder with certain rights on the board and quite regular interaction with the management.

and the chemicals sectors to make UCB a clear bipolar company, facilitating future strategic moves. As a result, the integration covered about EUR 1.5 billion in sales, 4,700 people, and 23 plants, more or less equally distributed among the three originating entities.

The integration had two objectives: (a) achieving a seamless integration of the three entities, and (b) building a high-performance organization. It was carried out in four phases:

1. Phase 1: Plan and get going (December 2002 – January 2003)
2. Phase 2: Design and implement the organization (February – May 2003)
3. Phase 3: Define improvement initiatives (May – September 2003)
4. Phase 4: Implement and follow up on results (October 2003 – December 2004)

The decision to separate the organizational design from the definition of the improvement initiatives was quite fundamental and unusual, but nevertheless feasible in an asset-heavy industry where site restructuring takes quite some time in any case. The reasons for doing so were to make absolutely sure that there would be minimum disruption to the ongoing business and that everybody would know his or her position in the new organization as quickly as possible. This was successfully achieved. Indeed, by June 1, six months after the announcement of the deal and four months after its approval by the anti-trust authorities, the new organization, called Surface Specialties, was fully in place and operational. This did not mean that no savings at all were identified and captured before Phase 3. In fact, all managers except for the head of manufacturing, when designing their organization, had to make sure they reduced their combined headcount costs by approximately ten percent, and the purchasing task force (30 percent of the synergies) had the task of capturing several quick wins as soon as the deal was approved. In parallel, a profit improvement plan for the films activity was launched.

Next to the above-listed four integration phases, there was also a so-called post-closure separation work stream, which covered the transition arrangements with Solutia (brand use, invoicing arrangements, building leases, etc.), as well as the restructuring of the legal entities.

26.2.1
Phase 1: Plan and Get Going (Dec. 2002 – Jan. 2003)

This first phase of the integration, between the announcement of the deal and approval by the anti-trust authorities, was probably the most important one, as it set the pace, governance, and objectives for the total process. Phase 1 consisted of the following actions, in chronological order:

- On the day of the announcement itself, both CEOs gave press conferences – at different locations, but well coordinated – and there was a cascade of information by the different management layers throughout the organization. This was an important undertaking for UCB in particular, since, as mentioned above, both the staff and the financial markets had been expecting a deal in the pharmaceuticals rather than in the chemicals sector. Staff on the chemicals side of UCB were very enthusiastic. People on the pharmaceuticals side, however, were initially somewhat concerned, but this faded quickly once they understood the acquisition would not jeopardize the financing of future growth in their sector. Finally, market reactions were positive in the sense that the stock – which had come down over recent months – remained stable;

- In the days following the announcement, both top management teams went around their respective sites and communicated extensively to their staff, suppliers and customers, using material that had been developed jointly by UCB and Solutia. Without quoting detailed figures and making commitments that would be difficult to meet later, managers explained to the staff that there would be synergies and hence headcount reductions but that these would be handled in a fair and transparent way with maximum respect for the individuals concerned. Looking back and having delivered on the promise of fairness, transparency and respect, this proved to be a good approach;
- Once that was done, the newly appointed head of Surface Specialties (the former head of UCB Chemical) started to reflect on the form and staffing of his management structure, the integration organization and schedule, as well as the aspiration for the new organization and its name. In parallel, the contractual agreements between seller and acquirer were finalized and the antitrust file was submitted:
  - Many models were evaluated for the new management structure. Eventually, the Group decided to move from a markedly regional model, in which the regional directors had responsibility for both sales and operations, to a model with regional responsibility for sales, but global responsibilities for manufacturing and logistics, in addition to existing global R&D and marketing functions. This re-organization offered the benefits of better sharing of best practices in manufacturing and purchasing, increased flexibility in logistics (e.g., swaps between regions depending on exchange rates and raw material costs), and hence lower overall costs;
  - The integration structure was pretty standard, with three types of bodies: the integration committee, which steers the integration, the integration office, which manages it, and the integration task forces, which design the detailed organization and processes and recommend improvement initiatives. The difference from most other integration programs, however, was the major involvement of human resources (HR) – the

corporate HR director was one of the four members of the integration committee and a mature and experienced HR manager led the integration office – reflecting the strong focus on the people aspect of the integration. The time frame for the integration was aggressive: four months to arrive at a single organization, and another four months to finalize the improvement plans;
  - In terms of aspiration-setting for the new entity, it was clear from the start that UCB wanted to go further than just capturing the integration synergies, and really take advantage of the "unfreezing" opportunity from the integration to rethink the organization and its way of working. Hence, the improvement aspiration (ten percent on sales), although still only roughly known in Phase 1 and not yet allocated to each of the taskforces, was set well above the pre-merger synergy estimates and consisted of a good mix of revenue and cost improvements;
- Less than two months after the announcement of the deal, on the day of the approval by the anti-trust authority, the newly selected management team – each member of it having committed to his/her new function and contractual terms – gathered for the first time. During the first day of this two-day off-site meeting, at which there was plenty of time for informal discussion, the head of the sector laid out his vision for the new entity ("a total solution provider for customers requiring high-end surface products and technologies"), the financial aspiration, the integration concept, and the structure and modus operandi of the new management team. Several sessions were interactive, using either breakout groups or consensor technology to surface and resolve misalignment within the team. On the second day, the integration office went through the integration process in more detail and started working with the members of the management team on the integration mandates (staffing of the task force, key objectives, main challenges, interfaces, etc.) for their respective areas of responsibility.

26.2.2
Phase 2: Design and Implement the Organization (February – May 2003)

In this second phase, the members of the management team (the "N-1s") designed the next level of the organization, developed new job descriptions, and together with HR appointed the N-2s. The N-2s, with the help of the task forces, then further developed the organization in all its detail, wrote new job descriptions, and again together with HR appointed the N-3s and below (Fig. 26.1).

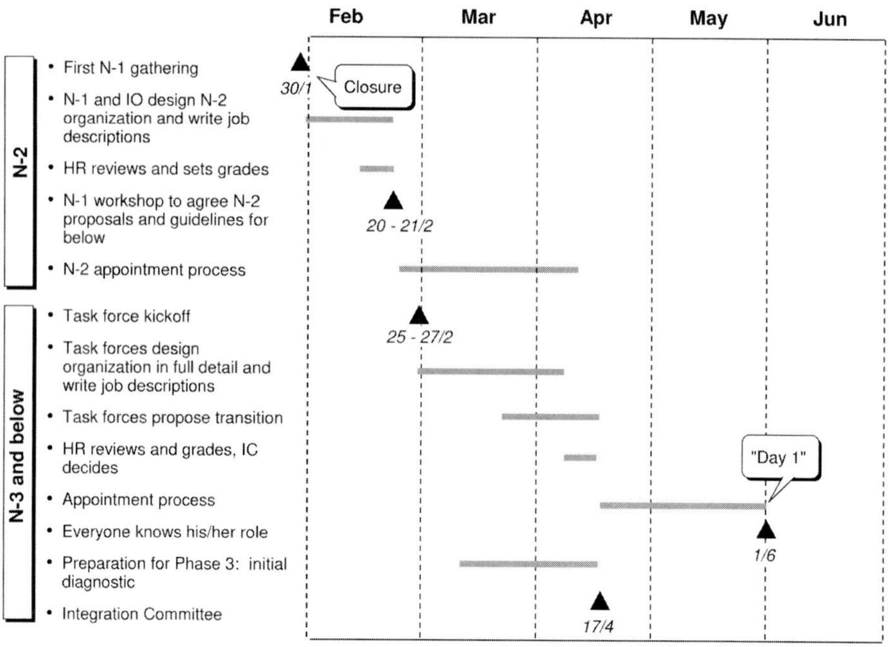

Fig. 26.1 Integration Phase 2 – timing for design of the new organization.

These activities were crucial in view of the time and delays that may be involved in staffing a new organization, and the anxiety that can reign in an organization when it is being fundamentally re-designed and management has expressed a clear desire to select the best (internal or external) candidate for each job. To avoid this anxiety and the consequent damage to the ongoing business, the appointment process had to be not only fast, but also absolutely fair, transparent, and thorough. Every manager therefore had to follow a number of guiding principles:

- The selection process for a function could not be started before approval by the integration committee of the organization design and its size, and before a job description and job grading were available;
- Cross-department contacts relating to appointments were strictly controlled, and no contact could be made to assess or attract a candidate without previously informing his/her superior manager and central HR;
- An HR group from the three originating entities facilitated the process and had a mandate to ensure fairness and diversity:
  - Together with the future superior for the position, it put together a list of candidates;
  - It coordinated the interviewing process by the future line manager (and any manager with whom the successful candidate would have a "dotted-line" reporting relationship) and

provided external support for candidate assessments or external searches;
- It made sure that no communications were made to the selected candidate before his or her current superior and other candidates, or people who had thought of themselves as candidates, were informed;
- It guaranteed that nobody was overlooked and that people left over from one round (e.g., N-3) were picked up in the next round (i.e., N-4), or taken care of in a different way.

At the end of this second phase – four months after the start of the integration – management had virtually fully finalized the re-organization, and on June 1 ("Day 1"), Surface Specialties was a reality with one face to the outside world.

26.2.3 Phase 3: Define Improvement Initiatives (May – September 2003)

About one month before the finalization of the previous phase, when the integration task forces had done most of their work but HR was still in the process of finalizing the last appointments, the integration task forces were renamed as project teams and refocused on developing improvement plans (Fig. 26.2). Some of these project teams differed slightly from the integration task forces, to reflect the new organization and appointments.

The target for the improvement plans was to reach a 14 percent EBIT margin by 2007. Taking into account a one percent price-cost squeeze, this meant an improvement of more than ten percent on sales (including the savings already implemented through Phase 2). Close to 50 percent of the improvement target was to come from manufacturing and purchasing, 30 percent from growth, product mix, and margin management initiatives, and the remaining 20 percent from non-FTE related economies in selling, general, and administrative expenses (SG&A) and from FTE savings.

Needless to say, these were stretch targets. Nevertheless, in four months, the project teams came up with 430 projects that made it possible to meet the aspiration. A total of 45 of these projects were "A" projects, which were to deliver 60 percent of the improvement. Implementation plans for "A" projects were very detailed with clear activities and milestones, as well as monthly resource plans and impact by activity. For the less important "B" and "C" projects, there were simpler planning requirements.

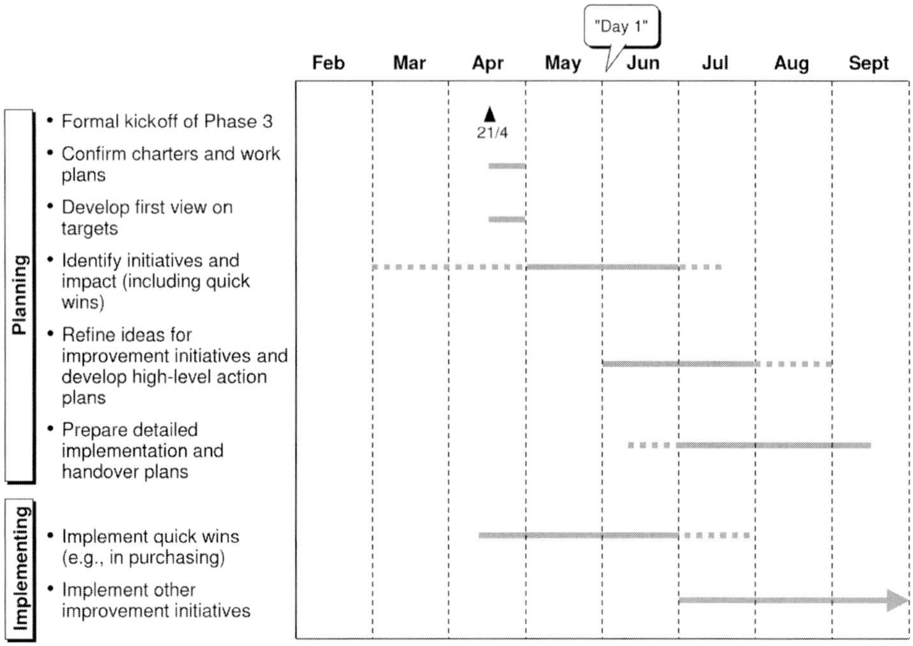

Fig. 26.2 Integration Phase 3 – timing for improvement initiatives.

26.2.4
Phase 4: Implement and Follow Up on Results (October 2003 – December 2004)

In business as elsewhere, what gets measured, gets done. Therefore, an explicit decision was taken in October 2003, when the project teams and integration infrastructure were disbanded, to install two managers dedicated to measuring the impact of the different improvement programs. Every month until the sale to Cytec, they reported to the management team on each of the "A" projects "by exception", and on the "B" and "C" projects in aggregate form.

It would be dishonest to say that these managers had an easy life and that the whole process of integration between a company that considered itself "entrepreneurial and pragmatic" and one that saw itself as having "better structure and processes" was an easy one. However, two years after the acquisition of the business from Solutia, at the time when the sale to Cytec was concluded, the company was clearly a very different one – having combined the best of both originating entities – and ready for the next big step. Thanks to the hard work of the joint workforce, UCB received several unsolicited bids for the business and eventually decided to divest to Cytec to create a EUR two billion focused specialty chemicals player at better conditions and faster than expected by the financial markets.

26.3
The Integration of UCB Pharma and Celltech

On the surface, the integration of UCB Pharma and Celltech was very similar to the integration of UCB Chemical and UCB Films with Solutia's resins, additives, and adhesives business. It covered about EUR 2.1 billion in revenue and 8,500 people; the aspiration was to achieve the same improvement target (five percent on sales before the merger, increased to around ten percent on sales in the planning phase); and the plan was to realize this through a similar four-phase integration approach (Fig. 26.3).

However, additional elements in terms of the aspiration, as well as the specifics of the pharmaceuticals business, made the challenge much bigger than in chemicals.

- First, the aspiration was to capture 65 percent of the improvement by 2005 and 100 percent by 2006 – much faster than in chemicals, but needed to finance the development costs of Cimzia, whilst still delivering double-digit earnings growth;
- Second, on top of the financial ambitions, and even more so than on the chemicals side, the new, externally-recruited Sector Head wanted to leverage the merger to fundamentally transform UCB Pharma. The new UCB Pharma had to:
  - Merge the proven commercial success of UCB Pharma with the innovation and R&D capability of Celltech;
  - Be better prepared to face new industry standards in terms of the commercial operating model (including market access, pricing, payer/provider relations), as well as regulatory affairs, quality, and drug safety;
  - Combine global standard processes for supply chain, R&D, and finance with locally-owned processes for the other functions, especially sales;
  - Thrive upon a set of strong values, which distinguish UCB Pharma from traditional biotech companies as well as from conventional pharmaceutical majors;
- Third, given that the stakes in the pharmaceutical industry are much higher, it was imperative to make sure that the integration would not jeopardize the successful growth of Keppra at UCB, or the development of Cimzia at Celltech.

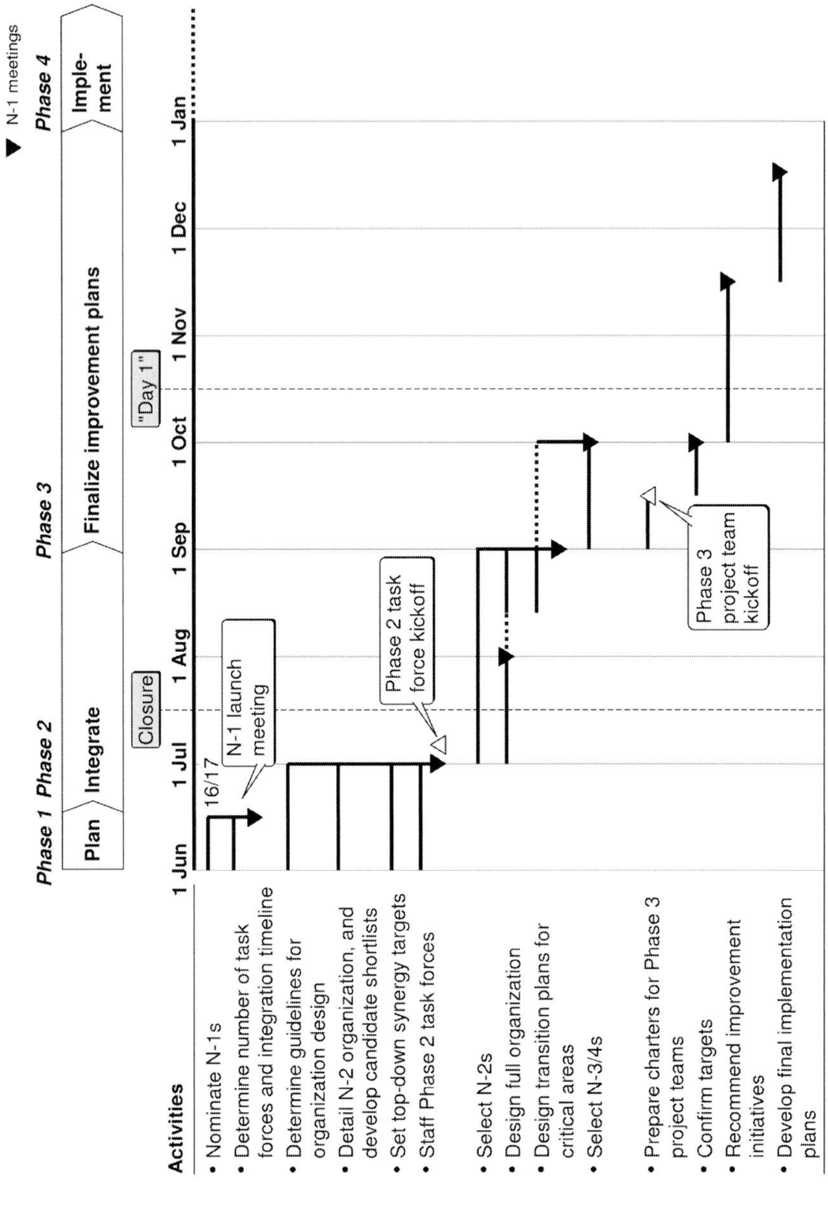

Fig. 26.3 Approach/timeline for the Celltech integration.

To address these additional challenges, we complemented the proven integration approach of Surface Specialties with the following elements:

- We encouraged the integration task forces – which again got an improvement target almost twice as large as that identified before the merger – to continuously update their improvement plans and to implement whatever was already possible as of the close of the deal (thereby de facto partially merging Phases 2 and 3, unavoidable in a business like pharmaceuticals, which is more people-intensive than chemicals). This resulted in 30 percent of the improvement plans already being implemented by October 15 ("Day 1"), three and a half months after the start of the integration, when the two organizations started to operate as one;
- To address the desire to fundamentally rethink the company, all major functions were reviewed following a zero-based approach, and the governance of major cross-functional processes such as portfolio management was completely redesigned. Also, a lot of attention was given in meetings and communications to the vision for the merged entity, its values ("aim high, act now"), and the short- and medium-term priorities. Whilst these values, for instance, were not new and did not differ widely between the merging entities, they were made much more explicit by the new Sector Head, the current CEO of UCB, to the extent that they have now become the explicit standard against which to measure our strategy and behaviors;
- Finally, we made sure the Keppra sales force and Cimzia teams were ring-fenced and remained unaffected by the integration process.

So far, the results of the integration have surpassed any of the original ambitions.

- The integration was fast and high-impact, but fair. As an example of this, the combined US primary care sales force was reduced in headcount by 25 percent, operational, and trained on the new products only three months after the start of the integration. This was achieved by removing the ten percent lowest performers and having all the other reps pass through an internal and external assessment. As a result of the latter, the current share of former Celltech and UCB Pharma reps is the same as what each of the former entities contributed in headcount;
- People feel involved and inspired by what the new UCB Pharma stands for. In a recent survey of the top 170 leaders, 63 percent of people described their engagement in the new entity as outstanding, 33 percent as very good, and four percent as good. Nobody rated it as average or poor;

- Last but not least, top-line and bottom-line results are exceeding expectations, and the extra profits UCB Pharma is making compensate for the profits lost due to the divestment of Surface Specialties.

26.4 Key Learnings

If 75 percent of mergers and acquisitions fail, as literature suggests, UCB has clearly done better than average, albeit the success of the Celltech acquisition will eventually be measured by the success of Cimzia. The key factors contributing to this success are as follows:

- We created value well beyond the pure integration;
- We set high aspirations – a ten percent improvement in return on sales in both the integrations;
- We made sure there was buy-in and commitment from the line for capturing the improvement potential;
- We recognized and dealt with anxiety by rapidly re-designing the organization and populating it, and making the people process the core element of the integration (e.g., senior HR representatives playing a very important role in the integration organization);
- We made sure all key managers and top performers were engaged in the integration – some, in fact, more than they would have liked.

Nobody is ever too old to learn, especially on such a complex matter as an integration of two companies. Dimensions where we could have done better and to which we would clearly pay particular attention next time are:

- *Identifying cultural changes more, and earlier on.* In the integration of Celltech in particular, we fell into the traditional trap of underestimating the differences in a "friendly" transaction. Whilst there was clear support for the transaction from both management teams, this did not take away the personal ambitions and the fundamental cultural differences between a Belgian and a UK company, a traditional pharmaceutical company and a biotech company, etc. – they were to be expected, but were not always planned for;
- *Communicating even more, and more openly.* Despite a lot of good intentions and initiatives, internal communications in particular has been an ongoing challenge throughout both integrations. It can be surprisingly difficult to pass on tough messages, to have information trickle down to all parts of the organization, includ-

ing in remote areas, and to make sure that managers play their role in information cascades or informal staff meetings – execution and having sufficient resources for follow-up are key in this matter;

- *Protecting current business momentum.* As already mentioned, we did particularly well on this in the pharmaceutical integration. In chemicals, however, it was more difficult to ring-fence certain parts of the organization, and the market environment at that time did not help either with rising raw material costs. It is crucial at such moments to make sure you still have the capability to quickly identify the issues through the accounting numbers and take action (hence, it is essential to maintain some form of parallel book-keeping during the transition, which we did not do consistently), whilst still being persistent on the integration.

26.5 Summary

- Over the last two years, UCB has gone through a tremendous transition from a hybrid chemical/pharmaceutical company to a pure bio-pharmaceutical player.
- The way it did this, i.e., by acquiring another chemical player, selling the thus reinforced non-pharmaceutical activities, and then strengthening the pharmaceutical branch by the acquisition of a biotech company, was quite unexpected by some observers, but came as a result of a long-time strategy – fully endorsed by a very engaged Board of Directors – and was highly value-creating.
- A significant share of the value created undoubtedly comes from the mechanistic, almost military, but very people-centric post-merger integration programs.

27
The Chemical Industry and Public Perception

Wilfried Sahm

License to Operate – today, an industry does not only achieve this vital goal because it is large and, therefore, important to a country. It must also be accepted or at least tolerated by as many members of the public as possible. Huge sales and hundreds of thousands of jobs remain strong arguments. For the main opinion-formers in our society, what the chemical industry manufactures and how much thought the industry gives to consumers and the environment count just as much. In other words: corporate and social responsibility are trumps.

Consequently, a modern industry association has two important tasks:

- First, it must try to assert the interests of the industry in contacts with politicians and public authorities – i.e., successful advocacy is called for,
- Second, it has to explain in a straightforward but effective manner that chemistry brings immense benefits to the individual and to the country as a whole.

Verband der Chemischen Industrie e.V. – VCI, the German chemical industry association – represents in Germany the economic policy interests of 1,700 German chemical companies and German subsidiaries of foreign enterprises. Thus, VCI stands for over 90 percent of the entire German chemical industry, which in 2004 employed some 446,000 persons and achieved sales of EUR 141 billion, making it the largest in Europe and the third-largest worldwide, after the USA and Japan.

27.1
Public Perception of the Chemical Industry – its Structure and Significance

Is there any such thing as a public image for the whole chemical industry? Does a positive industry image, assuming it exists, add value to individual companies? Under today's earnings and competitive pressures, the value of public image- and acceptance-building is sometimes questioned. After all, the important automotive, food, electronics, telecommunications, and other industries do not appear to

Value Creation: Strategies for the Chemical Industry. 2nd Edition. F. Budde, U.-H. Felcht, H. Frankemölle (Eds.)

ISBN: 3-527-31266-8

undertake any broad-based, public image-building for the sector as a whole. The chemical industry is quite different.

27.1.1
What Does the Public Associate with the "Chemical Industry"?

The "chemical industry" generally conjures up a small number of key images: research laboratories, chemical formulas, large production plants – but also people wearing protective clothing at scenes of accidents and polluted waters. Medicines, tablets, tinctures, and ointments are a few examples of clearly positive pictures (Fig. 27.1).

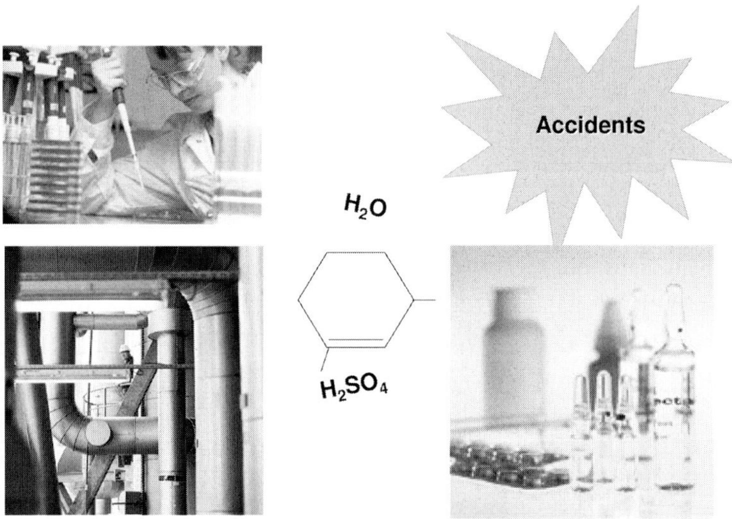

Source: Emnid survey commissioned by the "Chemie im dialog" ("The Chemical Industry in Dialog") initiative
May 2002, n = 1,570

Fig. 27.1 Key images in the perception of the chemical industry.

However, the industry's public image does not even come close to reflecting the great diversity of products and the many, mainly small and medium-sized, firms it contains:

- Around 50 percent of the German population is most likely to connect the "chemical industry" spontaneously with medicines. Only around ten percent think of textiles, cosmetics, toiletries, or insulating materials in this context. A good half only associates one or two product areas with it, or simply names "chemicals" in general.
- Partly because of globalization-driven restructuring, the public's spontaneous awareness of chemical firms is focused on only a

very few industrial heavyweights, some of which are no longer even in business.

Unlike end-consumer-oriented sectors, the chemical industry is, therefore, something of an *anonymous entity* to the man or woman in the street. The vast majority of the population has no grasp of the everyday benefits of most chemical products, because they are used as inputs for the manufacture and quality enhancement of cellphones, CDs, flat screens, or special plastics used in cars etc., and are therefore no longer recognizable to the everyday consumer as "chemicals". Against this background, the chemical industry is particularly susceptible to skeptical evaluations or even fears of "inescapable" risks. It is this fact that makes a good image a particular success factor for the chemical industry.

27.1.2 What Is the Value of a Good Image for the Chemical Industry?

Important decisions and behaviors in our market and political environment depend on the reputation of the industry rather than of individual firms or products: for example, in consumer markets, where values such as "ecological", "biological", and "chemical" lead underlying spending patterns from production to the grocery store shelves; in the market for new managerial talent, where graduates from disciplines not specific to chemistry deselect lackluster industries as potential employers (as borne out by European Graduate Surveys); and in regulatory initiatives in the political arena, where the backing of public opinion acts as a catalyst.

The sector's image impacts on market performance, recruitment costs, expenditure on the management of political issues – and, consequently, is hugely significant for the chemical industry.

27.1.3 How is the Image of the Chemical Industry Formed?

As a result of around 30 years of survey-based experience (the first major studies were conducted at the beginning of the 1970s), we in the VCI know that the reputation of the industry is influenced by many factors, which are only partly under our control.

The foundations of an image based on acceptance and trust are shaped and built over long years of experience. Factors determining fundamental attitudes can be described as follows: the influences of our social environment play an important role from our earliest infancy. For example, how is science regarded by parents and in the home generally? Are children enabled to absorb an optimistic view of the future and of technology?

How are these themes handled in kindergarten and school? Friends and acquaintances can also awaken an interest in science, including, of course, chemistry. Later on, superordinate social developments mould these attitudes further over the course of time, culminating in a collective consciousness: for example, a sense of the major economic importance of the chemical industry (with regard to business development and job creation, for example), memories of key events such as Seveso and Bhopal, or the knowledge of what the chemical industry actually does (both benefits and risks) and how it does it (responsibility). As a rule, these foundations are only changed gradually and within relatively narrow limits, or as a result of dramatic, far-reaching events, such as the fall of the Berlin Wall in Germany or a relative's or friend's life being saved by a medicine manufactured through genetic technology.

Attitudes towards the chemical industry then emerge on the surface of these foundations, and general opinions crystallize out. At times, these react very sensitively to developments within the industry and our industry's public presence, but also to events and changes in mood in the general economic and socio-political environment.

We can only increase the image value of the industry if we are aware of these paradigms and apply our levers at a point where we can really make a difference – with realistic objectives, continuity, consistency in word and deed, and with concrete, easily understood examples of the benefits of the industry. At the same time, we have to be very flexible in positioning our achievements and making our opinions felt effectively in every situation in the changing economic and socio-political environment.

27.2
Image and Acceptance in Germany

According to a representative survey carried out in November 2004, more than 60 percent of the total German population have a generally positive opinion of the chemical industry (Fig. 27.2). Among decision-makers in business and administration, this figure surges to 75 percent, and among better-educated young people – the decision-makers of the future – it is 65 percent. This places chemicals above the average in a cross-industry comparison, particularly among the core target group of decision-makers and opinion-formers.

Among the general public and better-educated young people, the chemical industry hits the average of all sectors covered in the survey, in a densely-populated midfield. The share of people who do not have a single good word to say for the chemical industry is a maximum of thirteen percent. There is therefore no reason for those involved in the industry to say that it has a generally poor image. After 20 years of sometimes fierce discussions about the safety of plants and environmental protection, this view seems so firmly established that many members of the chemical community do not yet fully realize that the trend changed some years ago.

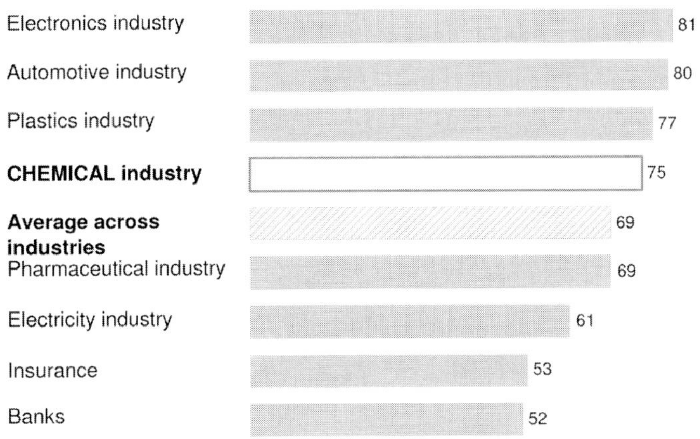

Source: Emnid survey commissioned by the "Chemie im Dialog" ("The Chemical Industry in Dialog") initiative
November 2004, n = 374 decision-makers in business and administration

Fig. 27.2 Generally positive opinion compared to other industries (decision-makers).

27.2.1 Credibility in Public Debate

Currently, 42 percent of the general public, 37 percent of better-educated young people, and 52 percent of decision-makers regard the chemical industry as having high or fair credibility with regard to current subjects of public debate (Fig. 27.3).

Environmental movements such as B.U.N.D. (the German branch of Friends of the Earth) and Greenpeace lead the credibility rankings, on account of their perceived commitment to public welfare. However, our credibility rating is extremely strong compared with other industries, the Federation of German Industries (BDI) or the German Confederation of Trade Unions (DGB), not to mention the Federal Government and Opposition – particularly among the opinion-forming core target group of decision-makers.

Fewer than ten percent of respondents attribute absolutely no credibility to the sector. That means the chemical industry has a better credibility base for its public statements than it is inclined to think – provided the arguments are clear and acceptable.

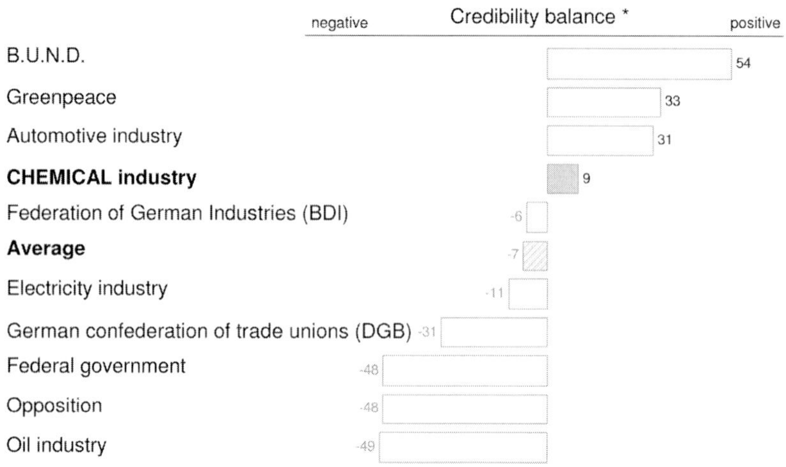

Source: Emnid survey commissioned by the "Chemie im Dialog" ("The Chemical Industry in Dialog") initiative
November 2004, n = 374 decision-makers in business and administration
* Credibility balance = % very/quite credible - % not very credible/not credible at all

Fig. 27.3 Credibility in public debate compared with other industries and organizations (decision-makers).

27.2.2
The Chemical Industry Viewed as an Economic Factor

The vast majority of Germans agree that the importance of our sector as an economic factor is indisputable: eight out of ten respondents, including those in the group of better-educated young people, believe that the chemical industry is important to Germany's economic strength and prosperity (Fig. 27.4). As many as 91 percent of decision-makers agree with this statement.

This attitude remains relatively stable across all levels of the population, even in times of high-profile problems. Ultimately, this is an expression of the chemical industry's traditionally very high significance to the German economy as a whole, and a consequence of the way it is reported in the media.

One image asset that has remained relatively constant, but could still do with some reinforcement, is the fact that the chemical industry is regarded as driving developments in other sectors due to its association with "research" and "basic products". However, respondents have only a vague knowledge of the type of products and services which bring advantages not only to other industries, but also to the consumer.

On the other hand, the public is more guarded in its opinions of other aspects of the economic image factor, partly due to events in the industry itself (for example, restructuring and jobs losses) or to general developments on the employment and capital market such as stock market slumps.

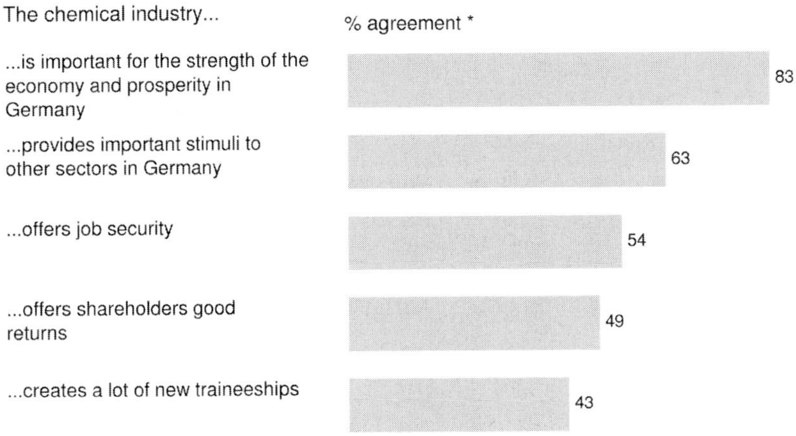

Source: Emnid survey commissioned by the "Chemie im Dialog" ("The Chemical Industry in Dialog") initiative
November 2004, n = 831 respondents, representative of the population

* % of responses ranging between 5, 6, and 7, on a scale of 1= totally disagree, to 7 = entirely agree

Fig. 27.4 Image of the chemical industry as an economic factor (general public).

27.2.3 Value and Responsibility Criteria

A robust climate of opinion and acceptance depends on people recognizing not only the macroeconomic value of the industry, but also the individual, personal benefits it brings to everyday life and the responsibility of its behavior (Fig 27.5).

The vast majority of the population – two thirds or more – are basically aware of the chemical industry's value to people and their quality of life. In the opinion-forming core target group of decision-makers, agreement is around ten percentage points higher.

Public opinion is more critical and controversial about aspects of our responsibility in terms of product and plant safety and, increasingly, social responsibility, in particular with regard to employees. However, only a maximum of 15 percent of the population categorically deny that the industry is aware of its responsibilities, though the figure goes up in times of high-profile problems.

We should, therefore, not allow ourselves to be thrown off track by a few loud voices uttering general fears of chemical products and chemical production. Nevertheless, particularly on such subjects as product and plant safety, we must make planned and realized improvements clear to the public – and we do so in our very detailed annual Responsible Care Report – to further increase their understanding and acceptance of the realities of our industry.

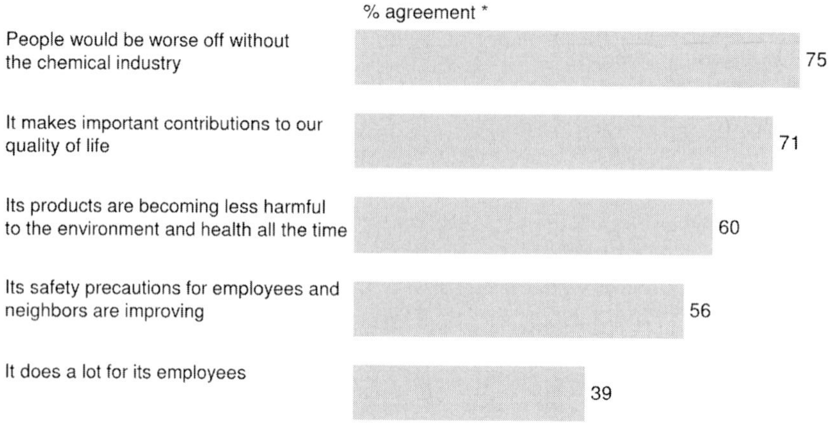

Source: Emnid survey commissioned by the "Chemie im Dialog" ("The Chemical Industry in Dialog") initiative
November 2004, n = 831 respondents, representative of the population

* % of responses ranging between 5, 6, and 7, on a scale of 1 = totally disagree, to 7 = entirely agree

Fig. 27.5 Image based on criteria of individual benefits and responsibility (general public).

27.2.4
Acceptance: Yes – But With Strict Controls

The climate of acceptance follows from the perception and image aspects discussed above, and directly defines the political and economic frameworks for the industry's success (Fig. 27.6).

Between 64 percent of the general public and 76 percent of the opinion-forming core target group of decision-makers accept the chemical industry – despite some reservations and objections. At the same time, however, 62 percent of decision-makers and 73 percent of the general public advocate stricter controls.

In-depth research into the results of the VCI surveys tells us that the call for tighter controls is an unconscious "precautionary reflex" resulting from the generally high level of skepticism towards the chemical sector. There are ways to soften the blow, however: for example, the reaction is lessened if the consequences of increased regulation are made clear. In addition, the underlying awareness of the sector's value and the progress that has been made on responsible behavior can be awakened.

In the present troubled economic and social climate, acceptance of the chemical industry is mainly founded on the simple fact that it is seen as an indispensable part of the economy, rather than on the – more robust – basis of a deep understanding of and trust in the industry. The existing readiness to accept the industry therefore has to be supported by entrepreneurial behavior and open communication, to tone down the call for tighter controls.

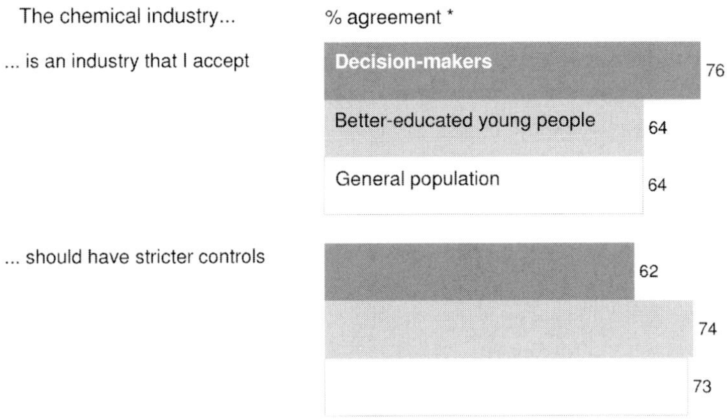

Source: Emnid survey commissioned by the "Chemie im Dialog" ("The Chemical Industry in Dialog") initiative
November 2004, n = 819 population, n = 372 better-educated young people, n = 374 decision-makers in business and administration

* % of responses ranging between 5, 6, and 7, on a scale of 1= totally disagree, to 7= entirely agree

Fig. 27.6 Acceptance of the industry.

We must, however, be careful not to pursue utopian visions of harmony or ambitious desires to be trusted. Comparisons with other industries have shown that the highest realistic level for "trustworthiness" is around 60 percent, and the realistic minimum when stricter controls are being called for is around 40 percent. These figures ultimately represent the bounds of the possible in a pluralist society with its inherent conflicts of interest. Against this background, we have to find a credible way of communicating the fact that companies need to make profits and, at the same time, that they do accept social responsibility.

27.2.5
Insights from the Long-term Trend

With many years of representative surveys behind it, the VCI can now draw on an experience base of decisive conclusions and insights. These make an important contribution to the organization's efforts to build public recognition and acceptance.

Despite highs and lows of varying degrees over the last 20 years, the climate of opinion on the chemical industry has shown a positive basic trend, ultimately heading towards the 60 percent mark (Fig. 27.7). No reason, therefore, to think that it has a poor overall image: the chemical industry can face the world confidently with its achievements and legitimate interests.

In the past, we have achieved particularly positive assessments above the basic trend, sometimes as high as 75 percent, during phases in which the chemical industry gives no cause for concern and, in some cases, the general public mood favors an upturn – because other actors are in the spotlight of public criticism.

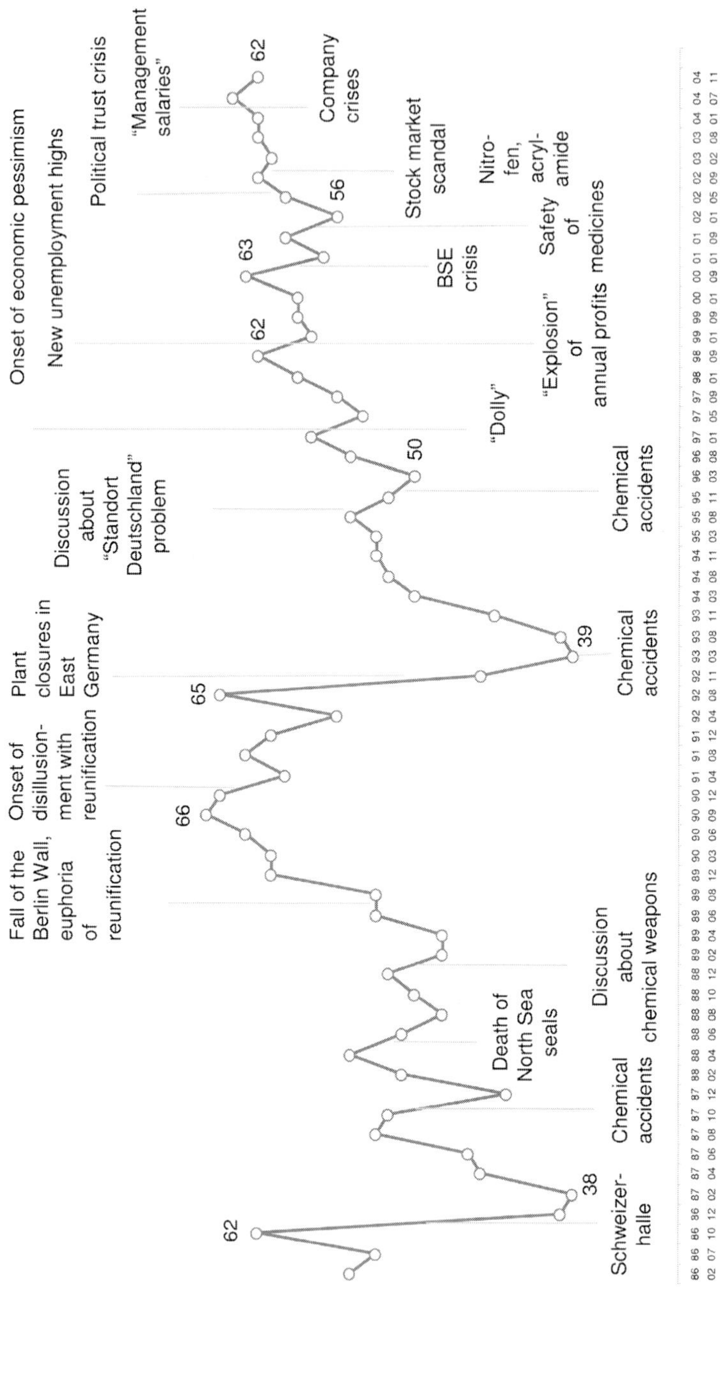

Source: Emnid survey commissioned by the "Chemie im Dialog" ("The Chemical Industry in Dialog") initiative
General assessment of the chemical industry - % predominantly positive from February 1986 to November 2004

Fig. 27.7 Long-term image development and influencing factors.

The biggest slumps – though the industry has never dropped lower than 38 percent – have occurred when problems arose in the industry (e.g., accidents) or in connection with products; that is to say, problems originating in the operational reality of the industry itself. These situations are catalyzed by intensive reporting in the media. A striking feature is that dips due to industry-specific problems have become smaller and less protracted over time, and the subsequent return to public favor has been faster.

This long-term image dynamic shows that most of the German public have a generally favorable and increasingly robust, positive basic attitude towards the chemical industry. However, this success story – a stable, positive basic trend – did not come about by chance. It was achieved by the timely, continuous, and successful implementation by the chemical industry of a wide range of measures to protect people and the environment. We have not only acted on that level, however, but also made a concerted effort to communicate our successes to our target groups and to the media.

27.3 Creating Awareness and Acceptance through Communication

As early as 1979, in response to the acceptance crisis following the industrial accident at Seveso, the German chemical industry set up an action group to represent the industry to the public. This group launched a multitude of measures such as joint advertising, political campaigns on current industry issues, target-group-specific dialog campaigns, and full-scale organized events.

27.3.1 Joint Advertising and Political Campaigns

The initiative launched the first joint advertising campaign "Chemie ist wenn ..." ("Chemistry is ...") to enhance public awareness of the ways in which chemistry adds value to daily life. Print adverts followed in the mid 1980s, then television commercials on the VCI environmental guidelines adopted in 1986. The intention was to demonstrate specific improvements in environmental problem areas, an increasingly hot topic at the time. The situation has changed dramatically since then. Economic and social challenges are much more pressing, there is a much broader range of important decision-making and opinion-leading target groups in politics and society, and the media have achieved a leap in status as a forum for political direction-setting and decision-making.

Our media advertising campaigns therefore now take a two-pronged approach. The new image campaign "Chemie. Element unseres Lebens" ("Chemistry. Elements of life") illustrates the everyday benefits of chemical products in examples such as computer technology or fuel cells, to put the spotlight on our anonymous up-front inputs into other industries' highly accepted end products, while our political campaign demonstrates to the interested general public that we are com-

Chemistry in Dialogue

Unbalanced EU proposals for chemicals legislation

Can we afford even less growth?

The German chemical industry wants efficient and practicable rules for the production and use of chemicals throughout the EU. The proposal for a Regulation developed by the Environment and Enterprise Directorates General of the EU-Commission does not meet these criteria.

The proposed provisions are excessively bureaucratic, remote from practice and too expensive. In particular small and medium-sized enterprises will be unable to fulfil such requirements.

These provisions do not only affect the chemical industry but also all downstream users, e.g. the automotive, mechanical engineering, electronics, textile and construction industries. The impacts on the overall economy would be devastating. French and German studies do confirm this.

An implementation of the proposals would result in a situation where many chemicals can be no longer manufactured and processed in Europe. Millions of jobs are at stake in the European Union.

The German chemical industry has made numerous proposals to improve the current system. We need practicable rules suitable for daily business and we are willing to cooperate to achieve such results.

The German Chemical Industry

For more information about the subject "Chemicals Policy" write to
Dr Gerd Romanowski, Verband der Chemischen Industrie e.V. (VCI),
Karlstrasse 21, 60329 Frankfurt am Main, Germany, fax +49 69 2556 - 1612.
Our full position is available in the internet: www.vci.de

VCI

Fig. 27.8 Examples from the political and current image campaigns.

mitted to finding sensible solutions for current political issues relating to the chemical industry (Fig. 27.8). Here, we use the weight of published opinion to support our lobbying of decision-makers and opinion-formers.

27.3.2 Dialog Programs and Events

Sustainable improvements in acceptance and trust depend on direct communication to reinforce the beginnings made by the image and political campaigns, and to open up dialog on controversial issues.

Under the title "Chemie im Dialog" ("The Chemical Industry in Dialog"), we incorporated these components into our communication strategy back in the early 1990s. The measures taken cast their net wide, and include:

- A nationwide theater tour of around 280 schools in 1989/90, with a play addressing current environmental subjects and followed by discussion
- Five nationwide open days since 1990, each with over 400,000 visitors at up to 250 locations
- Eight discussion events with representatives of the political, industrial, and societal spheres in respect of the EU chemicals policy
- More than 20 conferences and events for teachers to address current research and teaching topics relevant to chemistry, with almost 8,000 participants
- The "ChemiDrom" at the EXPO 2000 in Hanover, a multi-media presentation on chemistry which attracted 1.35 million visitors
- Sponsorship and a media program at the 36th International Chemistry Olympiad 2004 in Kiel.

All these projects have resulted in a wealth of associated activities at the regional level by the VCI regional associations and at the local level by individual firms.

The crucial factor for successful image-building in Germany lies in ensuring that the media campaigns, dialog programs, and events for target groups are integrated consistently and supported, continued, and thereby strengthened by the broad base of companies throughout the country.

27.4 Current Challenges

For the VCI, the issue of public perception is mainly focused on the image of the chemical industry in Germany. However, since Germany is the third largest chemical-producing nation in the world and exports over half its output in terms of value, and since German subsidiaries abroad manufacture products to a further value of EUR 125 billion, the reputation of the chemical industry in its main markets is also important.

It is also essential for German chemical companies to hold top positions in new technologies if they want to successfully face the challenges of the future. The acceptance needed for this new road is best achieved by giving a fair description of benefits and risks.

27.4.1
The Industry's Image in an International Comparison

Compared with other major European countries and the USA, the chemical industry in Germany has the best image. This is verified by the figures of the Pan-European Surveys carried out by the European Chemical Industry Council (CEFIC) since 1994 (Fig. 27.9).

The climate of opinion in Germany is also a positive factor in the country's attractiveness as a location for industry. However, the picture is not so rosy for our internationally operating member firms: the image and acceptance gap between countries has actually widened rather than narrowing over the course of the last few years. The VCI is thus faced with the obligation and the challenge of making its know-how and wide variety of communication activities available to other countries in Europe and further afield. Many of our measures are not directly transferable, but need to be adapted for local use. One specific aim in the development of our recent new image campaign was, however, to make sure that it could potentially be used in other European countries. Some neighboring countries have in fact already expressed an interest in it, while CEFIC has adopted the campaign's

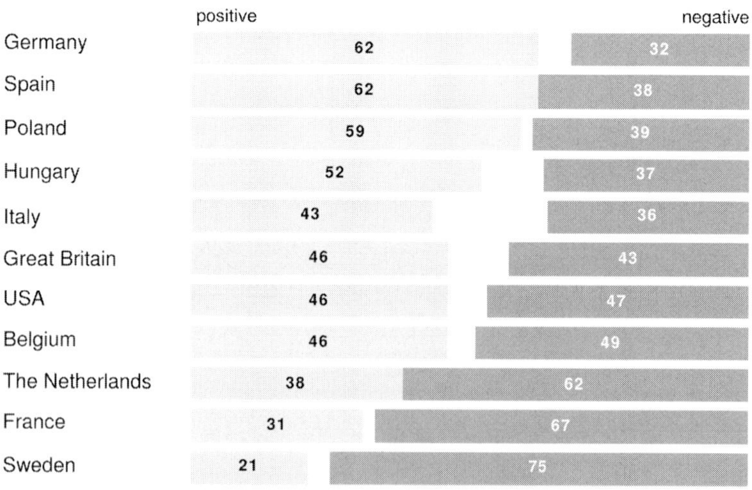

Sources: CEFIC Pan-European Survey Jan-Feb 2004,
Results for Hungary and Poland: APME Pan-European Survey May 2004,
Result for USA: Ri*QUESTA, Int'l Environmental Monitor June 2002

Fig. 27.9 Image of the industry in an international comparison.

website. These may be small steps, but we are heading in the right direction. After all, now more than ever, the whole of Europe is our home market.

27.4.2
Shaping the Future with New Technologies

New technologies can only be effective if they are accepted by society. Ideological barriers cost us a lot of time as well as market opportunities in bio- and gene technology in the 1980s and 1990s. We cannot let that happen again – in the case of nanomaterials, for example. German firms are international leaders in this field. If we want these technologies to prosper in Germany, we need to highlight the specific performance of the individual applications clearly and in a way that consumers can believe and accept, so that the public will realize and understand the benefits. Against this background, potential risks will be put in their true perspective and can be discussed properly.

Here, too, the "Chemistry. Elements of life" image campaign and the political campaign, as well as the dialogue and events programs, are important tools in successfully informing and convincing the public.

27.5
Summary

- The image of the chemical industry in society leads to consequences and reactions that have a very direct impact on its scope for action and chances of success. A good image for the chemical industry is, consequently, an important factor for any country in attracting industry.
- The German chemical industry recognized this early on. By steadily improving production and products, behaving responsibly towards people and the environment, and making continuous efforts to communicate it has achieved a respectable image position which actually makes it the international leader in this area.
- Knowledge of the ways in which public perception develops and more than 25 years of systematic, concerted reputation management form a sound basis for overcoming future challenges in a changing environment.

28
Industrial Biotech: From Promise to Profit

Rolf Bachmann and Jens Riese

Industrial ("white") biotechnology is already used to produce five percent of all chemicals today, and this share is expected to grow rapidly. Attracted by this promise, chemical companies have invested several billion dollars in exploring biotech opportunities over the past few years. However, the path from promise to profits has turned out to be long and arduous. Challenges in this new field are plentiful, from identifying the right opportunities to launching bio-based products and managing public acceptance. Due to the novelty of biotechnology and the uncertainties involved, achieving the best results depends on combining visionary thinking with critical and fact-based assessment, business judgment with advanced risk management, and internal expertise with challenging external views.

28.1
Time to Exploit the Potential

Following the inroads made by 'red' pharmaceutical and 'green' agricultural biotechnology, a third wave is beginning to spread: 'white' or industrial biotechnology, which uses renewable materials and copies tried-and-tested natural processes to produce industrial goods. This often saves resources and also enables companies to manufacture new, innovative products or existing products more effectively and efficiently than with conventional processes. Biotechnology lays a foundation for sustainable development reconciling social, ecological, and economic concerns.

The potential for industrial biotech has been broadly recognized and chemical and biotech companies are starting to move into this space and grow their presence. In fact, about five percent of the estimated USD 1.2 trillion total chemical sales already depend on biotech. The global market for bio-based ethanol alone is worth USD 15 billion; other basic organic molecules such as citric acid (USD two billion) and lactic acid are produced by fermentation, and so are all but three amino acids (approx. USD four billion); various basic, advanced and active pharmaceutical ingredients produced by the fine chemical industry are worth USD 7.5 billion; the attractive enzyme market has reached USD two billion in

Value Creation: Strategies for the Chemical Industry. 2nd Edition. F. Budde, U.-H. Felcht, H. Frankemölle (Eds.)

ISBN: 3-527-31266-8

sales and is growing by more than five percent per year, and specialty chemicals for flavors, fragrances, and other applications add several more billion dollars.

Biotech is changing industrial production in three specific ways. First, sugars, vegetable oils, and waste biomass are replacing fossil fuel feedstock (oil and natural gas). Second, bioprocesses such as fermentation and biocatalysis are replacing chemical syntheses; plant- and animal-based production may be added in the future. Last, new bioproducts are emerging including bio-based polymers, enzymes for use in textiles or feed, and innovative nutritional ingredients. Most remarkably, the combination of bio-based feedstock, bioprocesses, and new products could revolutionize the chemical industry's structures. In ten years' time, an integrated biorefinery could use row crops, energy crops, and agricultural waste as inputs to extract oil and starch for food, protein for feed, lignin for combustion, and cellulose for conversion into fermentable sugars and other byproducts. The sugars will be used to ferment not only ethanol for transportation fuel, but also a whole set of commodity and specialty chemicals and new biomaterials.

The results of a major market study suggest that biotech could affect at least ten percent of chemical sales by 2010, double today's figure (Fig. 28.1). Prospects are looking brighter than ever before thanks to three key drivers: advances in technology, environmental and economic benefits, and the need for innovation in the chemical industry.

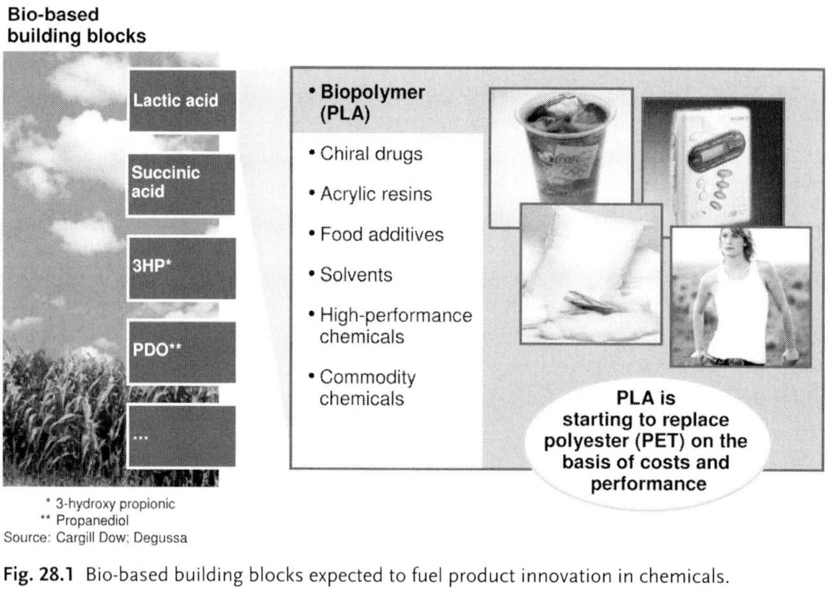

Fig. 28.1 Bio-based building blocks expected to fuel product innovation in chemicals.

28.1.1
Better Technology, Faster Results

A broad spectrum of enzymes and fermentation systems has already become available to the industry, and the number is increasing all the time. Biotech development has also picked up speed: it can now take a matter of weeks rather than years to develop new, highly specific and efficient enzymes. Until recently, slow development hindered the use of enzymes in pharmaceutical production. Now DSM's Pharmaceutical Product Unit, for example, is exploiting them systematically as a competitive advantage. Enzymes are also becoming more resistant to harsh environments such as heat and acidity, and are cheaper to produce, making inroads into other industrial production processes such as pulp and paper, oil exploration, and textile processing.

28.1.2
Environmentally and Balance-sheet Friendly

The increased pressure for sustainable production is also helping to spur the industry's prospects. Two reports – one by the OECD, the other by a consortium of companies, industry associations, the Öko-Institut, and McKinsey – clearly demonstrated that industrial biotech can help create jobs, boost profits, and benefit the environment.

Several studies of the environmental impact of replacing chemical synthesis with biotech routes have also demonstrated the benefits of industrial biotech. It is estimated that increasing energy efficiency by replacing fossil fuels could achieve seven percent of the Kyoto target on emissions (based on the case studies in the reports mentioned above, and ten percent of chemical product sales relying on biotech). When 20 percent of chemical products are produced using biotechnology, the contribution will increase from seven to 20 percent.

The economic benefits alone are also driving the adoption of biotechnology. BASF has reduced the production process for Vitamin B2 from eight steps to one through biotechnology, while DSM's bioroute for Cephalexin has also substantially reduced the number of process steps. These examples and those of dozens of pharmaceutical intermediates demonstrate that cost savings of 50 percent and more are not unlikely. The savings may come directly from lower variable costs, but also from reduced capital expenditures for simpler production assets, or from reduced scale and therefore lower risk, transportation costs, and/or overcapacity.

28.1.3
Rekindling Innovation

Finally, interest in biotech has increased recently thanks to its role in product innovation. At a time of increasing competition from Asia in established products and the subsequent commoditization and strong price decline, chemical companies are once again looking to innovation as a key source of differentiation. At the

2004 World Economic Forum in Davos, the leaders of the world's largest chemical companies placed biotechnology among the key drivers of change for the decade ahead.

The importance of stimulating innovation can be seen by looking at the introduction of new polymers. Over the course of the 20th century, the development of fossil fuel-based polymers increased steadily up to the post-war period, stimulated by the abundance and low cost of basic petrochemicals. However, it has declined dramatically since 1960. Innovation in the traditional polymer industry today is mainly related to the application and blending of existing polymers.

Just as low-cost petrochemical building blocks became available with the introduction of crackers in the 1930s, new bio-based ones are emerging today (Fig. 28.1). These include lactic acid, which can be polymerized to the biopolymer PLA. PLA has started to replace polyester thanks to its competitive cost and new applications. Lactic acid can also be processed into chiral drugs, acrylic acid, propylene glycol, food additives, and more. Other examples of innovation abound: Cargill is exploring the potential of 3-hydroxy proprionic acid; BASF is looking into new chemistry around the simple organic molecule succinic acid; and DuPont is to use cheap PDO as a monomer for its Sorona® polymer. Overall, a green chemistry is emerging that complements the traditional product trees and gives the industry more innovation headroom.

28.1.4
Increasing Corporate Action in all Segments

All these economic and environmental factors are encouraging companies in all major chemical market segments to make more definite moves. In fine chemicals, around 40 percent of all life science and nutritional products at DSM are based on biotech – a volume which is worth almost EUR 1.5 billion. BASF has committed more than EUR 500 million to exploring the potential of plant-based biotechnology. DuPont has invested more than USD 500 million in the development of Sorona®, and Cargill Dow's investments for PLA are of a similar scale. Ciba Specialty Chemicals has introduced an enzymatic process for acrylic acid, BP is exploring industrial biotech for oil exploration and basic chemicals, Degussa has recently opened its second project house dedicated to biotechnology, Givaudan is using biotech extensively for new flavors and aromas, and Novozymes has an annual R&D budget of more than EUR 100 million for new and enhanced enzymes. Finally, dedicated industrial biotechnology companies such as Codexis, Diversa, and Genencor have established themselves in the market, and have encouraged a new generation of startups to emerge.

28.2
Waste Biomass – a Feedstock with Mass Appeal

Early introductions of bio-based products have shown that few customers are willing to pay a high 'green' premium. For broad-based adoption, new products must be competitive with existing offerings. In this context, the use of alternative low-cost feedstock could give industrial biotech another boost.

28.2.1
How Waste Biomass Works

The most promising alternative feedstock is waste biomass, which comes primarily from agricultural sources such as straw and corn stover. Furthermore, dedicated energy crops such as switchgrass, grown in marginal areas that are not suitable for intensive farming, will open up additional sources of cheap biomass. These materials are abundant, cheap, and largely serve no other purpose. They are also obviously renewable and contain three useful raw materials:
- Cellulose and hemicellulose, which can be turned into sugars
- Proteins that can be used in animal feed and for industrial products such as hydrosylates
- Lignin that may be used as a combustible fuel: indeed, it can power the very biorefineries that process biomass, making them virtually self-sufficient.

In the longer term, biomass crops can also be genetically modified to serve as carriers for other bio-products, such as biopolymers.

Bioethanol is one of the first and the largest markets to profit from cheap biomass feedstock. Ethanol is usually produced from dextrose, which in the USA tends to derive from corn. The first ethanol biorefinery based on waste biomass is already online. It is a Canadian venture operated by Iogen and receiving investment from Shell, Petro Canada, and the Canadian government. With an annual capacity of 700,000 liters it is semi-commercial in scale and not cost-competitive with conventional ethanol refineries. However, the technology is expected to improve quickly.

28.2.2
Economic Benefits and Regulation

Biomass-based biorefineries have the potential to reduce sugar costs from about eight to nine ¢/lb in the USA today, to around four ¢/lb in three to four years, and lower still as the biorefineries get more integrated. The net cost of producing sugar in an integrated biorefinery may even hit zero, as byproducts such as lignin, proteins, and genetically introduced products generate value. The price of conventionally produced sugar will also fall, since genetic modification allows a variety of productivity improvements.

The technology for the cost-competitive, large-scale application of ethanol could be ready in the next few years and the value creation potential in shifting from corn to biomass is estimated to reach USD ten to 16 billion by 2010.[1)]

Legislation is in place to support production of fuel ethanol in Canada and Brazil, and the EU has committed to ensuring that 5.75 percent of all transportation fuel in the EU will be bio-based by the year 2010 – a target that cannot be achieved by biodiesel alone. Several European companies, such as Abengoa and Südzucker, have therefore announced aggressive plans for expansion in ethanol in the coming years.

Estimates for the US market suggest that 15 billion gallons (57 billion liters) of bioethanol will be produced from a combination of biomass and corn-based feedstock by 2020; and the cost of ethanol could drop from around USD 1.20/gallon (32 ¢/liter) today to just 40 ¢/gallon (11 ¢/liter) by 2010. This is substantially cheaper than gasoline from oil, which costs about 70 ¢/gallon at an oil price of USD 25/barrel. Experts estimate that the USA could derive up to 40 to 50 percent of its transportation fuel from biomass if energy crop farming in marginal areas provides additional sources of biomass.

28.2.3 Further Management Action Needed

To make biomass conversion and biorefining happen, all the players in the value chain have to contribute to improving efficiency and ensuring that the right investments are made.

All steps of the process need to undergo major efficiency improvements. The cost of enzymes, for example, needs to fall by 90 percent from its 2003 level. This might sound unrealistic, but it is not unusual to increase the effectiveness of enzyme production by factors of ten, 100, or even 1000. By modifying the amino acid sequences of the cellulase and hemicellulase enzymes, biochemists from Genencor and Novozymes have been able to make them dramatically more effective, and recent advances suggest that the cost target for the enzymes will be exceeded in due course.

The US Department of Energy predicts that two billion gallons (7.5 billion liters) of ethanol will be derived from biomass by 2010. To make this possible, however, companies have to invest. Despite major government support, especially in the USA, no major companies have thus far aggressively pursued this. The competitive threat is still perceived as low, and companies are afraid to risk being first movers. Only the technology companies, in particular Novozymes and Genencor as well as Spanish pioneer Abengoa, have publicly announced aggressive growth plans and commitments to biomass-based ethanol.

1) Derived from industry estimates for the increasing use of biomass as feedstock. After calculating marginal and overall cost savings, we multiplied the result by typical profit multiples for the industry. We assumed that the price for ethanol remains stable. The resulting change in market capitalization is the value creation.

The need to invest goes all the way down the value chain. Farmers need the right sort of equipment and storage facilities for corn stover, pre-treatment needs investment, and so on. No-one owns the full value chain, so collaborating with partners is essential. Today, the picture shows multiple players involved in diverse activities with very little explicit collaboration and without enough investment to launch a new value chain. Ultimately, this should shift to a structured network, in which multiple players from different parts of the value chain cooperate closely in a process that includes collecting, storing, and treating biomass, as well as constructing biorefineries. Possible governance and interface inefficiencies would have to be carefully managed here.

28.3 Turning the Promise into Profit

The path from awareness to profit for companies engaging in biotech is long and arduous. Most have already taken the first step and are aware of the chances and risks. Some have even committed to investments, but few have defined a proper biotech strategy. To become biotech players, chemical companies have to climb a long staircase of steps to build a new business, whilst also managing important external challenges (Fig. 28.2).

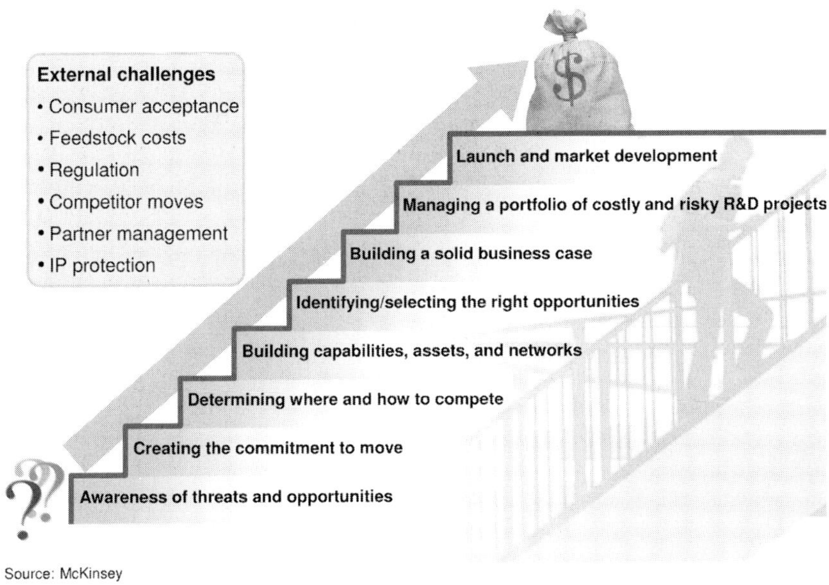

Fig. 28.2 Many challenges on the way to creating value in biotech.

28.3.1 Ascending the Staircase inside the Company

As we have said, most companies are now aware of the economic potential and benefits for stability that lie in biotechnology. Some have already taken the next step, and have committed to action. Investors and senior managers have recently shown that they will invest if presented with good opportunities, particularly for incremental investments. Those high risk/high return areas that require major investments, such as biomass conversion, still face high skepticism, however. Furthermore, very few companies have progressed beyond opportunism: it is important for them to formulate a true biotech strategy, with a clear focus on where and how to compete.

In order to execute a strategy, a company needs the right assets, capabilities, and networks, and no-one has all the requirements in-house. Identifying the gaps and understanding how best to fill them is not easy, and it is important to make partnerships work in practice.

The next step is to identify and select the right opportunities from the wealth of ideas available. This can prove difficult: as mentioned earlier, the list of biotech failures is much longer than the list of successes.

Failures cannot be avoided. The trick – as in the world of venture capital – is to manage the uncertainty and to build a portfolio of good prospects. And, as with venture capital, the starting point is a solid business case for the opportunities. It is important to ensure that the addressable market size and uptake are correctly estimated (by defining them narrowly enough) and that a sufficient budget is allowed for investments, in particular for market development and application development.

There remains the challenge of separating the biotech wheat from the chaff. Companies ultimately have to focus their scarce resources on the few most promising opportunities, and should not wait too long in doing so. In addition, ways must be found to accelerate the whole process: investors are no longer willing to tie their money up in projects that might take a decade to break even. The R&D work in biopolymers, for example, began in the 1980s.

In launch and market development, finally, the art is to focus on the right market segments and customers with the right value proposition, to align the value chain to adopt your innovation, and to use partners appropriately.

28.3.2 Handling External Pressures

Surrounding this set of internal challenges, there are external pressures that also need close attention.

Consumer acceptance is not a big issue yet, but is one to keep an eye on. Supermarket chains in the UK have refused a biopolymer because it was derived from genetically modified plants, despite the fact that it is an eco-friendly material. There is even a discussion as to whether vitamins produced by fermentation should be labeled GM. The environmental NGOs and pressure groups have been relatively quiet on industrial biotech thus far, but if it does come on the agenda it could cause lengthy delays. The industry therefore needs to invest resources in educating people on the benefits.

The cost differential between hydrocarbon and carbohydrate feedstock is clearly a moving target. In a climate of rising oil prices, carbohydrates look more appealing than ever before, but this situation may not last. Companies need to consider potential price changes, and analyze the sensitivity of biotech investment cases to various assumptions on future feedstock costs.

The regulatory situation is liable to change and is also subject to the influence of various interest groups. The proponents of industrial biotechnology, represented for example by the BIO and EuropaBio industry associations, need to maintain or even increase the level of activity on this front, which requires commitment from their members.

Finally, there are some important factors which lie outside the company but are more within the traditional management remit: company strategies and investment decisions also depend partly on competitors' moves, which should be watched and anticipated. Products need to be distinctive; the intellectual property needs to be in place to introduce a new biotech process; and companies should stay alert to opportunities to align interests and join forces.

28.4 Capturing the Value – How it is Done in Practice

A growing number of case studies demonstrate that the challenges can be tackled and the value from industrial biotech successfully captured. Here are some practical examples.

Case 1: Building a biotech strategy

After years of internal discussions and smaller investments, the board of a chemical company decided to take a major step in industrial biotech, which out of a number of fields had emerged as the one with the highest innovation potential and a good fit with the company's broader capabilities and position in the value chain. There were many potential entry points and business opportunities, but none had a clear rationale. The company approached the challenge from two sides (Fig. 28.3). It determined its distinctive skills and assets by benchmarking. At the same time, it assessed the relevant opportunities and threats – competitors' recent or announced moves, and also the potential for competitors if the company chose not to enter a specific area. It was key to think long-term and broad, and to consider the industry structure ten years out.

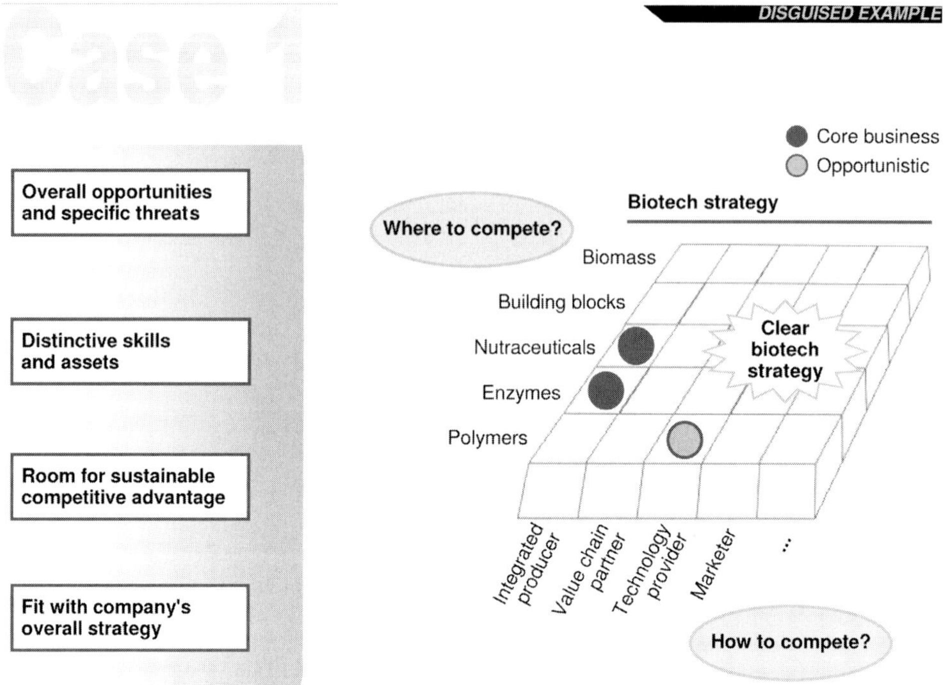

Source: McKinsey

Fig. 28.3 Building the biotech strategy for a chemical company.

This produced a list of strategic options: different business models and value chain positions that would match the most attractive areas. These options were then assessed by a set of criteria that included economic value, feasibility, risk, investments, and fit with the overall strategy and portfolio of initiatives. To test the economic viability of individual products, specific business ideas related to single products/markets were written and pressure-tested. The final decision was made after extensive discussions and more than 80 interviews with experts. While such interviews are essential, it is also important to remain skeptical and avoid the "common wisdom" trap. It is too early to determine the financial success, but the company has already achieved alignment between the overall strategy, the level of investment, and the organizational setup.

Case 2: Identifying the right opportunities

Another chemical company already had a clear biotech strategy in place and had built the capabilities, assets, and networks required to implement it. Execution was already successfully underway in several business units and new bio-based products and processes had started to generate healthy profits. However, the company was now seeking ways to change the old chemical production processes for

DISGUISED EXAMPLE

Case 2

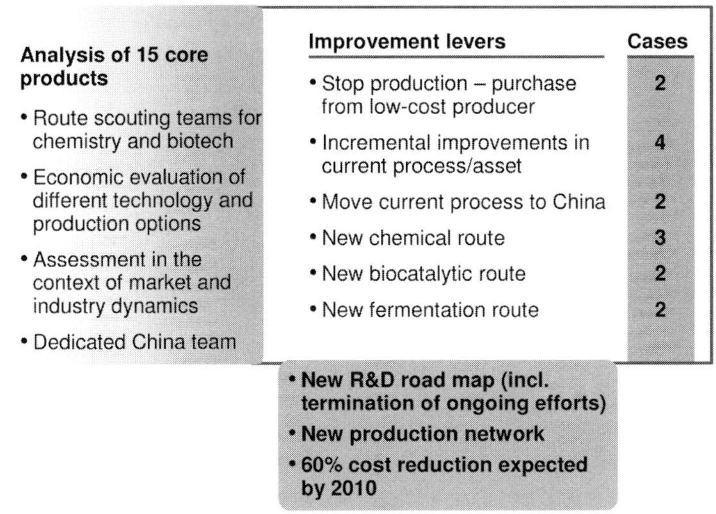

Source: McKinsey

Fig. 28.4 Systematic scan for process innovation opportunities.

its major products to new, more competitive synthesis routes in a recently acquired business (Fig. 28.4).

In the end, it expanded the scope of its project to include a complete review of core product strategies, including a detailed assessment of competitors' cost positions, anticipated moves, market trends, etc. This was important because a new biotech process can easily take five years or more to develop, so it is critical to know whether it will result in a distinctive cost position after that time. Since China was an emerging threat, one team was set up in Shanghai which focused on collecting intelligence on Chinese competitors for several months. Regulations and customer sensitivities also change – would there be a 'bio-based' premium or a 'genetically modified' discount for a product produced by fermentation?

The scope was also extended on the technology side. While one team investigated the potential for new biotech routes, a competing team tried to optimize the existing process, including analyzing different locations, and a third team searched for the best alternative chemical routes. In the end, biotech was just one of the solutions. Each potential solution was assessed; the final one was chosen based on the best risk/reward ratio and backed up by further technical feasibility studies.

Of the 15 products under investigation, biotech was the best solution for five; in four cases a new chemical process was found; incremental improvements to the existing one were made in three further cases; two products were moved to China, and one was outsourced to a low-cost producer. So even though the project was triggered by a desire to see how biotech might affect the company's processes, the outcome was an improvement of every process in a variety of ways. By 2010 it is predicted that costs will have fallen by 60 percent on average for these 15 processes.

Case 3: Preparing the launch and market development

This company was on the very last step of the staircase towards profitability in industrial biotech. It had a strategy, a highly capable organization, a great product, a new production facility, and some funding. However, the potential market applications for its new bio-product were extremely numerous, and it was finding that each market required a different product positioning. The company was looking for a go-to-market strategy for its new blockbuster that it could handle with its limited resources (Fig. 28.5).

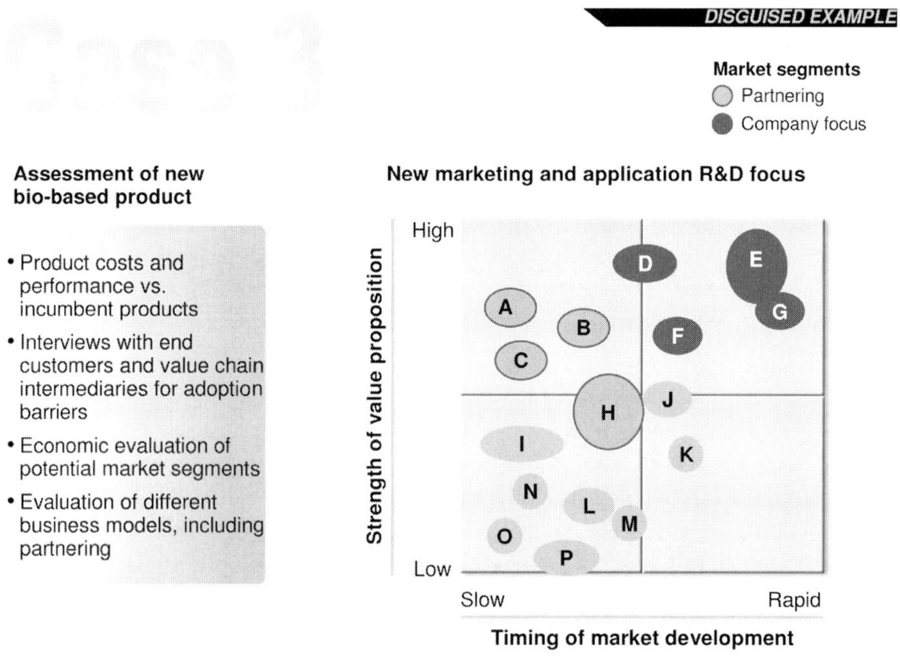

Source: McKinsey

Fig. 28.5 Defining the go-to-market strategy for a new bio-based product.

The full list of potential applications and addressable market segments was assessed against a) the relative strength of the new product's price and performance value proposition compared to existing offerings; b) the size and attractiveness of the addressable market, and c) the ease with which the value could be captured, i.e., the time and effort it would take to develop the product applications and markets and the hurdles for adoption of the product along the value chain. The last point required a lot of attention because in some market segments consumers had already expressed considerable interest in some of the applications and the economics looked attractive. However, interviews with companies along the value chain and supporting analyses showed that intermediaries in the value chain would be very unlikely to adopt it. In other segments, retailers were concerned about the brand risk and were not willing to proceed without a further demonstration of fitness-for-use and safety. These segments were therefore put on the back burner, but will get cooking again once results from other segments support the market case.

In the end, a handful of segments were chosen as top-priority targets for immediate focus, and specific targets, marketing strategies, and implementation plans were put in place. In some other market segments and geographies partnering was the preferred strategy, mostly because partners had better customer access or application technologies than the company. The remaining segments were put on hold.

The company has changed its strategy and its underlying assumptions fundamentally in the course of the project. For example, it has learned that it needs to price the product based on long-term positioning rather than current cost, that it needs to build a plant in a low-cost environment, that value chain intermediaries will need an incentive to adopt its product, and that it is sometimes necessary to work with smaller "attackers" as initial commercialization partners to demonstrate proof of principle, create market pull, and get the "big fish" to adopt a new product. The increased focus, clear product positioning, and partnering has already resulted in a number of major market introductions over the past year.

28.5 Finding the Right Answer

It does not take a miracle to find answers to the problems companies face in biotechnology, but it is by no means easy going. Often, the answers not only go against the company's core beliefs, but also against the experts' common wisdom. Due to the novelty of biotechnology and the uncertainties involved, classical project approaches that rely on fact-based analyses alone fail as much as gut feeling-based management decisions. The best results can be achieved when visionary thinking is combined in a meaningful way with critical and fact-based assessment, business judgment with sophisticated analyses such Monte Carlo simulations, internal expertise with challenging external views. A dedicated effort with senior management attention can be a catalyst in getting to the right answer in time – too many companies continue the same debates over many years, and waste valuable time and resources. This is a time of change and opportunities – companies that "wait and see" run the risk of missing out, pioneers with too much enthusiasm the risk of sinking big sums into the wrong investments. A targeted approach is needed to find the best way for each company to turn the promise of industrial biotechnology into profit.

28.6 Summary

The potential for industrial biotech has been broadly recognized, and chemical and biotech companies are starting to tap into it. However, transforming this potential into economic value still presents major challenges.

- The prospects for industrial biotech are looking brighter than ever before due to advances in technology, environmental and economic benefits, and biotech's role as a driver of innovation. Several chemical companies have already introduced biotech-based processes and products, and almost all are exploring the opportunities.
- The use of biomass as an alternative low-cost feedstock is still in its infancy, but could boost industrial biotech much further, and change the landscape of chemical and fuel production. To make biorefineries work, however, industry players and governments will need to invest and collaborate to create an entirely new value chain.
- Challenges in creating value in this promising but unfamiliar area can be surmounted. Visionary thinking needs to be combined in a meaningful way with critical and fact-based assessment, business judgment with sophisticated analyses, and internal expertise with challenging external views.

29
Industrial Biotech at DSM: From Concept to Customer

Colja Laane and Feike Sijbesma

Industrial (white) biotechnology is rapidly gaining momentum as a cost-effective and environmentally-friendly technology for producing bio-based chemicals, materials, and fuels in a safe and sustainable way from renewable resources. At DSM we are increasingly active in this growing business area. To that end we are optimizing a range of micro-organisms and cell lines for the contained production of a variety of specialties for our customers in the food, feed, pharmaceuticals, and fine chemicals industries. By combining our competences in fermentation, biocatalysis, biotransformation, advanced (gen)omics, and chemistry with our application and formulation skills, we have been able to replace several petro-based processes by bioroutes, and to find unique functionalities in very short development times. Typical innovations include the cost- and eco-efficient production of an antibiotic, the recent development of a peptide-based recovery sport drink, the replacement of an unnatural beer clarifying agent by a novel, cost-effective enzymatic solution, the microbial production of a lipid which is important in infant nerve development, as well as the manufacture of several novel chiral intermediates for the pharmaceutical industry. Promises are becoming reality and turning into profit, not only in the specialties area but increasingly in commodities. DSM, in partnership with relevant public and private stakeholders, will continue to play a leading role in shaping the third – industrial biotech – wave of biotechnology, which in our view will create a more sustainable society for the people of today and the generations of the future: a future where societal (People), environmental (Planet), and economic (Profit) benefits will go hand in hand.

29.1
From Petro to Bio

Over its 100-year history, DSM has transformed itself from a local coal-mining operation, through a predominantly chemical commodity producer, into one of the world's leading specialty companies, employing about 25,000 people and reaching almost EUR eight billion in sales. After the divestments of, among others, the petrochemical activities in 2002 and Bakery Ingredients in 2005, the

Value Creation: Strategies for the Chemical Industry. 2nd Edition. F. Budde, U.-H. Felcht, H. Frankemölle (Eds.)

ISBN: 3-527-31266-8

recent inclusion of DSM Nutritional Products (formerly Roche Vitamins and Fine Chemicals) in 2003, and the recent acquisition of NeoResins, our portfolio today comprises approximately 80 percent specialties. By substantially reducing the cyclic commodity character of our business we have achieved, and continue to strive for, a more stable and growing earning profile. Our life science activities (including nutritional products) have increased massively to today's level of about 50 percent of total turnover. DSM's involvement in life sciences and biotechnology was boosted through the acquisition of Gist-brocades in 1998, and strengthened further by the acquisition of Roche Vitamins & Fine Chemicals. Currently, about 40 percent of our life science/nutritional products for mainly the food, feed, and pharmaceuticals industries are biotech-derived, which is equivalent to almost EUR 1.5 billion in sales.

We realize that innovation will be the key to DSM's future success. At present we spend EUR 200 million on R&D in life sciences and nutrition, which is equivalent to more than four percent of sales. In selected areas, we spend eight percent of sales on R&D and more than 40 percent of the R&D budget on radical innovation. This percentage is bound to increase still further in the future.

29.2
From Principle to Product

Our proactive, customer-focused R&D activities form an integral part of our business process, as we translate our own ideas and developments in the world around us into commercial solutions, while simultaneously building on our expertise and knowledge.

DSM's activities in biotechnology are best placed in the field of industrial, white, or even environmental biotechnology. Industrial biotechnology can be defined as the modern use and application of biotechnology for the sustainable and competitive production of (bio)chemicals, biomaterials, and biofuels from (mainly) renewable resources, using living cells and/or their enzymes (Fig. 29.1). After red (health care) and green (agriculture) biotechnology, white biotechnology is increasingly being seen as the third wave, and as an important enabling technology towards a sustainable future. For a general introduction to industrial biotechnology the reader is referred to Chapter 28.

Industrial biotechnology is mainly based on fermentation technology and biocatalysis. In a contained environment, genetically modified (GM) or non-GM microorganisms (e.g., yeast, fungi, and bacteria), or cell lines of animal or human origin, are cultivated in closed bioreactors to produce a variety of goods. Likewise, enzymes derived from these (micro-)organisms may be applied to catalyze a conversion in order to generate the desired products. (Industrial) biotechnology is not new: its underlying processes have been used by mankind for thousands of years, for example in the production of bread, wine, and cheese. However, activities and opportunities in this field are growing rapidly at present due to recent breakthroughs in genomics, metabolic engineering, bioinformatics, systems biology, di-

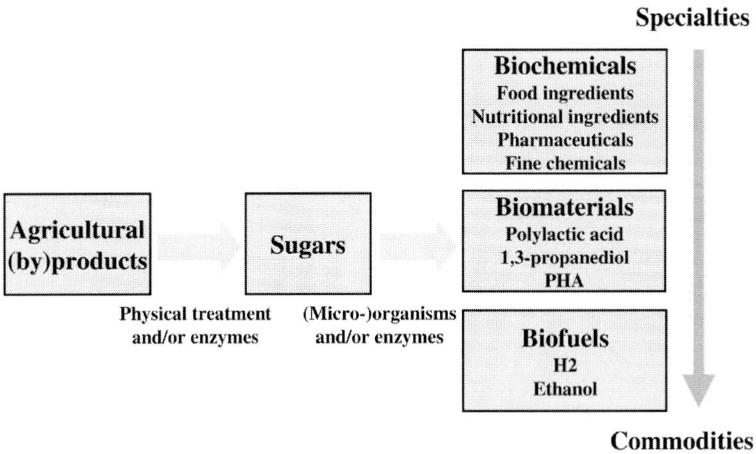

Fig. 29.1 Key steps in the industrial biotechnology value chain.

rected evolution, biocatalysis, and biotransformations, with wide applications in the fine chemical, pharmaceuticals, and food and feed industries. Promises are becoming reality and cells can now be used as tiny micro-factories that can be optimized for productivity, safety, and minimal environmental load. Processes are becoming more cost-effective, and novel products are emerging which could never have been made by any other means. Increasingly, environmental benefits are being realized by switching from petro- to bio-based processes.

Figure 29.1 shows the key steps in the industrial biotech value chain. Raw materials, including crops and organic byproducts from agricultural sources and households, are converted into (mainly) sugars, which can be readily converted by tailor-made (micro-) organisms and/or enzymes into the desired products. DSM is active in the second step of the white biotech value chain and focuses on the conversion of sugars and occasionally other feedstocks into value-added specialties, including a variety of food and nutritional ingredients, pharmaceuticals, and fine chemicals (Table 29.1). We also use enzymes derived from these and other sources to perform desired biocatalytic reactions for the production of a range of (chiral) intermediates.

Table 29.1 Examples of DSM's tool and product box. The (micro-) organisms of which we know the genome sequence are marked with an asterisk.

(Micro)organism / enzyme	Product (ingredient) example
	Food, feed, and nutritional ingredients
*Aspergillus niger**	Enzymes, citric acid
Selected *Bacillus strains**	Enzymes, vitamin B_2
*Kluyveromyces lactis**	(Intracellular) enzymes
Selected *Lactic acid bacteria**	Live cultures (starters and probiotics)
*Propionibacterium freundenreichii**	Vitamin B_{12}
*Saccharomyces cerevisiae**	Live cultures, yeast extracts, enzymes
Mortierella alpina	Arachidonic acid
Blakeslea trispora	Beta-carotene
Mucor miehei	Microbial rennets
Streptomyces natalensis	Natamycin (antifungal preservative)
Proteases from *A. niger**	Protein hydrolysates / peptides
Aspartase and thermolysin	Aspartame
	Pharmaceuticals and fine chemicals
Penicillium chrysogenum	Antibiotics
Selected *Streptomyces strains*	Proteins, antibiotics
Enzymes from *Escherichia coli**	Antibiotic / fine chemical synthesis
(e.g., aldolase)	Chiral intermediate for cholesterol lowering drug
Enzymes from *Pichia pastoris*	Fine chemicals synthesis
(e.g., hydroxynitrile lyase)	Intermediate for cardiovascular drug
CHO cells (animal)	Pharmaceutical proteins
PER.C6 cells (human)*	Pharmaceutical proteins

29.2.1
Food, Feed, and Nutritional Ingredients

The food and feed industry is one of the important fields of application of industrial biotechnology. Generally speaking, DSM produces three types of products in the food and nutritional ingredient arena:

- Biomass (e.g., living micro-organisms such as lactic acid bacteria, wine and beer yeast, as well as extracts of these);
- Enzymes (e.g., mainly hydrolases, some lyases and oxidoreductases);
- Nutritional ingredients (e.g., relatively small molecules, such as carotenoids, (pro)vitamins, anti-oxidants, preservatives, polyunsaturated fatty acids, aspartame, and tailored peptide mixes).

Throughout the third wave, and especially after the inclusion of DSM Nutritional Products, our portfolio has developed towards nutritional ingredients. Our technology base has developed in line with this, and now includes expertise in animal and human nutrition, complemented with advanced formulation skills. Furthermore, the traditional hypothesis-based screening of, for example, one enzyme after the other is currently – at least partly – being replaced by random screening of many enzymes from expression libraries for a desired or even unknown functionality. To that end, a range of miniaturized medium-to-high throughput application assays is being developed. Such facilities allow us to cope with the rapidly increasing availability of genome-based sequences and new enzymes created by directed evolution.

For all our little 'workhorses', we eventually want to have access to their genome sequence for two reasons. First, a high-quality and annotated genome offers new leads to improve our strains and processes, either by classical (non-GM) or GM means. We strive to offer our customers both options: classically-improved and classically-produced specialties as well as their GM equivalent, if feasible. Second, genomes – in general – contain many 'hidden' genes with unexpected enzymatic functionalities, which can be applied as such or used to generate other ingredients.

Take, for example, *Aspergillus niger,* which is our preferred host for the production of food enzymes. Typically, this generally-recognized-as-safe (GRAS) microorganism is used for the industrial production of several carbohydrases, such as pectinases and (hemi)cellulases for the beverages and bakery industries, but also for the production of proteases. Within DSM we have recently elucidated the complete genome sequence of this fungus. Out of about 14,000 genes more than 200 have been annotated as carbohydrases and another 200 appeared to encode for proteases. Of these, more than 100 genes have a signal sequence indicating that these enzymes are secreted. At present, we are systematically cloning and overexpressing these genes. Among these secreted enzymes we have recently identified a new protease, which shows superior performance in debittering casein hydrolysates. The enzyme was found to be a proline-specific endoprotease, which is a rare substrate specificity among proteases. The unique properties of this endoprotease have recently led to the development of a non-bitter, milk-based (casein-based) sport drink (Fig. 29.2), which has been shown to speed up the recovery after heavy and repeated exercise by stimulating the insulin response and glucose uptake in muscles (Edens, L. et al.; Dekker, P. et al.).

This debittering concept is now being further applied in the development of transparent milk-based beverages and clinical nutrition. Currently, we are expanding our protease toolbox with a range of di- and tripeptidyl-peptidases, which enables us to generate tailor-made hydrolysates rich in di- or tripeptides. Using these enzymes, we are now able to generate protein hydrolysates which exhibit discriminative advantages such as low allergenicity, low free amino acid content, and a high yield of small peptides for the health and nutrition market.

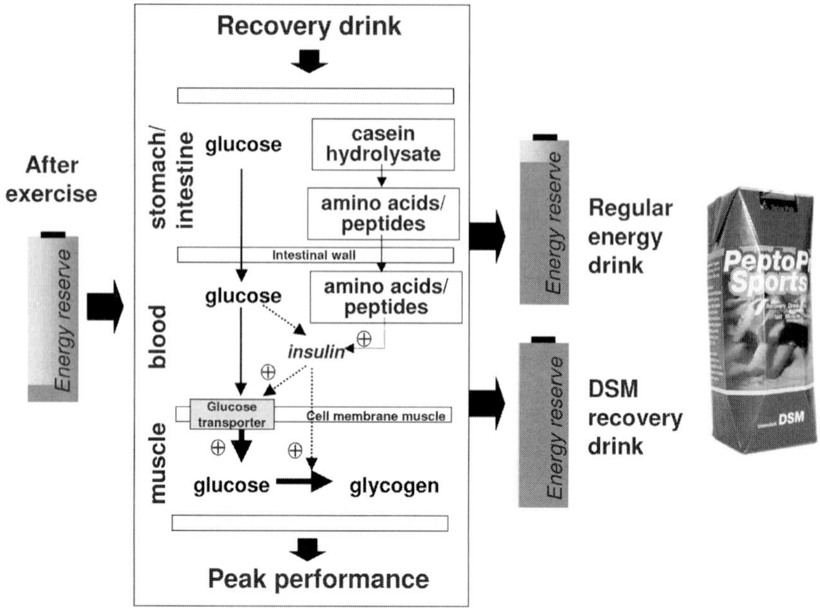

Fig. 29.2 PeptoPro™ for faster muscle refueling. Proposed mechanism.

The formation of chill-haze during beer production is known to involve a proline-rich fraction that is extracted from malt. The initial driving force of chill-haze formation is a hydrophobic interaction between this proline-rich protein fraction and polyphenols. Using one of our newly discovered proteases, we were able to show on a pilot production scale that even low levels of the enzyme can effectively prevent chill-haze in beer. Compared with the commonly employed polyvinylpolypyrrolidone (PVPP), the use of our enzyme, BrewersClarex™, results in improved beer stabilization, enhanced polyphenol (antioxidant) levels in the final beer, and considerably simplified processing requirements (Edens, L. and Lopez, M.). Our customers are now able to substitute an unnatural chemical clarifying agent by a simple, cost-effective biological solution.

Another important enzyme found in the genome of *A. niger* is asparaginase. This enzyme is capable of reducing the amount of the amino acid asparagine in foodstuffs, thereby preventing the undesired formation of acrylamide in baked and fried products (Plomp, P. et al.). The benefit to our business customers, and eventually the consumer, is clear: safer food.

Without the genome sequence, tracking down these (hidden) enzymes and scaling up the production process and application would have been much more time-consuming or even impossible. Clearly, (gen)omics has helped us enormously to shorten the time from concept to customer. The period from the identification of the endoprotease to the commercial production of the sport drink (PeptoPro™) was less than two years! Not so long ago, it would have taken at least five.

While genomic approaches are gaining momentum, we should realize that many traditional approaches are still extremely successful. One in-house example, which was co-developed with external partners, is the microbial production of arachidonic acid by *Mortierella alpina*. Arachidonic acid (AA) is a nutrient naturally present in breast milk in combination with docosahexaenoic acid (DHA). Together, these fatty acids are believed to be important building blocks in the development of the brain and visual system in infants. The fungus *M. alpina* is one of nature's surprises, accumulating a lipid fraction within its cells with an exceptionally high AA content. The AA may represent 50 percent or more of total fatty acids, even in wild isolates. This example shows that selecting and developing a classical high-yielding strain still pays off if the application is attractive enough and the volume is large. Furthermore, there are no realistic alternative processes for this product at this point in time.

Biotech may be gaining importance in the food and nutrition sector, but many nutritional ingredients are still produced by chemical synthesis or via extraction: for example, carotenoids are currently most competitively produced by chemical means. For vitamin B2, however, the situation has changed completely in the last five years. The traditional eight-step chemical synthesis has been replaced by one fermentation process. This biotech process, which is also practiced by BASF on a large scale, reduces overall cost by up to 40 percent and the overall environmental impact by 40 percent, as has been shown by detailed life cycle assessments. Similar trends have been described for other bio-based processes, indicating that economic and environmental benefits go hand in hand in today's white biotech practice (EuropaBio and McKinsey & Company, 2003, DSM position document, 2004).

29.2.2
Pharmaceuticals and Fine Chemicals

DSM's product portfolio in the pharmaceutical and fine chemical area covers a wide variety of specialties:

- Antibiotics (e.g., penicillins, cephalosporins, clavulanic acid)
- Chiral intermediates (e.g., non-proteinogenic amino acids, alcohols, amines, polyols, aminoalcohols, and acids)
- Recombinant proteins (e.g., antibodies)
- Gene therapy products and vaccines

Currently, DSM uses biocatalysis, biotransformation, and fermentation technologies in addition to chemical methods to produce these specialties. To keep ahead of the competition and provide the best service to our customers, we use classical and/or advanced (gen)omics tools such as proteomics and metabolomics to make continuous improvements in the productivity and quality of our industrial workhorses. In the many cases where it is difficult – or impossible – to apply fermentation, we use (multi-step) biocatalysis to achieve our goal. Other in-house competences that are of growing relevance to the pharmaceutical area include the formulation of such products into various sterile and oral dosage forms under stringent cGMP conditions.

A good example of a white biotech process in the pharmaceutical area is our route to the antibiotic cephalexin, practiced on a large industrial scale for several years now. By advanced enzyme and metabolic engineering we were able to replace the traditional 10-step, mainly chemical synthesis (Fig. 29.3A) by a fermentative route followed by two mild enzymatic steps (Fig. 29.3B). The white biotech process has been shown recently to use far less energy (–65%), less input of (harsh) chemicals (–65%), is water-based, generates less waste, and is very cost-effective (–50%) (Bruggink, A.). Again, this is a bio-based process where environmental and economic benefits go hand in hand and for which there is no competitive chemical alternative (EuropaBio and McKinsey & Company, 2003; DSM position document, 2004).

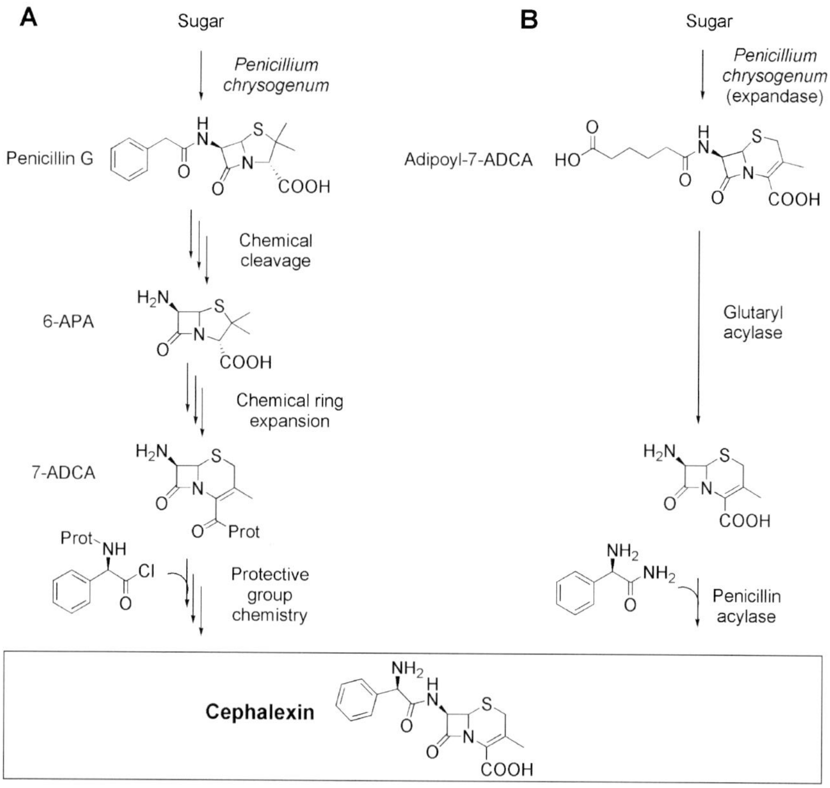

Fig. 29.3 The DSM cephalexin case. Changing from a chemical route with ten difficult chemical steps (A) to a white biotech route based on metabolic engineering and biocatalysis (B).

A very recent example, which clearly shows the power and benefit of biocatalysis, is the synthesis of a chiral intermediate for the production of cholesterol-lowering drugs, also known as statins (Fig. 29.4A). The enzyme deoxyribose aldolase (DERA) from *E. coli* is used to catalyze two consecutive aldol condensations between two molecules of acetaldehyde and one molecule of chloroacetaldehyde (Panke, S. et al.). These starting materials are readily accessible at low cost and the enzyme makes two chiral centers in one step at over 96 percent diastereomeric excess. Traditional chemical methods are far more complex and laborious. Interestingly, many routes to the same intermediate have been developed and almost all of them have a biocatalytic step integrated in the overall process (Müller, M.). This is just the second aldolase-based process on an industrial scale since GlaxoSmithKline developed an aldolase route to the antiflu drug Relenza™ (Liese, A. et al.). In this case the time span between discovery and commercialization was less than two years, and it is a clear example of a process where biocatalysis beats chemistry.

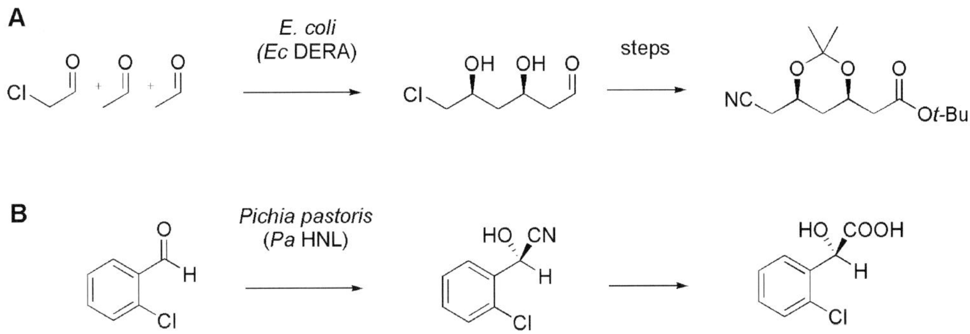

Fig. 29.4 Recently developed biocatalytic processes at DSM. A) Aldolase catalyzed route to chiral statin intermediate, and B) hydroxynitrile lyase catalyzed route to intermediate for a cardiovascular drug.

Yet another recent in-house example exhibiting the synthetic power of enzymes is the use of hydroxynitrile lyase (HNL) from the almond tree *Prunus amygdalus* (*Pa*HNL) for the production of *(R)*-2-chloromandelic acid (Fig. 29.4B), which is produced on an industrial scale for the synthesis of a number of biologically-active compounds, including a cardiovascular drug (Effenberger, F.). Here again, a carbon-carbon bond-forming enzyme is used to generate chiral products at high carbon efficiency. The enzyme was cloned from the almond tree and transferred into the yeast *Pichia pastoris* (Glieder, A. et al.). The process is very efficient and yields products at an enantiomeric excess of over 98 percent at close to 100 percent yield. Alternative processes based on resolution have a maximal yield of 50 percent. Thus, the bio-based HNL technology provides a more carbon-efficient route than chemical alternatives to this important class of compounds and other α-hydroxy acids (Pöchlauer, P. et al.).

The success of biocatalysis depends ultimately on the economics of specific processes. It provides enormous opportunities, and with the introduction of each new process, experience and confidence accumulate. It thus becomes easier to develop and justify the next biocatalytic step (Schmid, A. et al.; Schoemaker, H. E.). The challenge is now to develop a range of off-the-shelf biocatalysts that can be used to rapidly screen their potential in organic synthesis. In principle the screening and genetic tools are available to select the desired functionality from the vast biodiversity nature offers us, and to tailor or even create those activities which suit our purposes best. The future for industrial biotech processes involving biocatalysts is bright.

Besides biocatalytic and microbial systems, we are increasingly employing animal and human cell lines for the production of advanced pharmaceutical products, including recombinant proteins such as antibodies. In collaboration with the Dutch-based biotech company Crucell, we are currently improving their PER.C6-based production platform for a range of human pharmaceutical proteins.

Since pharmaceuticals are becoming increasingly complex molecules, we will keep strengthening our science base in this area too, by combining our internal skills with relevant external expertise in fermentation, biocatalysis, biotransformation, and chemical synthesis.

29.3
From Specialties to Commodities

The examples above are just a few selected DSM cases, and represent only a handful of the white biotech processes currently being practiced and developed. Not only chemical companies like BASF, Bayer, Dow, DuPont, Degussa, and Lonza are active in industrial biotechnology, but also enzyme providers such as Novozymes, Genencor, Diversa, and Maxygen/Codexis, not to mention energy and raw material providers including Shell, Total, and Cargill (see Chapter 28).

These and a growing number of other chemical and biotech (startup) companies are active somewhere along the industrial biotech value chain (Fig. 29.1). For broad-based adoption, new products must be competitive with existing offerings. The use of alternative low-cost agricultural (by)products such as biomass is bound to give industrial biotech another boost, facilitating the shift from specialties to commodities. According to Riese and Bachmann (Chapter 28), biomass-based biorefineries have the potential to reduce sugar costs substantially as they become more integrated. The net cost of producing sugar in an integrated biorefinery may even hit zero, when waste products or byproducts such as lignin and proteins generate value.

The shift from bio-based specialties to commodities is already visible in the marketplace with biopolymers made from corn. The first example is NatureWorks™ from Cargill, which is made from corn sugar-derived lactic acid. As in the biochemicals examples described above, the environmental benefits are eye-opening: NatureWorks™ already requires 25 to 55 percent less fossil resources, and it is planned to replace fossil resources completely in the next four to six years (EuropaBio and McKinsey & Company, 2003). Other high-potential biomaterials are a polymer based on 1,3-propanediol from DuPont and Genencor (Sorona®) and polyhydroxyalkanoates from Metabolix and ADM. Nexia is developing even more sophisticated protein-based materials which combine the strength and flexibility of spider polymers. Once they are being more widely produced on a global scale, bio-based polymers may eventually become cost-competitive with traditional fossil-based mass polymers such as polyethylene and polypropylene.

A different field with its own dynamics is the bioethanol market. Huge volumes are required to meet, for example, the EU target of 5.75 percent ethanol to fuel our mobility in 2010. To meet this ambitious objective, 9.3 million tons of bioethanol will have to be produced or imported by 2010. Given the limited government support in Europe, the artificially high sugar prices, and the fact that companies are hesitating to enter this risky market, it is doubtful whether this target will be met in time. The impact of cheap bioethanol on the chemical industry would be very high, since ethanol can not only serve as a potential energy source for (bio)-chemical processing, but also as a cheap building block for (bio)chemical synthesis. A typical example here would be polyethylene.

Different chemical markets are introducing and using biotechnology at different rates. In general, the development time for biomaterials is longer than for biochemicals. As our examples have shown, the development time for biochemicals is steadily decreasing and can be as short as two years. For biomaterials, a time span of five to as much as 15 years between concept and customer seems more realistic at present, but this period is bound to decrease as well in light of the rapid technological advances and the growing need to replace fossil resources.

29.4
From Innovation to Impact

The global impact of industrial biotechnology on the chemical industry is currently relatively low. About five percent of all chemicals today are produced by industrial biotechnology. Riese and Bachmann estimate that ten percent of all chemicals sold by the year 2010 will be biotech-derived using renewable resources. The impact will be greatest in the fine chemical sector, with an estimated 60 percent share of white biotech products.

As DSM shapes its life-science activities through in-house developments, acquisitions, venturing, joint ventures, and public-private partnerships, we see white biotechnology gaining momentum by the day and being recognized as a potential gateway to a more sustainable future. While the United States and Japan have already endorsed a very progressive position towards industrial biotechnology, Europe is shaping its own future at the moment by establishing – among other things – a Technology Platform on Sustainable Chemistry. Industrial biotechnology forms one of the main sections in this platform, together with materials technology and process & reaction design. The Technology Platform can be regarded as a multi-stakeholder public-private partnership between industry, the research community, agriculture, and society, and it is bound to set the strategic research and policy agenda for the European chemical industry (the biggest in the world!)

in the next decades. In accordance with the European vision for white biotechnology (EuropaBio and Cefic, 2005):

- An increasing number of chemicals and materials will be produced using biotechnology in one of their processing steps;
- Biotechnology will allow increasingly eco-efficient use of renewable resources as raw materials for the (chemical) industry;
- Industrial biotechnology will enable a range of industries to manufacture products in an economically and environmentally sustainable way;
- Biomass-derived energy, together with other renewable energy sources, is expected to cover an increasing amount of our energy consumption.

To meet these targets, the main tasks of this Technology Platform will be to develop a strategic research agenda and road map, remove technical, economic, regulatory, and implementation barriers, and involve society in the decision-making process via a dialogue with stakeholders. By technical barriers we mean that, despite the undisputed progress in (gen)omics, the technology – in its broadest sense – is still relatively immature compared with, for example, conventional chemistry. Economic barriers include the requirement of high capital investments to replace traditional processes, or to start up new ones. Incentives such as corporate tax advantages, or investment tax credits, are highly desirable to encourage the switch from petro to bio. Another important economic barrier is the high European sugar price discussed above. Developments such as the announced reform of the sugar regime will be helpful in making European-grown feedstock available at world market prices. Regulations can often be a barrier, either when new products have to comply with existing legislation, or when stringent, precautionary regulation makes compliance unfairly costly. A more balanced risk/benefit approach to regulation rather than a purely risk-centered philosophy would be beneficial for both industry and society. On the intellectual property front, the lack of an affordable European patent system is a severe disadvantage for European science and innovation. In addition, encouragement should be given to supporting policies which will allow the increasingly stringent environmental criteria to be met (EuropaBio and McKinsey & Company, 2003, EuropaBio and Cefic, 2005). Key to the successful exploitation of white biotechnology is that market push and market pull are properly attuned by combining strong industry leadership with broad-based stakeholder commitment.

The worldwide prospects of industrial biotechnology are looking brighter than ever before due, amongst other things, to the momentum generated by the agenda of the Technology Platform on Sustainable Chemistry, the rapid advances in (gen)omic technologies, the proven environmental and economic benefits it offers, increasing oil prices, and biotech's role as an important driver of innovation. Promises are increasingly becoming reality and turning into profit. DSM is highly committed to playing a leading role in shaping the third – industrial – wave of biotechnology which, in our view, will create a more sustainable society for the people of today and the generations of the future: a future where societal (People), environmental (Planet), and economic (Profit) benefits will go hand in hand.

29.5 Summary

- Industrial (white) biotechnology is rapidly gaining momentum as a cost-effective and environmentally-friendly technology for producing bio-based chemicals, materials, and fuels in a safe and sustainable way from renewable resources.
- DSM is increasingly active in this growing business area, with about 40 percent of its life science/nutritional products currently biotech-derived. The company realizes that innovation will be key to its future success.
- This chapter outlines a number of examples from DSM's key biotech product areas, demonstrating their advantages and improvements over time:
  - Food, feed, and nutritional ingredients, where it focuses on biomass, enzymes, and nutritional ingredients.
  - Pharmaceuticals and fine chemicals, where its product portfolio covers a wide variety of specialties including antibiotics, chiral intermediates, and recombinant proteins.
- The authors point out that an increasing number of companies are becoming involved in white biotech, and that biotechnology is spreading from specialties to commodities.
- Finally, they outline the role played by governments and government bodies such as the European Technology Platform on Sustainable Chemistry as well as other stakeholders.
- In the authors' eyes, industrial biotechnology will create a more sustainable society for the people of today and the generations of the future.

References

Bruggink, A. (Ed.) **2001** Synthesis of β-lactam antibiotics, Kluwer Academic Publishers, Dordrecht

Dekker, P., van der Hoeven, R., Edens, L., and de Lange, L. **2002**, Patent WO 02/45524

DSM Position Document, *Industrial (White) Biotechnology: An Effective Route to Increase EU Innovation and Sustainable Growth*, **2004**

Edens, L., van der Hoeven, R., and Delest, V. **2002A**, Patent WO-02/45523

Edens, L., and Lopez, M. **2002B**, Patent WO-02/46381

Effenberger, F. Hydroxynitrile lyase in stereoselective synthesis in *Stereoselective Biocatalysis*, R.N. Patel (Ed.), Marcel Dekker, New York, **2000**, 321–342

EuropaBio and Cefic, *A Vision for 2025: Industrial or White Biotechnology, A Driver of Sustainable Growth in Europe*, **2005**, http://www.europabio.org

EuropaBio and McKinsey & Company Position Document, *White Biotechnology: Gateway to a more Sustainable Future*, **2003**, htttp://www.europabio.org/upload/documents/wb_100403/Innenseiten_final_screen.pdf

Glieder, A., Weis, R., Skranc, W., Pöchlauer, P., Dreveny, I., Majer, S., Wubbolts, M., Schwab, H., and Gruber, K. *Angewandte Chemie Int. Ed.* **2003**, *42*, 4815–4818

Liese, A., Seelbach, K., Buchholz, A., Haberland, J. *Processes*, in A. Liese, K. Seelbach, C. Wandrey (Eds.) *Industrial Biotransformations* Wiley-VCH, Weinheim, **2000**, 95–392

Müller, M. *Angewandte Chemie Int. Ed.* **2004**, *43*, 2–5

OECD Report, *The Application of Biotechnology to Industrial Sustainability*, **2001**, http://www.1.oecd.org/publications/e-book/9301061e.pdf

Panke, S., Held, M., and Wubbolts, M. *Curr. Op. Biotech.* **2004**, *15*, 272–279

Plomp, P., de Boer, L., van Rooijen, R., and Meima, R. **2004**, Patent WO 04/030468

Pöchlauer, P., Skranc, W., and Wubbolts, M. The large-scale biocatalytic synthesis of enantiopure cyanohydrins in *Asymmetric Catalysis on Industrial Scale*, H. U. Blaser, and E. Schmidt (Eds.), Wiley-VCH, Weinheim, **2004**, 151–164

Schmid, A., Dordick, J. S., Hauer, B., Kiener, A., Wubbolts, M., and Witholt, B. *Nature*, **2001**, 409, 258–268

Schoemaker, H. E., Mink, D. and Wubbolts, M. *Science* **2003**, *299*, 1694–1697.

30
Leveraged Buyout Transactions – Challenges and Learnings

Achim Berg, Florian Budde, and Bernd Heinemann

In the mid 1990s, buyout firms began to conduct leveraged transactions in the chemical industry. A slew of multi-billion dollar LBOs suggest the trend has intensified since the start of the new millennium. This chapter explores the relevance of private equity investors for the chemical sector and describes the value generation levers they apply; it examines what traditional chemical corporations can learn from their financial competitors; and concludes with a description of buyout firms' challenges in the chemical industry and an outlook.

30.1
Chemical Sector LBOs Show No Sign of Abating

In a series of articles in 2001, Butler, Samdani, and McNish described the increase in LBOs in the chemical industry over the second half of the 1990s (Butler, P.; Samdani, G. S. et al.). They labeled this phenomenon the "Alchemy of Leveraged Buyouts" and predicted a natural convergence between traditional chemical corporations and their new financial competitors in terms of management procedures and skills as well as how they created value.

LBOs – the purchase of a controlling stake in a company or division from its owners to be held for a limited time, financed through a combination of equity and a large amount of debt and with strong involvement of a financial investor – certainly gained prominence in the late 1990s. However, the real boom started around the turn of the century, especially in Europe. Over the past four years, buyout firms have emerged as formidable competitors in many segments of the industry, often outbidding publicly traded chemical companies for acquisitions. As major players in the sector, they compete for capital, corporate control, and management talent. However, the fierce competition within the buyout industry continues to force them to differentiate themselves in order to stay competitive.

At the time of writing, financial investors own majority stakes in more than 100 European chemical companies. A report published by the Chemical Industries Association and Cogency very conservatively estimates that at least EUR ten billion of assets remain available for acquisition in Europe, and buyout investors

Value Creation: Strategies for the Chemical Industry. 2nd Edition. F. Budde, U.-H. Felcht, H. Frankemölle (Eds.)

ISBN: 3-527-31266-8

expect to take a respectable share of that cake (Smith, P. & Butler, S.). Globally, financial investors account for 25 to 30 percent of all chemical sector M&A transactions by value, rising to 40 percent in Europe.

The deal wave seems set to continue. A handful of major deals have dominated the news: in 2003, for example, the German chemicals distributor Brenntag was bought by Bain Capital (USD 1.5 billion); former Hoechst subsidiary Celanese was acquired and delisted by The Blackstone Group in the largest buyout transaction in chemicals to date (USD 3.1 billion); Dynamit Nobel was sold by MG Technologies to Rockwood/KKR (USD 2.3 billion); Apollo acquired Borden Chemical from KKR (USD 1.2 billion); Platinum Equity became the new owner of DyStar (acquired from Aventis, BASF, and Bayer for USD 700 million); while the Texas Pacific Group purchased British Vita for GBP 668 million (USD 1.28 billion) in 2005. An analysis of chemical buyout transactions[1] proves the increase in activity (Fig. 30.1). The value of disclosed transactions conducted between 2001 and the end of 2004 was more than double that in the previous four-year period. In total 179 transactions were carried out between 1985 and 2004, of which the 108 that disclosed their deal value accounted for more than USD 41 billion. Furthermore, the analysis showed that specialties are the focus of transactions, and that most targets are European (Fig. 30.2).

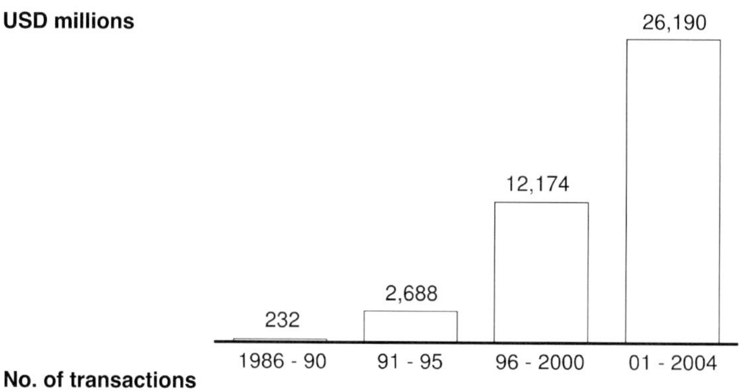

	1986 - 90	91 - 95	96 - 2000	01 - 2004
No. of transactions with disclosed transaction value	3	29	32	44
Total transactions	7	48	62	62

Source: McKinsey analysis

Fig. 30.1 Value of chemical buyout transactions involving public companies with disclosed transaction value, 1986–2004.

1) Chemical buyout transactions where public companies have been involved as seller or acquirer.

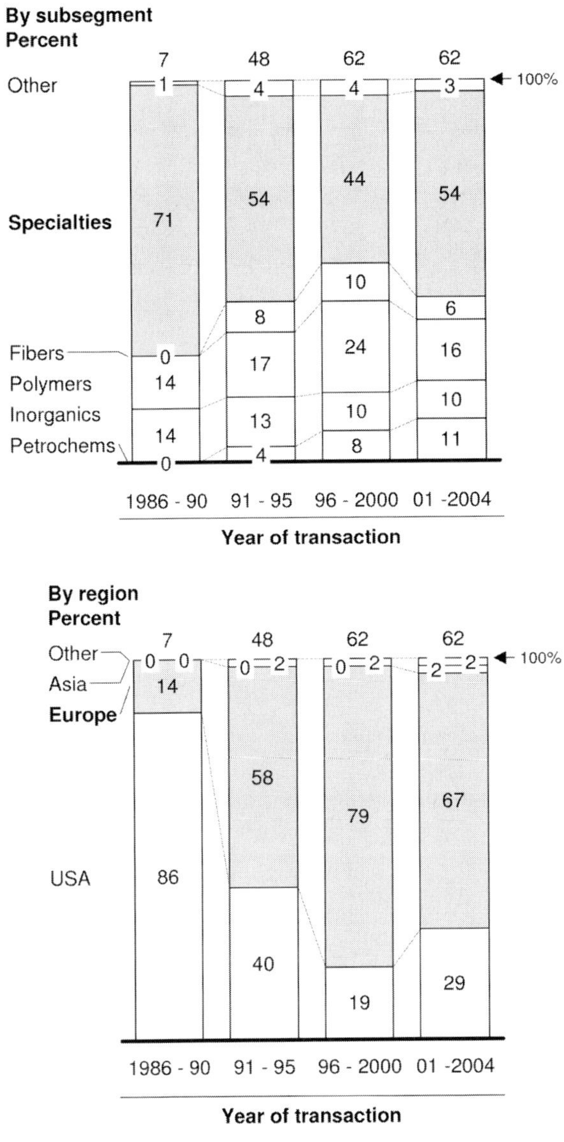

Source: McKinsey analysis

Fig. 30.2 Chemical buyout transactions involving public companies by subsegment and by region, 1986–2004.

At first glance, the chemical sector appears unattractive for investors. It tends to have relatively low growth, high capital intensity, rather high volatility of raw material cost, currency risk in a highly internationalized market, and numerous environmental and regulatory issues. In addition, revenues and returns are notably more cyclical and volatile than in some other sectors, and different chemical subsectors may be at completely different points of the business cycle at any given time.

However, these intrinsic drawbacks are matched by some major benefits. Chemicals is a safer industry than the darlings of the 1990s, such as telecoms, because it has well developed international markets and can generate cash even in depressed economic environments. There are plenty of low-risk targets that have physical assets and reliably high cash flows perfectly suited for investors looking to acquire companies that can quickly pay down their high piles of debt. Also, many chemical companies have an attractive and tradable asset base – an important starting point for the portfolio company's exit story. In addition, many have invested in skills and in operations, resulting in limited or non-existent capital expenditure requirements for the next few years. Products and basic processes are also often in good shape, and improving efficiency is nearly all that buyers need to do to make a stable and profitable company.

In fact, some of the sector's very problems also create opportunities for shrewd investors. Its cyclical nature, combined with the debt burden of many global players, means that the industry consolidates and restructures more frequently than many others. The consolidation wave of the 1990s forced many trade buyers to pay high multiples in the race to build larger, more integrated businesses. Recent times have shown that many of these players are overstretched in the face of the market downturn and the increased competition from Asia, thereby creating a raft of opportunities for financial investors. In addition, the absence of industry buyers and the capital markets' apparent lack of interest are resulting in lower multiples than in the 1990s. Low multiples combined with low earnings, caused by depressed performance results, means bargains to be had for private equity investors.

Buyout firms, in turn, are keen to conduct transactions as they face pressure to employ the money raised from investors in recent years. According to estimates, the private equity industry as a whole raised USD 900 billion between 1998 and 2003 (3i/PwC 2000 to 2004). However, the industry has invested only about USD 800 billion, leading to a capital overhang of around USD 100 billion by 2003. Part of this immense capital wave is flooding into the chemical sector. European markets in particular are profiting from the capital inflow as many large and experienced US buyout firms have started only recently to build up resources in Europe in order to prepare for high profile LBOs (Lerner, J. et al.; Dixit, A. & Jayaraman, N.).

In addition, many of the buyout firms that have already invested still monitor the deal pipeline for further acquisitions in order to strengthen the competitive position of their portfolio companies and/or to consolidate market segments. For example, in late 2004, The Blackstone Group announced it had bought Acetex for USD 490 million and Vinamul Polymers, ICI's emulsion polymer business, for USD 208 million in order to strengthen the competitive position of its portfolio company Celanese.

The increased buyout activity in the chemical sector is a consequence of all the effects described above. However, this does not explain how top-tier buyout firms actually generate value through these transactions. Studies have revealed that, on average, buyout investments slightly outperform public investments (Wright, M. & Robbie, K.; Doran, A.; Raicher, S.; Kaplan, S. & Schoar, A.), and McKinsey analyses have shown that this holds true for the chemicals sector (Butler, P.; Samdani, G. S. et al.). Nevertheless, this does not mean that all buyout funds outperform this benchmark. On the contrary, McKinsey studies (Kehoe, C.) as well as academic research have revealed that the bulk of returns are earned only in the top quartile of funds. Those funds heavily outperform the public market benchmark, but the other three-quarters barely come close, or end up destroying value (see also Lamm, R. M. & Ghaleb-Harter, T.; Sood, V.).

30.2 Understanding Value Generation in Chemical Buyouts

Due to these companies' inherently private nature, there is still a lack of empirical data linking outperformance by leading buyout firms to specific value generation levers. However, it is still possible to describe the levers applied by buyout firms operating in the chemical sector. This is particularly useful for chemical company executives as it helps to demystify some of the myths that surround the operations of their new financial competitors.

Berg, A. and Gottschalg, O. (2004; 2005) identified six general groups of value generation levers, which can be further divided into seventeen sublevers (Fig. 30.3).

Financial arbitrage is a well known lever for value generation in chemical buyout transactions. Buying low and selling high has always been a goal of financial investors and skills in financial arbitrage are obviously beneficial in a highly cyclical market such as chemicals. But it seems today that financial arbitrage has become as much a commodity in chemical buyout transactions as *financial engineering*. Although the ability to optimize the capital structure of the portfolio company and to reduce corporate tax may have resulted in a competitive edge in the past, it seems that today financial engineering alone may not be sufficient to ensure a successful investment.

Improving operational effectiveness through, for example, cost cutting and margin improvement programs and the *strategic repositioning* of the acquired company are much more effective levers. Many leading buyout firms have accepted that it is generally not sufficient to capture value through wheeling and dealing, and that they need to make significant changes to the organization, processes, and strategy of the acquired company. In addition, it is crucial to understand that all of the levers mentioned above can have a real impact only if the buyout firm is able to *reduce the agency cost* involved. The interests of owners and managers of the portfolio company need to be aligned through high incentives and close, professional monitoring and controlling of management's activities. Furthermore, many transactions show that it makes a difference which buyout firm conducts the transaction and how able it is to exercise the *parenting effect* and restore the entrepreneurial spirit of managers as well as to act as a valuable adviser to the management of the portfolio company.

Lever	Sublever	Description
A. Financial arbitrage	A-1. .... based on changes in market valuation A-2. ... based on private information about the portfolio company A-3. ... through superior market information A-4. ... through superior deal-making capabilities A-5. ... through an optimization of corporate scope	Generating returns from differences in the valuation applied to a portfolio company between acquisition and divestment, independent of changes in the underlying financial performance of the business (buy low/sell high).
B. Financial engineering	B-1. Optimizing the capital structure B-2. Reducing corporate tax	Optimizing capital structure and minimizing after-tax cost of capital of the portfolio company as a consequence of financial knowledge and experience
C. Increasing operational effectiveness	C-1. Cost cutting and margin improvements C-2. Reducing capital requirements C-3. Removing managerial inefficiencies	Implementing measures that enhance overall productivity and effectiveness of operations. The configuration of a company's resources, i.e., how they are deployed/readjusted
D. Increasing strategic distinctiveness	D-1. Corporate refocusing D-2. Buy and build strategies	Adjusting the strategic objectives, programs and processes of the portfolio company
E. Reducing agency costs	E-1. Reducing agency cost of free cash flows E-2. Improving incentive alignment E-3. Improving monitoring and controlling	Reducing the agency costs that arise from the owner/manager conflict in the portfolio company
F. Parenting effect	F-1. Restoring entrepreneurial spirit F-2. Advising and enabling	Increasing revenues or cutting costs as a result of the benefits the portfolio company accrues from its association with the buyout firm

Fig. 30.3 Levers of value generation in buyouts.

We find it helpful to consider the value creation process for leveraged buyouts along three dimensions: transaction timing, value generation mechanisms, and the sources of value, linked to investor characteristics. This approach offers an entry point to understanding the interdependencies between the individual levers described above, and the overall process of buyout value generation.

30.2.1
Timing: Three Phases of a Buyout

The *acquisition phase* starts with the negotiation and due diligence process, during which investors familiarize themselves with the company and develop a business plan for the buyout. In a management buyout, where existing managers participate with financiers, this is when the foundations are laid for the relationship between the two parties. The valuation of the target company and the corresponding acquisition price is probably the most important factor in determining the value of the deal, as it sets the hurdle for all future valuations. At the end of the acquisition phase, the buyout structure is defined, including the degree of financial leverage, the distribution of management equity stakes, the design of incentive systems, etc. Much of the buyout value generation is determined by decisions taken during this phase.

During the subsequent *holding period*, the investors implement the strategic, organizational, and operational changes prescribed in the initial business plan, and realize any operational improvements. In practice, this is often an iterative rather than a linear process, with the business plan undergoing constant updates.

The *divestment phase* marks the end of the buyout. Here, decisions are made on the type of divestment mode (trade sale, IPO, etc.) as well as the valuation.

30.2.2
Value Generation Mechanisms

A business's equity value can be split into valuation multiple, revenues, margin, and net debt. Any changes in equity value must therefore be linked to a change in at least one of these four components.

The first type of increase in the equity value of a company is *value capture*. This is linked to changes in the valuation of the business without any underlying change in financial performance. The valuation may be partly influenced by changes in the financial performance, but there are several other possible factors, such as changes in the market valuation multiples for comparable companies or updated expectations of the financial performance of a business or an entire industry. Naturally, capturing any increase in a company's equity value will depend on the valuations at the time of acquisition and divestment. Returning to the six levers of value generation we identified earlier, only financial arbitrage is linked to value capture.

The second type of value generation – *value creation* – is directly linked to a fundamental change in the financial performance of the organization through improvements in operating performance (revenue growth, improved operating margin, etc.), reduced cost of capital (optimization of capital structure, better financing terms, etc.), or the freeing-up of resources by reducing the necessary fixed or variable assets. The remaining five levers discussed above all have a direct impact on the financial performance of the business. They can be split into primary and secondary levers.

Primary levers have a direct impact on the bottom line. They include financial engineering, operational effectiveness, and strategic distinctiveness. The secondary levers are reducing agency costs and the parenting effect from the new owner. These typically enhance the effect of the primary levers. For example, while an increased incentive alignment between the target company's management and new shareholders may have no impact on profits or cash flow generation by itself, it may help to remove operational inefficiencies or to achieve a better strategic positioning.

30.2.3
Sources of Value Generation

Buyout value generation can also be differentiated by the degree to which it depends on specific characteristics of the private equity investors. At one extreme, value generation occurs entirely within the boundaries of the portfolio company and would occur in any buyout context irrespective of any specific characteristics of the investors. This could happen, for example, through improvements in operational efficiency that were achieved without any knowledge transfer from the investors. At the other extreme is value generation that is inherently linked to the investors' own network, expertise, experience, capability, strategy, etc. This is comparable with the parenting effect advantage that business units may enjoy as part of a conglomerate company. Clearly, these two cases mark a continuum and many types of value generation fall between these two extremes.

30.3 Learnings for the Chemical Industry

A detailed review of the levers applied by private equity players makes it clear that they are neither new to the chemicals business, nor ignorant of its particularities. Nevertheless, top-tier buyout firms perform so well that it seems that applying the right levers and aligning the interests of all parties can add significant value.

However, understanding how buyout firms generate value is only one part of gaining an insight into the reasons for their success. It is also helpful to understand the motives of financial investors in chemicals as they have proved to be quick to buy divisional spinoffs and subsidiaries of chemical conglomerates as well as whole companies. Shareholders are demanding more value, and competitive pressure is increasing in the industry. Many over-diversified companies therefore feel the need to restructure in order to achieve greater focus. They put in place aggressive business unit disposal programs to raise much-needed cash. Especially when markets are weak and traditional chemical companies have limited interest in acquisitions, buyout firms have emerged as important players in the market. Buyouts also seem to be the primary option for assets that are difficult to sell, either because of a lack of growth potential or low profitability levels. Additionally, private equity firms often seem to be the only acquirers interested when direct chemical competitors cannot enter the fray because of antitrust concerns.

By accumulating significant assets in chemicals, buyout firms have not only become acquirers of disposed entities but also respectable competitors for traditional players. They compete against each other in the market for capital, corporate control, and talent. And in many respects it seems that buyout firms could build a competitive edge.

Executives therefore have to accept that their performance and value generation approach will be benchmarked against those of buyout firms. A better understanding of the 'buyout recipe' should help them stay competitive.

30.4 Learnings for Buyout Firms

Despite some buyout firms' very visible successes in the chemical industry, they have found the macroeconomic environment challenging in recent years. At the beginning of the new century they had to cope with a general downturn and low share prices, and financial turbulence has meant that both banks and bondholders have been more cautious about lending for leveraged transactions. While this situation has generally improved since the beginning of 2005, buyout firms still have to invest much time and effort to secure financing for their transactions.

30.4.1 Good Deals Are Getting Harder to Find

Buyout firms have to put more equity to work for transactions to succeed. Back in the late 1980s, a fund would put up only ten percent of the deal value in equity; the other 90 per cent would be debt-financed in the form of bank loans and junk bonds. Ten years later, the required equity stake had increased to 20 percent of the deal's value; and at the turn of the century the buyout specialists had to put up around 30 to 40 percent, given the stage of the chemical and capital markets cycle and the degree to which the big banks' balance sheets were stretched as a result (Clow, R. & Smith, P.; Hardymon, F. et al.; The Economist). This translated into lower expected returns from successful investments, forcing buyout firms to pay lower prices for portfolio companies due to the higher cost of financing and reluctance on the part of financiers to cater for a highly leveraged capital structure. Putting up more equity and suffering restrictions in paying appropriate prices for portfolio companies puts even the successful managers of buyout associations under pressure if they are to match the returns on equity earned in the 1980s and 1990s. The situation has now improved and buyout firms are benefiting from relatively low interest rates. However, it seems unlikely that the days of 90 percent debt funding will return any time soon.

The challenge is heightened by the fact that sellers of chemical assets are becoming more sophisticated and experienced in negotiating with buyout companies. Many request assistance from M&A advisors and are generally better informed than in the past, with the result that price negotiations are more evenly balanced. Sellers and management teams willing to conduct a buyout therefore usually negotiate with a number of financial sponsors. In fact, vendor auctions are so commonplace that it is rare for a fairly large deal to occur without some form of open or sealed auction (Wright, M. & Robbie, K.; De Leenheer, L. & Wong, M.; Harper, N. W. C. & Schneider, A.). This leaves little scope for buyout associations to pick up firms at a discount. Furthermore, corporate buyers are expected to return as competitors in many transactions after the economic downturn ends.

30.4.2
Exiting Through the Right Door

The second real danger lies in exiting the acquired assets. Buyout firms are forced to sell portfolio companies at the end of the predefined limited lifetime of their funds. It is no secret that this has become more problematic in recent years and that the average retention time of assets has increased. In fact, relatively few have managed to successfully divest in the past few years and exiting portfolio companies is one of the top issues for private equity players operating in the chemical sector.

Our own analysis of public-to-private chemical buyout transactions conducted between 1986 and 2004 shows this in impressive fashion (Fig. 30.4). Only a third of total deals conducted have reached any exit, and about 20 percent of the companies acquired between 1991 and 1995 are still owned by their financial investors.

In fact, *trade sales* to non-financial/non-private equity companies have been the dominant route for exiting a portfolio company. The dominance of trade sales as an exit channel is in line with the assumption that the chemical industry offers, in theory at least, a greater number of potential trade exits thanks to its position as a truly global industry with market segment players in most regions of the world. There has always been a lot of corporate turbulence, which usually spawns more M&A transactions. But it also seems that, at times, even trade sales become challenging, reflecting the cash flow and profitability constraints that face would-be acquirers. Many traditional chemical companies prefer to concentrate on paying down their debts and creating operational efficiencies rather than growing through acquisitions. However, this trend may be just about to change again.

The need for timely and profitable exits for buyout firms has given rise to a growing secondary market for buyout transactions. Although they are historically less lucrative than both IPOs and trade sales, *secondary buyouts* have increased in importance and acceptance in recent years (Davison, C.). For example, Ripplewood sold Kraton Polymers to the Texas Pacific Group in 2003 for USD 770 million and in 2004 Borden Chemical, held for nine years by KKR, scrapped its IPO plans and was sold instead to Apollo for USD 1.2 billion.

It might seem counterintuitive, but the returns achieved by buyout firms depend crucially on public markets (e.g., Wright, M. & Robbie, K.; Gompers, P. & Lerner, J.; Clow, R. & Smith, P.). Weak capital markets and the decline in valuations have made it extremely difficult to use IPOs as an exit channel. Structural impediments have also arisen. For example, an organization needs to be much larger now before it attracts analyst attention. This means that only businesses with financial critical mass are likely to have any chance of floating in Europe. The situation is different in North America where the market for mid-caps has remained popular, keeping the threshold for public businesses considerably lower.

We do not expect the stock market exit option to pick up in the near future, at least in Europe (Davison, C.; Smith, P.; Maier, A.), but some of the large buyout firms have started to revitalize this channel. Compass Minerals, for example, completed its IPO in December 2003 raising USD 217 million; Apollo, Blackstone, and Goldman Sachs conducted a partial IPO of Nalco Holding at the end of 2004; and Blackstone also managed to sell off part of its stake in Celanese via an IPO at the beginning of 2005. On the other hand, Polypore, a portfolio company of Warburg Pincus, withdrew its IPO offer in August 2004 because of unfavorable market conditions.

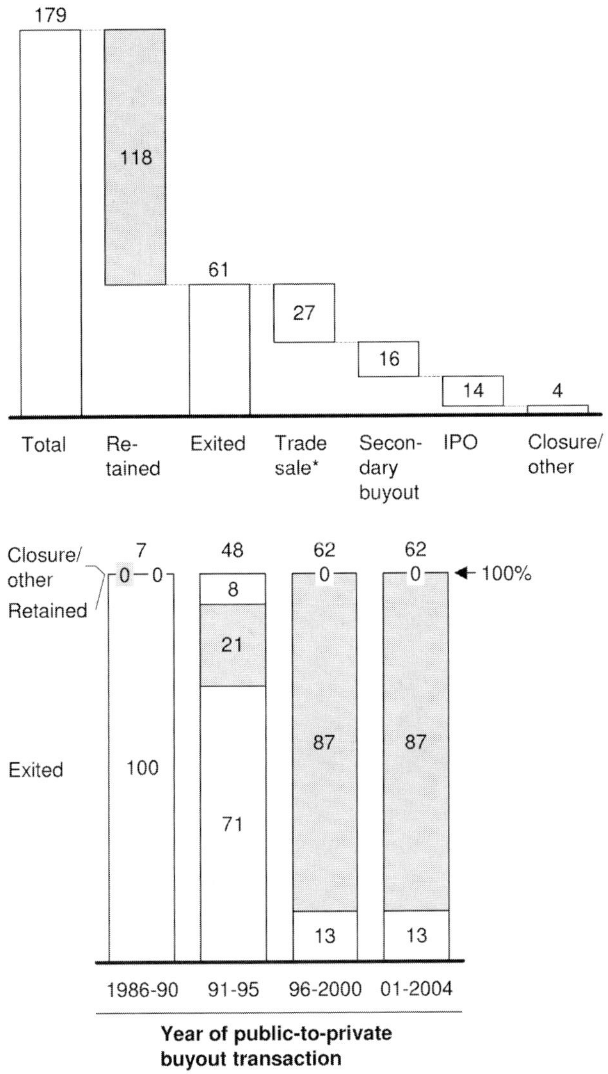

* Sale to a non-financial/non-PE quoted/private company
Source: McKinsey analysis

Fig. 30.4 Outcomes of chemical buyout transactions involving public companies,1986–2004 (companies, percent).

30.4.3
Time to Stand Out From the Crowd

Taking all these challenges into account, it is surprising to realize that some buyout teams operate in the chemicals sector with little or no experience. Furthermore, one would expect financial investors to be aware that highly sophisticated commercial due diligence regarding both deal selection and execution plays an important role in such a cyclical and volatile industry. However, it seems that this is not obvious to some, despite the lack of evidence that a purely opportunistic investment strategy can lead to sustainable above-average performance. Indeed, it is clear that only buyout firms with a consistent strategy will achieve and maintain superior performance compared to their financial and non-financial competitors.

In general, we perceive that a number of buyout firms pay insufficient attention to their own strategy[2]. They seem to underestimate the importance of the choices they make in generating and maintaining a long-term competitive advantage. In addition, they do not spend enough time articulating their goals or configuring activities to establish and then defend an attractive strategic position relative to their competitors. In consequence, they fail to deploy resources and to design organizational structures that are aligned with the strategic positioning in order to generate a reinforcing pattern of activities and choices.

Although there is still only limited differentiation among European buyout firms and their funds, their longer standing US counterparts have evolved further (Lerner, J. et al.; De Leenheer, L. & Wong, M.). It is fair to assume that buyout firms competing in Europe will develop distinctive strategies in order to acquire a source of competitive advantage.

On the one hand, buyout firms have to make clear to potential investors why investments in their funds are worthwhile and why they are best positioned to satisfy investors' needs and expectations. Buyout firms are placing increasing emphasis on enhancing the service they provide to their fund investors in order to differentiate themselves, and they are increasingly seeking ways in which they can make themselves attractive to fund providers (cf. Wright, M. & Robbie, K.).

On the other hand, buyout associations have to compete for portfolio companies and therefore have to convince potential vendors as well as their management teams of their strategy. The increase in competition for deals is therefore also leading to differentiation and specialization.

All this highlights the relevance and importance of strategy for buyout firms; or, to quote Markides, C. C. "Behind every successful company, there is superior strategy. The company may have developed this strategy through formal analysis, trial and error, intuition, or even pure luck. No matter how it was developed, it is the strategy that underpins the success of the company. To understand corporate success, we must therefore understand the logic of successful strategies. Even the 'luckiest' of strategies has an underlying logic – a set of underpinning principles."

2) For a more detailed discussion of strategy for buyout firms see Berg, A. (2005).

30.5 Summary

- Leveraged buyouts have become an important part of the chemical industry and it seems that the spate of deals in the late 1990s will pick up speed. Financial investors have good reasons to invest in chemical companies and the top players have developed a set of value generation mechanisms that let them achieve above average results.
- Executives of traditional chemical companies can benefit from these developments. Buyout firms have become grateful acquirers of non-core businesses and have brought liquidity to some segments of the market.
- However, they also represent a competitive threat that needs close observation. By examining buyout firms, chemical executives can gain deep insights for their own strategies and the set of value generation mechanisms that they themselves should apply.
- Buyout firms also face important challenges that they may only overcome if they manage to achieve differentiation. They need to understand what will make them distinctive in the future in order to become or remain superior performers. Crucially, they need to understand their sources of competitive advantage.

References

3i/PwC. 2000. Global Private Equity 2000: A Review on the Global Private Equity and Venture Capital Markets. *3i Group & PricewaterhouseCoopers*: pp. 1–16. London, UK.

3i/PwC. 2001. Global Private Equity 2001: A Review on the Global Private Equity and Venture Capital Markets. *3i Group & PricewaterhouseCoopers*: pp. 1–20. London, UK.

3i/PwC. 2002. Global Private Equity 2002: A Review on the Global Private Equity and Venture Capital Markets. *3i Group & PricewaterhouseCoopers*, (September): pp. 1–16. London, UK.

3i/PwC. 2003. Global Private Equity 2003: Report Charts. *3i Group & PricewaterhouseCoopers*: pp. 1–37. London, UK.

3i/PwC. 2004. Global Private Equity 2004: Charts for Press Release. *3i Group & PricewaterhouseCoopers*: pp. 1–27. London, UK.

Berg, Achim. 2005. *What Is Strategy for Buyout Associations?* (1st ed.). Berlin, Germany: Academic Readings on Private Equity, Verlag für Wissenschaft und Forschung.

Berg, Achim & Gottschalg, Oliver. 2004. Understanding Value Generation in Buyouts. *European Venture Capital Journal*, no. 114 (July – August): pp. 22–24.

Berg, Achim & Gottschalg, Oliver. 2005. Understanding Value Generation in Buyouts. *Journal of Restructuring Finance*, Vol. 2, No. 1, pp. 9–37.

Butler, Paul A. 2001. The Alchemy of LBOs. *The McKinsey Quarterly*, no. 2: pp. 140–151.

Clow, Robert & Smith, Peter. 2002. Scandals Help Break the Deal-Drought: Life Has Come Back to a Moribund Sector. *Financial Times*, FT Report on Private Equity, December 12: p. 1.

Davison, Chris. 2002. European Buy-out Firms: Robust and Realistic. *AltAssets*, November 27, http://www.altassets. com/features/arc/2002/nz1410.php.

De Leenheer, Lien & Wong, Melissa. 2000. Battle of the Atlantic: U.S. Private Equity Groups' Invasion of Europe. *Journal of Private Equity*, vol. 3, no. 3 (Summer): pp. 15–26.

Dixit, Amit & Jayaraman, Nithya. 2001. Internationalization Strategies of Private Equity Firms. *Journal of Private Equity*, vol. 5, no. 1 (Winter): pp. 40–54.

Doran, Alan. 2000. Historical Performance of European Private Equity. In R. Lake & R. Lake (Eds.), *Private Equity and Venture Capital: A Guide for Investors and Practitioners*. pp. 241–248. London UK: Euromoney Books.

Gompers, Paul A. & Lerner, Joshua. 2001. The Venture Capital Revolution. *Journal of Economic Perspectives*, vol. 15, no. 2 (Spring): pp. 145–168.

Hardymon, Felda, Lerner, Joshua, & Leamon, Ann. 2003. Between a Rock and a Hard Place: Valuation and Distribution in Private Equity. *Harvard Business School note*, no. #9-803-161: pp. 1–28. Boston, MA.

Harper, Neil W. C. & Schneider, Antoon. 2004. Private Equity's New Challenge. *McKinsey on Finance*, (Summer): pp. 1–4.

Kaplan, Steven N. & Schoar, Antoinette. 2003. Private Equity Performance: Returns, Persistence and Capital Flows. *MIT Sloan School of Management working paper*, no. 4446-03, (November): pp. 1–45. Fontainebleau, France.

Kehoe, Conor. 2002. Top Performance in European Private Equity. *Speech given at the EVCA Annual Symposium*, June 19. Athens, Greece.

Lamm, R. McFall & Ghaleb-Harter, Tanya E. 2001. Private Equity as an Asset Class: Its Role in Investment Portfolios. *Journal of Private Equity*, vol. 4, no. 4 (Fall): pp. 68–79.

Lerner, Joshua, Indahl, Reynir, & Zinterhofer, Eric. 1998. A Note on European Private Equity. *Harvard Business School note*, no.# 9-299-017: pp. 1–11. Cambridge, MA, USA.

Maier, Angela. 2004. Private-Equity-Börsengänge oft schwierig. *Frankfurter Allgemeine Zeitung*, no. 134, June 12: p. 19.

Markides, Constantinos C. 1999. In Search of Strategy. *MIT Sloan Management Review*, vol. 40, no. 3 (Spring): pp. 6–7.

Raicher, Serge. 2000. European Private Equity: Reaching Adulthood. In R. Lake & R. Lake (Eds.). *Private Equity and Venture Capital: A Guide for Investors and Practitioners*. pp. 27–33. London, UK: Euromoney Books.

Samdani, G. Sam, Butler, Paul, & McNish, Rob. 2001. The Alchemy of Leveraged Buyouts. In F. Budde, G. A. Farha, H. Frankemölle, D. F. Hoffmeister, & K. Krämer (Eds.), *Value Creation: Strategies for the Chemical Industry*. pp. 93–107. Weinheim, Germany: Wiley-VCH Verlag.

Smith, Paul L. & Butler, Steve. 2004. Private Equity in Chemicals. *Chemical Industries Association & Cogency Chemical Consultants Ltd.*, January 22: pp. 0–56. London, UK.

Smith, Peter. 2003. Looking to Better Times on the Horizon. *Financial Times*, Special Report European Private Equity, June 11: pp. I–II.

Sood, Varun. 2003. Investment Strategies in Private Equity. *Journal of Private Equity*, vol. 6, no. 3 (Summer): pp. 45–47.

The Economist. 2003. The Charms of the Discreet Deal. Special Report Private Equity, July 5: pp. 59–61.

Wright, Mike & Robbie, Ken. 1999. The Venture Capital and Buy-out Markets in the UK and Europe. In M. Wright & K. Robbie (Eds.), *Management Buy-outs and Venture Capital: Into the Next Millennium*. pp. 57–80. Cheltenham, UK: Edward Elgar.

31
What Attracts Private Equity Firms to the Chemical Industry?

Thomas Jetter

European chemicals managers today are asking themselves the same question as their US counterparts did a decade or two ago: how can private equity funds succeed, given their limited knowledge of the chemicals business itself? Conventional wisdom says that such limited knowledge of engineering and chemical markets and a lack of operational synergies for such M&A transactions should be obstacles to successful private equity investments.

An analysis of the strategies of private equity firms must start with an understanding of their investor base: the world's largest and smartest institutional investors. Big pension funds and other professional asset managers usually allocate a small part of their huge portfolios to private equity. Private equity investments are not publicly traded, the timing of the exit is uncertain, and the risks are considerable, which means that the returns need to be higher than for most other asset classes. Any investment nowadays will be judged and undertaken against a fairly aggressive business plan, the yardsticks are aspirational when private equity comes into play. Anyone working with financial sponsors should be aware that investors in private equity funds expect a 25 percent annual return (IRR) and a money multiple that reflects fairly long holding periods. The best and largest private equity firms have historically surpassed those return levels.

One of the key themes in financial markets is the large amount of private equity available. During the first quarter of 2005, approximately EUR 50 billion was available for investment. Some critics rightly say that such amounts create considerable pressure to execute transactions, and caution against deteriorating return levels. But the biggest asset managers in the world continue to provide the funding for the private equity sector because they believe that private equity can generate superior returns.

In an ideal world, a successful private equity investment would follow a clear trajectory: identify a hidden, undervalued enterprise; have a solid and actionable hypothesis on how to add strategic value; finance it smartly; develop a clear implementation path to improve the business; strengthen the management to execute the strategy; and find the best owner for the business at the exit point. In reality, of course, the situation is rarely so clear-cut, but some of these factors always play a role. In the real world, private equity firms also need to be involved in the evolv-

Value Creation: Strategies for the Chemical Industry. 2nd Edition. F. Budde, U.-H. Felcht, H. Frankemölle (Eds.)

ISBN: 3-527-31266-8

ing industry patterns, for example, taking the growing Asian economies into consideration. The holding period needed by financial investors to reach the results mentioned above is usually between three and seven years, after which time the required IRRs on the initial equity investment are diluted (Fig. 31.1).

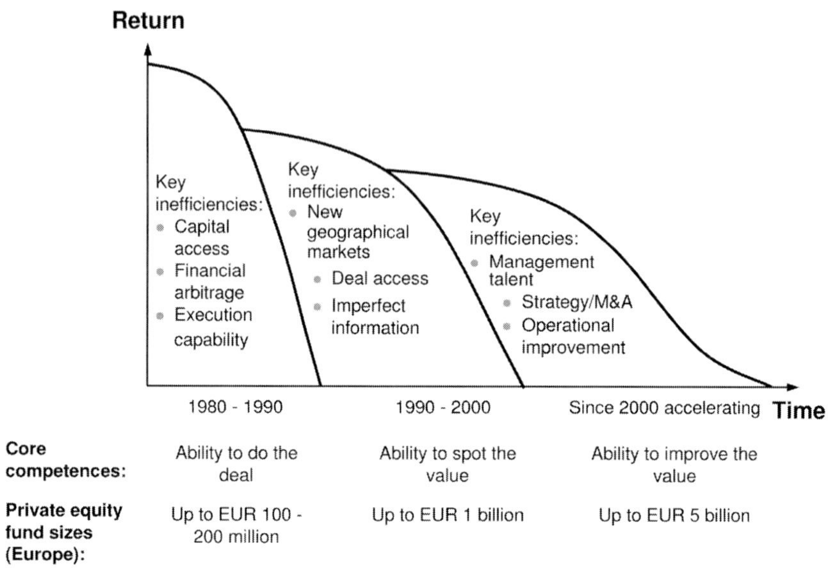

Source : Permira, McKinsey (adapted)

Fig. 31.1 Private equity key success factors.

Finding the right target and improving the value of an enterprise during the investment period requires a mix of industrial, strategic, financial, and soft skills. Big financial sponsors are key accounts of investment banks, consulting firms, executive search firms, and accounting and tax practices.

This chapter will explore the routes to value for private equity practitioners, focusing on how their pragmatic perspective leads them to act differently from the corporate world to achieve returns.

31.1 Chemical Industry Restructuring

Private equity funds are attracted to the chemical industry for the same reasons as they like any sector – they believe they can generate good returns. The investment opportunities are driven by the industry's structural challenges and evolution. Therefore, continuous analysis of equities markets, consolidation trends, and related M&A activity is essential.

Capital-raising through the equity markets, either through IPOs or capital increases, has been fairly low key over recent years. However, it picked up again in 2004 and that growth is expected to continue. Analysts are now shifting their focus away from cost management, streamlining, and restructuring, towards support for larger, visible acquisitions.

Global M&A activity in the chemical industry peaked during 1998 and 1999 at about EUR 100 billion p.a., reflecting the equity markets' support of such transactions. The market's focus on cost restructuring later led to a fall in M&A to an annual volume of about EUR 15-25 billion. Recent industrial buyer-driven transactions in chemicals include Lubrizol's acquisition of Noveon in June 2004, Cytec's acquisition of Surface Specialties from UCB in early 2005, and Albemarle's acquisition of Akzo Refinery Catalysts in August 2004.

The chemical industry has the potential for substantial further consolidation over the next few years. The precise timing will be influenced by the equity market cycles and general M&A activity. However, as in the past and in other industry sectors, consolidation will provide investment opportunities for financial sponsors.

31.2 Private Equity Transactions in Chemicals – Success Stories, Mostly

There were more than 30 sizable private equity chemicals transactions in the USA and Europe between 1999 and 2004, with a total valuation of about EUR 40 billion. Twenty of these deals were in Europe, and the average deal size was greater here as well. Two of the ten largest European chemical LBOs were public-to-private cases, the remaining eight were divestments of conglomerates to financial sponsors (Fig. 31.2).

Of the ten biggest European chemical LBOs since 1999, one was a full realization (100 percent sale), three were partial realizations (partial sale during IPO and/or recapitalization), one was a credit default, and five are still fully invested. Exit opportunities were limited until 2004 due to the equity market dynamics in Europe and the USA.

The exit by disposal of Messer Griesheim to strategic buyers and the partial Celanese IPO in the USA are successes with above-average returns. Recapitalizations due to strong operational cash flows and the availability of debt have also been successful: for example, the recapitalization of Cognis has returned more than the money originally invested, although the financial sponsors still own it entirely.

There has been only one equity write-off, as Vantico underperformed and the debt financing covenants went into breach. Matlin Patterson acquired the debt at a discount and took control via a debt-equity swap.

Taken together, these transactions confirm that chemical LBOs in Europe have been successful. Estimates suggest that the total returns after realizing the investments will be viewed as positive by investors in private equity funds.

Announcement date	Financial sponsor	Target	EV (m)	EBITDA multiple	Investment realized?
Dec. 2003	Blackstone	Celanese	$ 3,809	6.2x	Partial IPO
Apr. 2004	KKR/CSFB	Dynamit Nobel	$ 2,700	8.5x	No
Apr. 2001	Goldman Sachs/ Allianz Capital	Messer Griesheim	$ 2,468	7.4x	Yes
Sep. 2001	Permira/ Goldman Sachs	Cognis	$ 2,248	6.6x	Recapitalization
May 1999	Investcorp/ Cinven	Avecia	$ 2,106	9.4x	Partial sale
Dec. 2003	Bain Capital	Brenntag	$ 1,700	N.A.	No
Jul. 2002	EQT	Haarmann & Reimer/Dragoco	$ 1,680	11.9x	No
Oct. 2000	KKR/CSFB	Rockwood	$ 1,175	7.2x	No
Oct. 1999	Industri Kapital	Dyno	$ 1,130	7.4x	No
Dec. 1999	Morgan Grenfell	Vantico (Ciba)	$ 1,013	9.5x	No

Source: CSFB, Goldman Sachs

Fig. 31.2 Biggest European chemical LBO transactions 1999–2004.

A review of the situation in the USA leads to a similar conclusion. Fewer large deals were concluded between 1999–2004 than in Europe, but all have been successful. Noveon (formerly BFGoodrich Performance Materials) was acquired by an AEA-led consortium in February 2001 and sold in June 2004 to Lubrizol with a substantially higher valuation. Kraton Polymers, an LBO transaction concluded by Ripplewood in March 2001, was sold to the Texas Pacific Group in December 2003 with a good value uplift.

31.3 The Value Drivers of Private Equity Investments

From the perspective of financial investors, there are three factors that influence the degree of success: the business's strategy and operations, conditions on the financial markets, and the aligned interest of management teams and financial sponsors.

31.3.1 Business Strategy and Operations

Financial sponsors place heavy emphasis on strategy and operations, and getting these two right can lead to performance improvements and cash flow acceleration. These are usually rewarded by incentive systems for management teams and are supported by powerful implementation principles.

31.3.1.1 Performance Improvement

Bottom-line performance improvement is the key to value creation for a private equity investor. Even before doing due diligence for a potential acquisition, private equity practitioners establish hypotheses on how to create value. They will usually focus on three factors: organic growth potential, portfolio streamlining/add-on acquisitions, and cost improvement.

Organic growth potential is the most important factor. Companies with limited growth potential because of the nature of their markets are naturally seen as less attractive. Growth from targeted capital expenditure, internationalization, and investments into the sales and marketing infrastructure provides scale and often opportunities to increase margins. Only companies that can capture growth opportunities will remain competitive, thrive on innovation, and maintain leadership positions in their markets. Thus, "the top private-equity firms focus all their energies on accelerating the growth of their business." (Rogers, P. et al.). A recent empirical study of LBOs over a five-year timeframe shows that private equity-led companies grow substantially faster than their peers (Lüdke, U.).

Portfolio streamlining and add-on acquisitions are the second source of performance improvement. Many companies have complex portfolios of products and production infrastructure, which need to be simplified through divestments. The businesses to be sold are often small, but nevertheless add substantial complexity to the entire portfolio. Add-on acquisitions that improve the scale and scope of businesses are also a focus of chemical companies owned by private equity funds. Conventional wisdom suggests that leveraged transactions are not suitable for acquisitions because of the high debt burden, but there is ample empirical evidence that financial sponsors play the M&A market actively for acquisitions. The leveraged finance contracts that support LBOs usually provide room for such deals. Assuming that an acquisition generates value, financial sponsors will certainly consider an equity increase to strengthen the business.

Cost improvement and productivity gains are a continuous process in any enterprise. Financial sponsors focus heavily on cost, particularly in the early phase of an investment, and especially when an LBO has been a divestment by larger conglomerates. In chemical companies, cost improvement projects often focus on overhead cost and plant productivity. Such approaches are usually implemented in tandem with projects that strive to make the organization more market-driven and functional for the business. They may also require investment in plant automation to improve plant productivity.

31.3.1.2 Cash Flow Acceleration

LBO financings require a healthy cash flow for interest payments and repayments of debt tranches. The net debt position also determines the equity value that can be realized for financial sponsors on exit. The two major levers for improved cash flow management are working capital management and a disciplined capital expenditure program.

Working capital consists mainly of accounts receivable, physical stocks of raw materials, and semi-finished/finished goods; accounts payable are subtracted from this position as they are a source of finance until paid. For a typical chemical enterprise, the balance of these three positions will be about 25 percent of sales. Thus a chemical company with sales of EUR four billion will have working capital of about EUR one billion – an amount that requires careful management. Transparent and centrally monitored information on working capital is highly important, especially in complex international organizations. It helps initiate and follow up action plans to manage working capital, and increase cash-consciousness within an enterprise.

Capital expenditure is, of course, required to maintain and expand the company's asset base. It should be analyzed with the same close scrutiny as any other cash expense and must yield a return on investment that strengthens the economic value of the asset base. Specifically, capital expenditure for maintenance, removing bottlenecks, and major investments require different analytical approaches.

Flawed decisions can occur even in companies with a solid set of rules on capital expenditure. Best-practice guidelines for minimum hurdle return rates, defined technical evaluations, and established information and decision processes do not necessarily generate the optimum capital productivity. A sustained improvement in capital productivity requires embedding a capital ownership mindset into the organization. Beyond a good methodology toolset and organizational prerequisites for better capex decisions, it is the people who make the difference – i.e., the entire management team that prepares, takes, and executes such decisions needs to handle the information required for them with great commercial skill. In many corporations, however, financial incentives and informal rules can impede rational decision-making. In companies where cash flow is not measured and where the size of assets signifies the importance of managers, such decisions are often distorted.

31.3.1.3 **Implementation Principles**

Critics of financial sponsors may ask what they bring to the table that is different. The strategies for creating value are well known. The difference between most traditional ownership situations and financial sponsors lies in the rigor of managerial discipline in reaching goals once they are defined, and in putting together strong management teams (Rogers, P. et al.).

The private equity practitioners' approach differs in three ways from that of a traditional corporate organization. First, they concentrate on specific metrics, second, they institutionalize change projects with a lasting impact on performance early on during the investment period, and third, they take great care over the choice of management teams.

Limiting the financial metrics is a key ingredient in energizing a company for change. Corporations tend to focus on net earnings (per share), EBIT, sales growth, and market share. Financial sponsors, on the other hand, use a value-based reporting system that emphasizes EBITDA (resulting from sales and margin growth) and cash flow (resulting from net working capital and capital expenditure management). Business line managers also need to work with those financial metrics.

After concluding an LBO, early change projects are vital to signal a fresh start. Such change programs demonstrate different ways to operate, a rigorous strategic review and direction setting, and action plans that are logically derived from the strategic review and followed through with focus and energy. In many cases such change programs mark a sharp contrast to the traditional ownership. Performance-oriented and ambitious people enjoy this.

Superior management teams are, beyond all analysis and financial modeling, the most important factor in laying the foundations for good investor returns. Financial sponsors systematically seek out 'star managers' who can implement changes in organizations and businesses. Such people are also usually able to attract and retain other top talent – this brings speed and greater dynamics to idea generation, decisions, actions, and results. The compensation systems of such management teams are linked to the value-based reporting systems mentioned above.

31.3.2 Financial Market Conditions

Financial markets value companies. The cycle of equity markets measured as a multiple of the current EBITDA performance of a company is important, as it influences the value and the timing of exits for financial sponsors. The availability of debt markets is equally important because LBOs require sophisticated debt financing arrangements. Conditions in both equity and debt markets have an impact on private equity investments.

Understanding these cycles is an important ingredient of financial sponsors' success. In the debt markets, the risk attitude of investors in such instruments is changing. Debt leverages on last twelve months' (LTM) EBITDA have historically fluctuated from ranges of 3.5–4.5 times EBITDA to 5.5–6.5 times EBITDA. These differences reflect the supply/demand balance in capital markets for such industry risks, and also reflect the overall economic risk. The valuations on equity markets are equally volatile. The enterprise valuation measured as a multiple of the EBITDA can fluctuate by two to three times EBITDA through the chemical and capital markets cycle.

Beyond the capital market analysis, a financial sponsor needs to judge the sustainable value of a business over the medium to long term. This needs to be adaptable, as equity markets may not be available for IPOs, equity increases, or larger acquisitions. From the financial sponsor's viewpoint this means that the private equity fund has limited exit opportunities during such periods.

Another characteristic of international equity markets, especially for chemical companies, is that valuations on US markets are high in by comparison with those on European markets (Fig. 31.3). US enterprise valuations are at least 1.7

times EBITDA higher than in Europe. Global financial investors will try to arbitrage this valuation differential. The recent Celanese public-to-private transaction on the German equity market and its re-IPO on the US market is a good example.

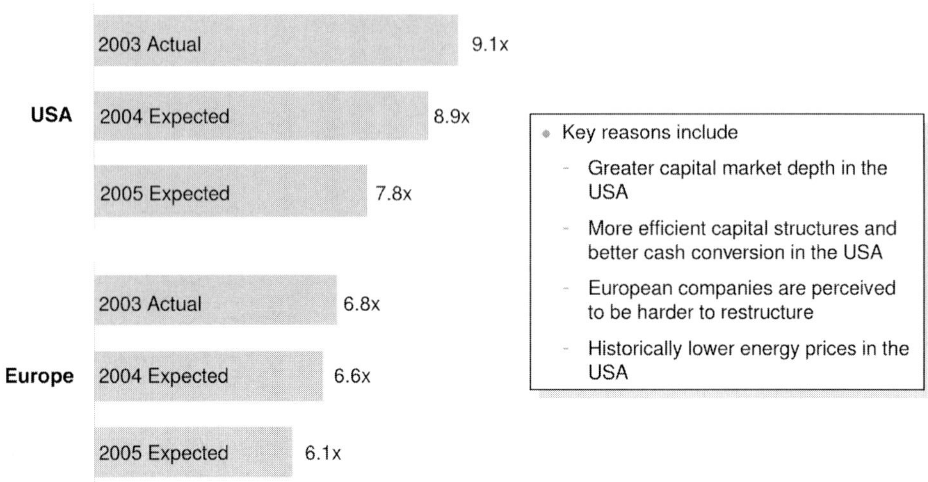

Source: CSFB and Lehman Brothers Analysis Summer/Autumn 2004

Fig. 31.3 Specialty chemical companies' valuation comparison USA and Europe.

Two fundamental aspects need to be considered when assessing the impact of the financial markets on the success of private equity investments. First, all financial markets are cyclical, and the supply and demand of financial products drive the availability, volumes, and pricing of equity and debt financing. Second, European and US capital markets provide access to different investors and therefore different market conditions. In almost all cases, the US market provides higher volumes and more favorable conditions for equity and debt transactions.

31.3.3
Aligned Interest of Management Teams and Financial Sponsors

However thorough the financial analysis and however rigorous the operational and organizational improvements are, every success story is underpinned by dedicated and competent people. Highly skilled and motivated management teams lead companies that have been acquired by financial sponsors. Such management teams seek entrepreneurial freedom, but they also seek entrepreneurial success.

The increase in value of an LBO-led company at exit is shared between the financial sponsors and management teams. This is why LBOs are often synonymous with management buyouts. This participation in the success is important as it aligns financial sponsors and management teams in their own economic best interest. Joint financial goals help create a solid platform for cooperation in two ways.

1) First, they are a common denominator when discussing annual budgets, cash flow statements, and performance reviews. Any of those sets of information has a measurable impact on the value of the invested equity.
2) Second, when the financial sponsor exits the business, the joint financial success is equally measurable for the private equity investors and the management team sharing the equity participation. This is important, as it demonstrates fairness to both those who have worked hard on the financial sponsors' side and those who have worked hard as responsible managers of LBOs.

Managers who have worked with financial sponsors greatly appreciate the entrepreneurial freedom. Although the limitations of LBO financing suggest limitations on spending money, a stand-alone, private-equity-led company offers greater entrepreneurial freedom than any other ownership situation. Good CEOs of such companies find unusual but effective ways to energize their organizations. Corporate rules in large conglomerates that are not necessarily functional for a specific business can be eliminated and 'politics' within the top ranks of a company play a much smaller role once financial sponsors set the pace. Simpler goals, rules, and means to generate results can also strengthen leadership.

31.4 Summary

- Private equity investors have been successful in the chemical sector; and have helped strengthen chemical businesses that have been carved out of larger conglomerates.
- There is substantial consolidation and de-merger potential in the industry. Private equity investors are ready to cope with the challenges of this industrial evolution.
- A limited number of value-oriented metrics and visible change projects have energized businesses acquired by private equity investors.
- Their skilled and highly incentivized management teams have benefited from entrepreneurial freedom.

References

Butler, Paul, Samdani, G. Sam, McNish, Rob. The Alchemy of Leveraged Buyouts. In: Budde F., Farha G.A., Frankemölle H., Hoffmeister D.F, Krämer K. (Eds.), Value Creation: Strategies for the Chemical Industry. Pp. 93–107. Weinheim, Germany: Wiley-VCH Verlag. 2001.

Lüdke, Ulrike. Wachstumsturbo Private Equity. in: Finance. Dezember 2004/ Januar 2005. pp. 44–49.

Rogers, Paul, Holland, Tom, Haas, Dan. Lessons from Private Equity Masters. in: Harvard Business Review. June 2004.

32
Facing China

Sönke Bästlein, Ralf Dingeldein, Tomas Koch, and Karsten Neuffer

The numbers are mesmerizing. Since 1980, China's economy has grown by more than nine percent a year and the country now has a GDP of approximately EUR 1.3 trillion. It is the world's seventh largest economy (second if purchasing power parity is used), and annual growth is expected to be around double-digit rates.

The chemical industry is a pillar of this booming economy. It employs 3.5 million people and contributed 2.5 percent to the total value-added in China in 2004 – roughly four times the figure in the USA and about twice as much as in Western Europe.

Consumer demand will be a key growth driver: China's 1.3 billion people have both an overwhelming demand for a more modern lifestyle and rising annual per capita disposable incomes (from less than EUR 500 in 2004 to almost 700 by 2010). This makes China increasingly important for major end-user industries. Significant government infrastructure expenditure as well as high levels of foreign direct investment also fuel this growth.

Multinational chemical companies have long been aware of China's potential, but what was once viewed as a market only for the brave must now become a cornerstone of any ambitious CEO's agenda. Nevertheless, seizing the opportunity without encountering problems is not easy.

The Chinese market may be highly attractive, but it also differs greatly from that in Europe or North America. It is extremely dynamic, and the speed of change combined with a high degree of uncertainty means that companies appear and vanish at rates unknown in the West. This makes industry and market data hard to obtain – and the validity of available data questionable.

Trust-based relationships are still very important for doing business in China. They open doors and influence decisions: a contract may be worth only as much as the relationship behind it.

The third big difference in the Chinese market is that competition is based on cut-throat efficiency. Chinese entrepreneurs tend to go for the maximum cash return and focus on scale rather than margins, with the goal of becoming market leaders as fast as possible.

Chemical companies cannot afford to ignore this market. Here we give an overview of the importance of the chemical industry in China, analyze why multina-

Value Creation: Strategies for the Chemical Industry. 2nd Edition. F. Budde, U.-H. Felcht, H. Frankemölle (Eds.)

ISBN: 3-527-31266-8

tional chemical corporations have been hesitant about China for so long, and then suggest how to succeed in this attractive arena.

32.1
China Is No Longer 'Optional'

The Chinese chemical industry is still at an early stage of development and is highly fragmented. Almost 14,000 companies compete, with state-owned Sinopec and PetroChina the most dominant players. State-owned enterprises are, however, being replaced increasingly by private enterprises. In only four years, the share of private companies in the sector has grown from 23 to 33 percent.

China accounts for thirteen percent of the world chemical market, with sales of EUR 141 billion. Consumption is expected to grow at between eight and nine percent per annum over the next ten years – already substantially lower than in the relatively recent past (Fig. 32.1). Consumption growth is driven mainly by textile and clothing end-user industries, followed by automotive, electronic goods, and industrial machinery. By 2015, the Chinese market will have a volume of more than EUR 370 billion and account for more than a fifth of the world's market share. In other words, it will be on a par with the US or Western European markets.

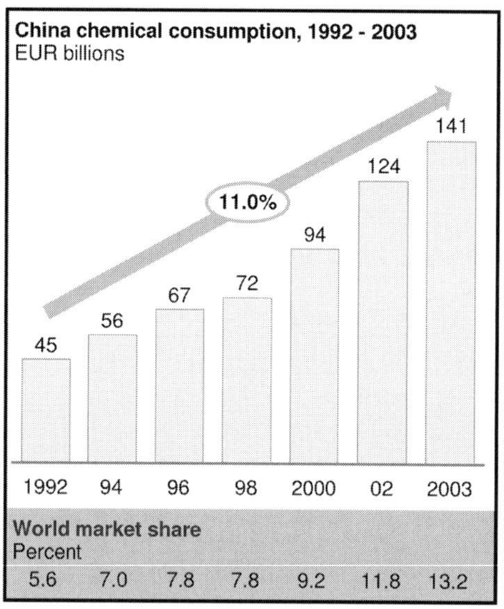

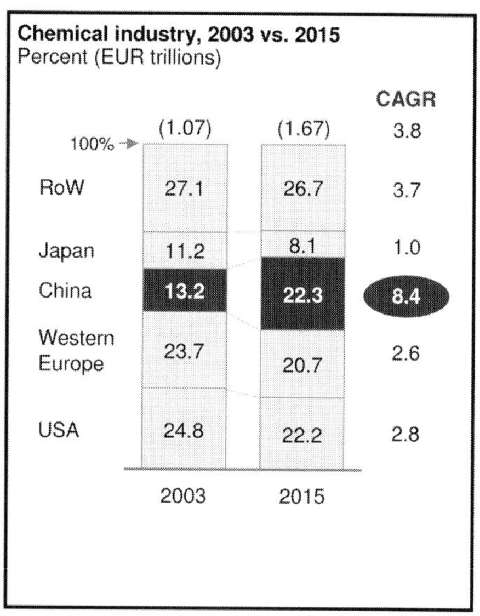

Source: ACC, Import and Export Database, Global Insight, China Chemical Companies Database (CCCD), McKinsey analysis

Fig. 32.1 Strong growth of the Chinese chemical industry.

The Chinese chemical industry is already as profitable as that in the West, with reported EBIT margins of between six and seven percent. There is, however, a great disparity between the best and worst performers. Wholly foreign-owned enterprises and joint ventures are typically the most profitable, at least in specialties (Fig. 32.2). Commodities average an EBIT of six percent, with little variance between sub-segments. Specialties have an average EBIT of eight percent, with a much wider variance between sub-segments, electronic chemicals show the highest returns.

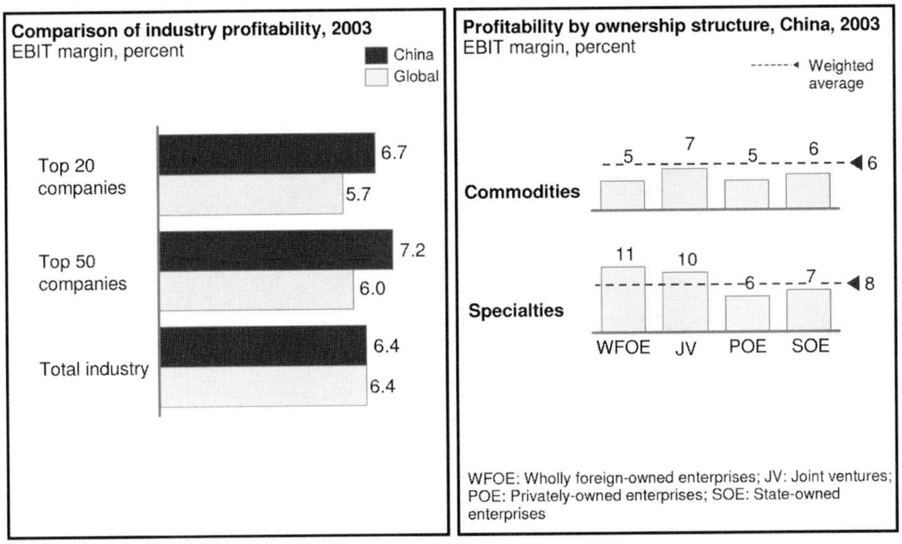

Source: C&EN analysis, annual reports, China Chemical Companies Database (CCCD), McKinsey analysis

Fig. 32.2 China's chemical industry is as profitable as that of the USA or Europe.

The margins are partly supported by a sustainable share of imports, allowing for import parity pricing. In 2003, approximately 21 percent of China's total chemicals consumption was covered by net imports, mainly commodities. For example, at least half of the consumption of synthetic rubber (69%), plastics (55%), and organic chemicals (50%) was met by net imports. Even at production growth forecasts of 9.6 percent p.a., Chinese capacity levels will not meet demand in the foreseeable future and the country is expected to remain a chemicals net importer beyond 2020.

Nevertheless, some chemical companies have started expanding abroad, using the domestic market as a springboard to international success. Indeed, for some specialty chemicals, for example in the agrochemicals/pharmaceuticals segment, China has already become a net exporter, profiting from landed cost advantages of up to 40 percent when exporting to Europe or the USA. These cost advantages are based primarily on significantly lower labor, land, and investment costs. The new competition is likely to have a noticeable effect on price and market share in these segments, adding to the profitability pressure on multinational companies in their home markets.

To take one example, in the late 1990s a Chinese entrepreneur noticed the rising imports of a specialty intermediary chemical and opened a small plant to meet local demand. Prompted by its domestic success, he rapidly increased capacity and began exporting. The company's entry into the global market brought down prevailing prices by about 40 percent, and by the end of 2003 it was one of the leading global players in the niche. Another striking example is the nutritional and health supplement sector. Chinese companies produce these supplements at about a quarter below the cost of US manufacturers. As a result, almost three quarters of such supplements sold in the USA come from China. The Chinese dragon is starting to show its teeth.

32.2 Most Chemical Companies Are Not up to Speed

The Chinese chemical market is attracting huge foreign direct investments. Between 1993 and 2003, investment projects amounted to around EUR 20 billion of contractual FDI, with Bayer, BASF, Shell, and BP the four biggest investors. Over the last ten years, five percent of foreign direct investment was into the chemicals sector. Multinational corporations (MNCs) tend to enter via joint ventures partially driven by the need for feedstock access, but wholly foreign-owned enterprises will become more common following China's membership of the WTO.

There is still a long way to go before MNCs' presence in China is equal to the country's importance. Only a few have managed to reach top-ten positions in any segment: BASF and Dow in synthetic rubber/thermoplastics, BASF and DuPont in paints and coatings, and Ciba Specialty Chemicals and Clariant in organic pigments. Not one MNC generates a share of sales in China that is even close to China's global market share of thirteen percent (Fig. 32.3). Based on today's level of investment and activity, this discrepancy is not likely to be overcome soon. Moreover, MNCs in China still import approximately 80 percent of what they sell, forgoing profit opportunities from local production.

We see two reasons for MNCs' lack of enthusiasm for investing in China: a lack of strategic foresight and a lack of determination and persistence in implementation.

From a strategic point of view, MNCs have tended to overestimate the risks of the Chinese market and underestimate the opportunities. In the aftermath of Asia's economic crisis in the late 1990s, and with the bursting of the Internet bubble still fresh in people's minds, companies were too afraid of taking on what they saw as yet another risky business strategy. At the same time, they did not anticipate the speed with which the Chinese market and potential competitors developed. Between 2000 and 2003 almost one in three Chinese chemical companies grew faster than 20 percent p.a., for example Zhejiang Longsheng Group at 32 percent p.a., or Dalian Luminglight S&T at 28 percent p.a. MNCs also underestimated customer migration. Manufacturers of textiles, electronics, and automotive vehicles, for example, have migrated to China faster than expected in order to capture the factor cost advantages and gain access to the big domestic market. In addition, they have started sourcing locally faster than expected.

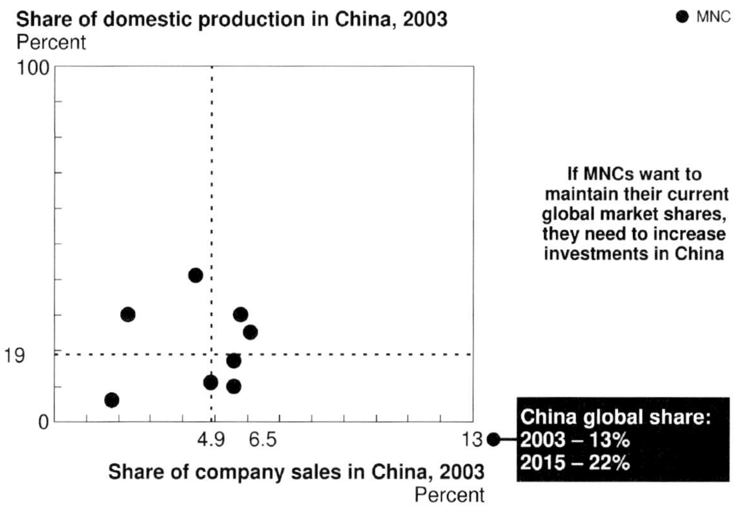

Source: SRI, press clippings, web research, McKinsey analysis

Fig. 32.3 MNCs have a limited presence in China – for now.

For those companies getting to grips with the strategy, implementation still proves problematic. Chemical companies have often lacked the rigor, the determination, and the persistence needed for success in such a difficult market. Four stumbling blocks in particular have impeded their development.

Insufficient corporate support and encouragement. Chemical companies have a long tradition of focusing on their home markets and dealing with the restructuring pressure from the continuing price-cost squeeze. Emerging markets were on the radar of most companies, but senior management typically did not pay China sufficient attention. This lack of support and encouragement from the corporate center often left business units venturing into China feeling isolated. The financial resources and management capacity provided were not sufficient to develop the new operations successfully. And, while there was too little corporate push, there was also too little corporate pull. Corporate incentive systems and processes are typically geared towards short-term profit improvements with low rewards for higher risk-taking – such as achieving top-line growth in the Chinese market.

Market dynamics create uncertainty and complicate data gathering. The highly dynamic development of the Chinese chemical market has made it hard for foreign players to develop up-to-date market knowledge, which has often left business plans for market entry without a solid foundation. The sheer size of the chemical market alone is awesome. Companies have to get to grips with how to build a database of potential customers when there are thousands and thousands of geographically widespread companies, vanishing, emerging, and growing. They have to think about how to assess the risk of such customers. They need to consider how to screen potential acquisition targets in a market where 20 percent of the companies did not even exist three years ago and where ROS ranges from less than –15 percent to over +15 percent with no clear correlation between performance and segment, size, or ownership type. And then foreign entrants have to think about how to find the right location for direct investments, when 21 different chemical zones each offer different tax advantages, segment foci, infrastructures, and so on.

Not enough friends in the right places. Relationships with different levels of government and local authorities are essential for doing business successfully in China. Many chemical corporations, however, lack the kind of network needed, for example, to tap into the complex world of health, safety, and the environment (HSE) legislation. The absence of these relationships slows down the processes for M&A negotiations or for obtaining production permissions and product certifications. Moreover, a limited level of integration in industry-specific networks inhibits word-of-mouth recommendations and thus slows the expansion of both the customer and the supplier base. Even when relationships are established, it can take considerable time before they yield their full benefit.

Unprepared for cut-throat competition. As we mentioned earlier, Chinese companies are ruthless in their efficiency, and foreign players trying to emulate this can run into major hurdles. In particular they are at threat from a weak intellectual property environment, disappointing supplier quality, and unreliable local distributors.

Intellectual property (IP) is technically protected by Chinese law, but enforcement is a major problem. Uncooperative local authorities often tolerate counterfeiting because it can benefit local economies, and Western chemical companies may find their drums with counterfeit labels and products. The authorities have insufficient IP knowledge, especially outside the major business centers, and enforcement bodies lack the funding – and often the willpower – to implement the laws; few counterfeiters have yet been prosecuted – the most effective way of deterring forgery. This may lead MNCs to feel they are stuck between a rock and a hard place. If they enter the Chinese market with state-of-the-art technology, IP drainage is likely. If they enter with anything less than cutting-edge technology, they must start at a disadvantage.

Finding local chemical suppliers that can meet price, product quality, and reliability standards is another major hurdle. Chemical companies have often learned the hard way that the quality of the final products can be very different from that of the original samples. Extensive screenings, checks, and quality tests are indispensable.

Infrastructure and distribution raise similar problems. Capable third-party chemical distributors offering consistently high levels of service are not universally available. According to a 2002 study published by the State Council's China Development Research Centre, 70 percent of China's commercial enterprises have their own fleet of vehicles and 80 percent own their warehouse facilities. MNCs thinking about setting up their own (Western-style) nationwide chemical distribution network have soon been discouraged by the high risks and costs involved.

Taking all these barriers into account, MNCs that want to maintain their world market share have only one option: they must put their concerns to one side and boost their investments in the fast growing Chinese chemical market.

32.3 The Lessons of Confucius

Most MNCs acknowledge that increasing their Chinese activities is a strategic necessity. Paying heed to four Chinese proverbs may help them maximize the opportunity without taking undue risks.

China is not an easy market to enter; a deep understanding of the country and culture is critical to success. Chemical companies that want to thrive must base their strategy on the Chinese way of doing things. They must be flexible and pragmatic, and accept that traditional Western approaches to doing business may have limited, or worse, no success in this particular context.

Companies therefore need to work on two fronts: at home they must develop a corporate strategy for China and adjust the organization and processes accordingly, and in China they need to achieve excellence in setting up and managing the local operations.

32.3.1 A Good Strategy at Home Wins the Battle a Thousand Miles Away 运筹帷幄，决胜千里

A company's success in China is, of course, decided in the Chinese market. However, the corporate center must lay the foundations for this success. Moving into China is not the same as entering just another market: there will be significant consequences for the entire organization. In addition to the top management stewardship which will be vital in steering the company in the right direction, three areas in particular merit close attention: the top-down corporate perspective, the organizational structures, and key corporate processes.

Management must develop a clear and consistent *corporate perspective* for China. This should start with a clear vision on strategic priorities based on the opportunities and threats, and should include growth and sales share targets. The vision then translates into an investment strategy – specifically deciding on whether to build or whether to acquire.

The company's *organizational structures* then need to be aligned with the strategic vision. Establishing a China council at headquarters, consisting of board members and other senior managers, can provide clear and visible support and demonstrate the seriousness of a company's engagement in China. This needs to be supported by other operational functions.

Chinese R&D activities should eventually be integrated into the global R&D network of the MNC to make best use of the Chinese factor cost advantages and technological innovations. China is slowly but surely becoming a scientific research powerhouse. The number of patents granted in chemistry and metallurgy, for example, climbed from 1,572 in 1992 to 8,974 in 2003. However, the potential of leading universities and R&D facilities, such as the Chinese Academy of Sciences or the Chinese Chemical Industry Research Institution, has barely been assessed and they are therefore underused. Tapping into Chinese R&D networks early on will help companies participate in the latest technology developments. BASF, for instance, has established a Sino-German R&D fund to support scientific co-operation, which has led to the establishment of 39 scientific and research cooperation projects with 15 colleges and research organizations. Meanwhile, Henkel Technologies cooperates with six research groups from five universities.

During the early stages, however, companies must keep in mind the problem of IP theft. GE Plastics' R&D in China manages this by focusing more on development than on research, producing those parts with the most strategic and sensitive research knowledge only in the USA and Europe. Another chemical company has successfully bundled its product with testing devices to identify counterfeits. Now that requirements for joint ventures are diminishing, many MNCs are moving to protect their IP by setting up wholly foreign-owned enterprises.

Chinese sourcing and production functions should also be integrated into the company's global setup. This may require rethinking both networks, but defining an optimal global footprint from the outset helps avoid internal battles for capacity utilization as Chinese-produced sales cannibalize their western counterparts.

Implications for production include a re-allocation of volume and mix to optimize re-import opportunities from China. This will have consequences for existing plants in other locations – some may need to be closed, others upgraded or refurbished. Companies also need to consider how to transfer best practices between plants in different countries.

Corporate functions such as M&A and human resources need to be replicated in China to accommodate local particularities. In M&A, for example, the deal volume in the Chinese chemical industry quadrupled between 2003 and 2004. Keeping track of opportunities, evaluating and structuring deals, as well as managing the multitude of stakeholders involved is too complex a job for a remote headquarters to do alone. Using local experts not only increases transaction efficiency but also secures better deals and thus lower multiples. Nevertheless, in the wake of a thorough cost and risk analysis, some chemical companies have established a completely Western 'shadow management' to limit the risks involved in mega-investments.

To capture economies of scale and scope, some large MNCs have set up multi-user service platforms in China where functions such as legal affairs, financing, tax, and IT are used jointly by several business units. Smaller companies must at least ensure that their China back-office functions have critical mass from the outset.

Finally, companies need to adapt key *corporate processes* that influence the entrepreneurial behavior of those divisions and business units active in China. One example of this is adjusting the budgeting and planning processes: scenario planning with a two-year timeframe can help companies handle the level of dynamism in China. Companies should also, in the growth period, put more emphasis on volume than on profitability targets in the Chinese business world. Nevertheless, minimum profitability targets must be both set and met. This change in key performance indicators will need to be reflected in incentive schemes.

32.3.2
He Who is Close to Water is First to See Reflections
近水楼台先得月

Gaining proximity to customers and markets is vital for success in the fragmented Chinese chemical market. Market knowledge – hard to gain from outside – must be collected and consolidated systematically from the bottom up. Analysts must examine how customer needs and buying factors differ, both from China to the rest of the world and within China by region and segment. The profit potential of each customer needs evaluating, considering not just size and growth but also risk and relationship. The company must understand how products should be adapted to meet local customer preferences, raw material availability and/or low Chinese price levels. To take one example, Ciba Specialty Chemicals is developing a UVA sun blocker that meets the special healthcare and skin color requirements of many Asian women.

Customer and product portfolios must be reviewed continuously given the speed of change and rapid maturation of both the demand pattern and the supplier base. In polyolefins, for example, MNCs will typically sell base grades before they focus on special grades for high-end applications. This strategy will work only for a limited period and will need then to be adapted very quickly given the speed at which market conditions change.

Despite the demanding nature of the Chinese market, the marketing and sales skills of local staff are usually not as well developed as they are in the West. MNCs therefore need to set up special training programs for these staff. A marketing and sales academy that systematically covers topics such as customer portfolio management, pricing excellence, or multichannel management can give companies a head start.

Getting the products to the customer also needs to be tackled systematically. It is unlikely that a reliable nationwide distributor will be available, so companies must establish and manage a portfolio of distributors to achieve national coverage. In addition, companies may find it beneficial to establish in-house distribution centers in some regions to put them in close proximity to key customers. Regularly asking these customers for feedback on product delivery and service level will help align the service centers with customer needs. Third-party distributors can be used to serve smaller customers in core regions or for developing non-core regions thereby keeping costs and risks low, but this needs to be evaluated regularly. Distributors should be chosen based on their growth potential and their capabilities, including the willingness to share market intelligence.

CNOOC-SHELL, the joint venture between Shell and China National Offshore Oil Corporation, differentiates its distribution channels by products and customers. Core products (e.g., mono-ethylene glycol and styrene monomer) as well as non-core products for core customers are sold through its own sales channel, with proprietary distribution centers in key regions near ports. Other products (such as polyethylene) are sold through Sinochem and smaller distributors/logistics providers – thus drawing on existing assets and customer relationships.

32.3.3
When Away from Home, Seek Help from Friends
出门靠朋友

The importance of well-established relationships is slowly fading, but they will always have a significant impact on a company's ability to influence the regulatory environment. Governments and regulatory bodies at all levels (central, provincial, municipal) still have substantial degrees of freedom in decision-making as written legislation (e.g., on plant expansion) is often quite vague.

The best approach is first to conduct a comprehensive screening to identify all the relevant bodies and then to prioritize them. Companies must approach stakeholders in order to understand their individual agendas following a key account management approach. This requires having dedicated resources to establish, maintain, and update relationships – be it with the central government, the local Chinese Communist Party, or potential business partner networks. It is almost impossible for a foreign company to invest too much management time in giving *mianzi* (face). BASF, for example, reportedly spent four years negotiating with partners and multi-level governments, addressing different stakeholders' interests and China's broader goals in the petrochemical industry, to implement the Yangtze-BASF ethylene project in Nanjing.

Considering how different the Chinese market is, local professionals must be employed from the start. The number of expatriates – typically high at the time of market entry – is likely to decrease over time. However, recruiting and retaining local talent has emerged as one of the most significant problems for Western companies. China has an abundant supply of both low-skilled labor and graduates, but the number of eligible candidates falls rapidly when foreign companies look for the talent required. Most graduates have little familiarity with Western management styles and have foreign language problems. Furthermore, many graduates do not see the chemical industry as very exciting compared with consumer goods companies such as Procter & Gamble or Unilever, which makes it even harder to attract talent. Also, job switching in China is comparatively high at around 15 percent.

The best way to attract and retain local talent is to become thought of as a profitable, strong, and trustworthy company. Cooperating with top universities (e.g., Peking University, Tsinghua University, or Fudan University) and research institutions (e.g., Chinese Academy of Science) by sponsoring projects or by involving them in the recruiting process is one way of fostering this kind of image. Another is to establish scholarships or internships for students. Providing a long-term perspective will help retain high-caliber people. Compensation packages often include attractive loans, housing, company options and stocks, as well as tenure-based bonuses. Some MNCs have built on-site staff training facilities while others offer tailored multi-year development programs and mentoring programs, overseas postings and opportunities for quick promotion, like McKinsey with its European Asia House.

32.3.4 The Sharper Saw Does the Better Job 工欲善其事，必先利其器

Operational excellence in production and sourcing is a must for any MNC investing in China. Given the multitude of chemical parks in the country, choosing the right production location is not easy (Fig. 32.4). Site selection should be based on functional factors including proximity to customers (particularly important for specialties), availability of a skilled workforce, land price, tax relief, and investment regulation; as well as on operational factors such as raw materials supply (particularly important for commodities), energy supply, and infrastructure. Evaluating candidate sites against these criteria will usually produce a shortlist of just two or three locations. Integrated sites obviously offer substantial synergy opportunities as well as other benefits such as preferential energy prices and a higher priority during energy shortage seasons. The presence of top global utility providers ensures high service levels and reduces supply risks. BASF-YPC, a joint venture between BASF and Sinopec, operates an integrated petrochemical site in Nanjing featuring a streamcracker and nine downstream plants. Degussa, meanwhile, is investing in a multi-user site in the Shanghai Chemical Industry Park, with a polyesters and a colorants plant as the first projects.

Investment levels for plants must be adapted to local standards without compromising on HSE standards. Costs average around a third lower than in the West (Fig. 32.5). The equipment and technologies used must take into account the high price sensitivity of the Chinese chemical market: Chinese plants tend to be less automated given the low cost of labor and the high cost of imported machines. More and more foreign companies have also started using local design institutes to reduce their costs, and although the Chinese chemical industry is still dominated by older, smaller plants, the number of state-of-the-art plants, such as BP's in the Shanghai Chemical Industry Park, will increase significantly within the next couple of years.

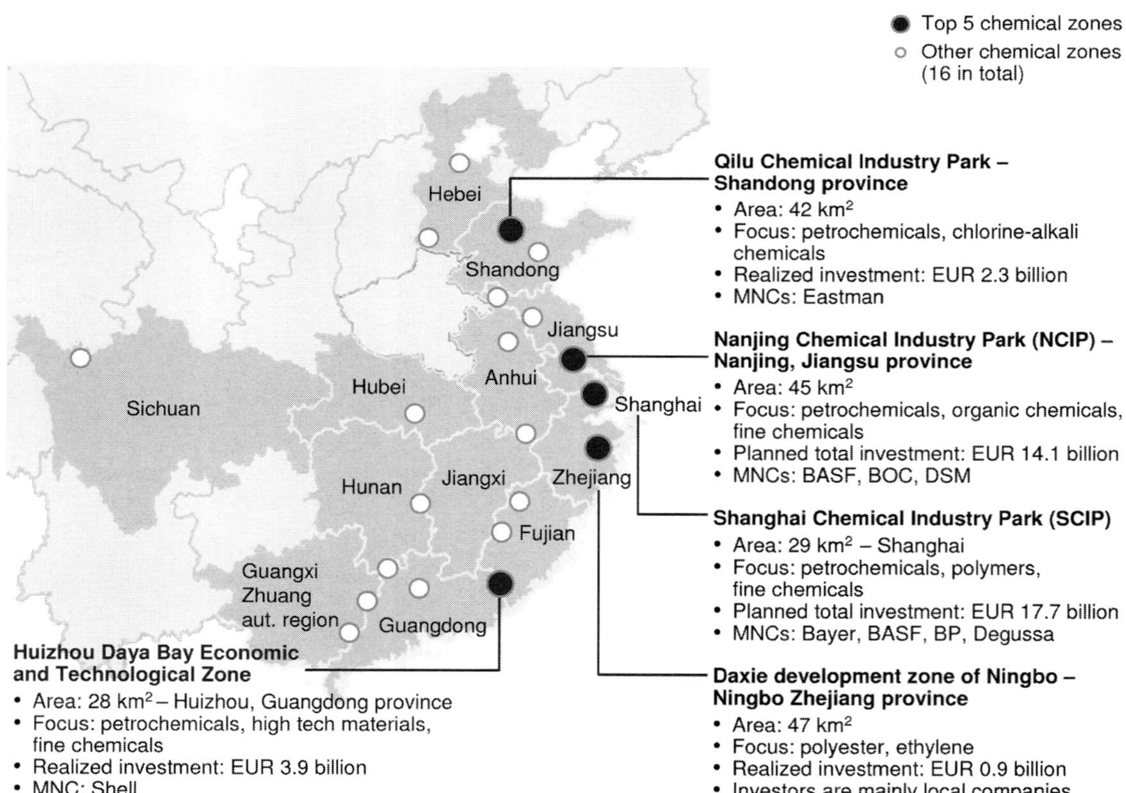

Source: SCIP research report, web research, McKinsey analysis

Fig. 32.4 Overview of top five chemical zones.

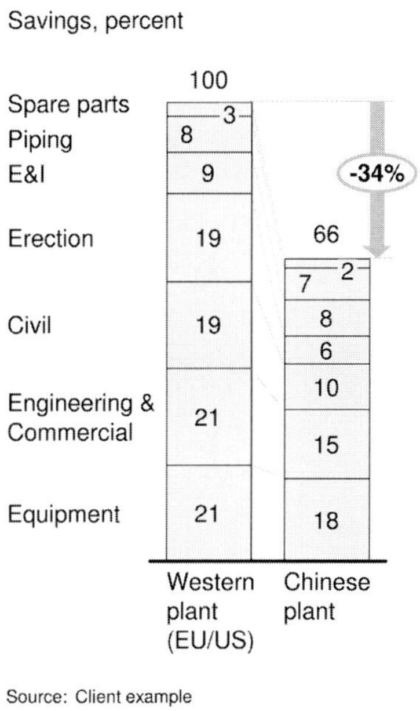

Fig. 32.5 Chinese chemical plants offer potential savings of a third compared with the West.

Sourcing in China builds the basis for a competitive advantage beyond pure cost savings. Developing a reliable supplier network is the first challenge. The local purchasing group will need guidance from and close links to its counterparts at headquarters; but shifting decision-making to local buyers will help them respond quickly to market developments. Any skills gap can be overcome with coaching, for example in using advanced purchasing tools such as clean-sheet costing and linear performance pricing.

The supplier base is likely to be fragmented to start with as MNCs seek to mitigate the risks, they might even rely on Western suppliers for critical materials. Eventually they will amass enough data to assess local suppliers and can reduce the number used. The most capable ones should then be developed through knowledge sharing and consulting. Introducing concepts such as lean manufacturing will lead to further cost savings, and joint application development will improve quality and speed up the development processes. Once a close relationship with a supplier is established, exclusive contracts should be signed.

Two questions nevertheless remains open: can China provide the raw materials at the competitive price levels necessary for commodity production; and can MNCs secure access to these raw materials?

32.4 Summary

- China's chemical market is growing rapidly and by 2015 will be on a par with the USA's and Europe's. Multinational chemical companies have long underestimated the opportunities of this booming market and have been too tentative about the risks.
- The time for pondering is over – if these companies want to maintain their global market shares they must boost their Chinese investments and adhere to the wise Chinese sayings.
- A top-down corporate strategy for China is an essential start. Thinking about how to get close to customers will help put the company on the right trajectory, and its path will be eased by building the right relationships and hiring and training the right people.
- However, none of this will matter if the company loses sight of operational excellence – paramount in the highly competitive Chinese chemicals market.

33
China – Key for Success in Asia

Jürgen Hambrecht

In 2005, BASF will celebrate its 140th anniversary and can look back on 120 years of trade with China. This year BASF also starts up the largest investment project in the history of the company: its new “Verbund” site in Nanjing, China. This article analyzes China’s key role in achieving BASF’s strategic goals in Asia and reviews BASF’s progress in the country.

33.1
Why China?

According to current estimates, approximately 17 percent of the world population, in other words 1.1 billion people, are reckoned to be potential consumers of products from the chemical industry[1]. By 2015, the total number of consumers is expected to almost double to two billion due to rapid economic growth, mainly in the emerging economies. Almost half of these consumers will then live in Asia, and the number in China is likely to climb nearly tenfold to 700 million.

Together with the trend toward offshoring of production to low-cost countries, these demographic developments will lead to a dramatic shift in chemical markets in the coming years. With a 31 percent share of the global chemical market, valued by BASF at EUR 1,200 billion, the Asian chemical market is already as large as that of North America. Within Asia, Japan is the largest market, closely followed by China. In the period up to 2015, BASF expects the global chemical market to grow on average by 2.5 to three percent per year, whereas Asia is forecast to grow by 3.5 to four percent per year. As a result, Asia will become the world’s largest chemical market in only a few years. Within Asia, the Chinese and Indian markets are anticipated to be the most dynamic, with annual growth of six to seven percent. China will therefore overtake Japan as the world’s second largest national chemical market by 2010 and will become by far the largest chemical market in Asia.

1) Consumers of chemical products are defined as consumers with an annual income of more than USD 10,000 in terms of purchasing power parity.

Value Creation: Strategies for the Chemical Industry. 2nd Edition. F. Budde, U.-H. Felcht, H. Frankemölle (Eds.)

ISBN: 3-527-31266-8

BASF recognized the enormous potential of the market early, and had already defined a long-term strategy with ambitious goals for Asia in 1996. By 2010 in Asia Pacific, BASF wants to:

- Contribute 20 percent of the BASF Group's sales and earnings in its chemical businesses
- Strengthen its position as one of the top five suppliers in strategically relevant markets
- Form the best team in industry
- Build a local manufacturing base to secure 70 percent of domestic sales

And as the biggest and fastest growing Asian chemical market of the future, China plays a key role in our strategy.

33.2
BASF and its Long Relationship with China

In trying to achieve these goals, BASF can build on a long tradition and excellent starting position in Asia, and in China in particular.

33.2.1
BASF's First Steps in China

The business with synthetic dyes forms the starting point of BASF's success story in China. Expansion of the company's sales organization prompted the company's first activities in China in the mid 1880s, when it sold and marketed the magenta dye fuchsin to the flourishing cotton cloth market (Sung, J. K.).

Between 1890 and 1904, there was a significant improvement in sales of BASF products to Asia, which until then had accounted for only four percent of turnover. The surge was mainly due to the fact that large quantities of synthetic indigo were being purchased as a substitute for the inferior natural product. Performance in China mirrored that of Japan, and each country accounted for around seven percent of BASF's total worldwide turnover by 1904 (Abelshauser, W. et al.).

33.2.2
The Post-Second World War Era

After the Second World War, the difficult years in the early 1950s did not drive BASF away from China, but rather provided it with time to collect insight into market trends and customers' wishes. While still accumulating its local wisdom, BASF had appointed Hong Kong-based Jebsen & Co. as its sole agent for trade in China. Jebsen had managed to maintain its offices in Shanghai and Tianjin during the early years of the People's Republic, and the firm continued to represent

BASF from Hong Kong until BASF began its own direct engagement with China in 1982.

BASF's first direct investment in Taiwan came in 1969, when it established Cheng Kuang Chemical Industrial Co. Ltd. and Teh Hsin Dyes and Chemicals Co. Ltd. Cheng Kuang was the manufacturing arm, while Teh Hsin was a trading company. Cheng Kuang became BASF Taiwan in 1984, and merged with Teh Hsin four years later, after Taiwan eased its restrictions on foreign companies.

By the 1980s, China was becoming more open in the wake of the ambitious reform program instituted by Deng Xiaoping. The door to investment had been reopened, and BASF was quick to seize the opportunity.

Well aware of the fact that to get into a market you have to be local, BASF took its first step to strengthen its presence in China in 1982 by establishing a subsidiary in Hong Kong. This company, BASF China Ltd., had responsibility for selling, marketing, and distributing BASF products in Hong Kong and on the mainland. A representative office was opened in Beijing in 1986, followed by offices in Shanghai, Nanjing, Guangzhou, Qingdao, and Chengdu.

In 1994, China relaxed controls on foreign participation in several industries, including the chemical sector, in the hope of attracting USD ten billion in investment and technology by the end of the decade. While other players in the industry were skeptical about the profit potential of such investments, BASF had already established three production facilities in China, making it the second-largest German investor in the country. The first of BASF's joint ventures – Shanghai Gaoqiao BASF Dispersions Co. Ltd. (SGBD) – was set up in 1988 in conjunction with the Shanghai Gao Qiao Petrochemical Corporation to produce styrene-butadiene dispersions for coating paper and carpets to supply rapidly expanding customer markets. Since then, BASF has successfully established other joint ventures, expanding its local production base in Greater China (the People's Republic, Taiwan, and Hong Kong).

By the mid 1990s, around 700 staff people were employed in BASF's China operations, and the company was selling products worth about USD 460 million, of which almost ten percent were manufactured in China itself. Since the early 1980s, chemical consumption in the People's Republic had grown at the rate of eight percent a year, and the market, worth USD 48 billion, was projected to more than triple by 2010. In 1995, China's importance as a market was instrumental in the decision to base BASF's East Asia regional headquarters in Hong Kong. For the first time, a member of the Board of Executive Directors was based in Asia, underlining BASF's commitment to the region. The primary function of the new headquarters was to oversee operations in Greater China and South Korea and develop new projects for the BASF Group. The following year, a holding company, BASF (China) Co. Ltd., was formed in Beijing with responsibility for integrating BASF's operations in mainland China and providing corporate services – human resources, accounting, financing, logistics, reporting, IT, marketing, and sales – to joint ventures.

The three words central to BASF's "Vision 2010" – change, focus, and speed – were the success factors for BASF's approach to the market. Investing in East

Asia, and China especially, needs a great deal of flexibility to respond to the challenges of working across cultures and meet the specific needs of the market. It is important to change from an opportunistic market approach to systematic and sustainable market penetration based on domestic manufacturing facilities.

"Change" was the rule. In 1996, BASF therefore established the targets quoted above, which define where the company wants to be in Asia Pacific by the end of the first decade of the twenty-first century. The targets were ambitious by any standards: local production, for example, would have to double. As part of this ongoing process, BASF aims, where possible, to delegate responsibility for activities and operations from its traditional base in Germany to the regions. Establishing the regional headquarters in Hong Kong and a holding company in the People's Republic played an important part, since such decentralization enables a quicker and more efficient approach to meeting customers' requirements.

1997 was something of a watershed year in the region, marked most prominently by the Asian economic crisis, which was precipitated by currency devaluations in Thailand, Indonesia, and Korea. By 1998, the shockwaves had spread throughout Asia. Two countries, though, seemed miraculously unscathed: India and China, whose economy – according to official statistics – continued to grow at an impressive rate, while Taiwan and Hong Kong both languished. Despite the slump, BASF pushed ahead with ambitious expansion plans for the region, including a vast petrochemical project in Nanjing. BASF saw the crisis in Asia Pacific as a chance to enforce necessary structural reforms and refocus on sound economic development. It stood by its ambitious targets for the region.

33.2.3
Toward the Future

China was the main focus in the region. Two joint ventures had just come on stream: Shanghai BASF Colorants and Auxiliaries Co. had begun production of textile dyestuffs and pigments in 1996, and BASF Shanghai Coatings Co. had started manufacturing and marketing coatings and paints in early 1997. Yangzi-BASF Styrenics Co. Ltd. became operational by autumn 1997. In addition, there was the Nanjing project on the drawing board.

Together with China Petroleum & Chemical Corp. (Sinopec) in a 50-50 partnership, BASF planned to build its first "Verbund" project in East Asia – an integrated petrochemical site (IPS) on 220 hectares of land. The core of the project was an ethylene cracker with a capacity of 600,000 tons per year. Nine new plants downstream would be supplied by the cracker, producing 1.7 million metric tons of chemical products for local consumption, including ethylene, aromatics, polyethylenes, ethylene oxide and ethylene glycols, acrylic acid, acrylates, oxo alcohols, formic acid, propionic acid, methylamine, and dimethylformamide.

Integration, or "Verbund" as it is known in German, is a core competence at BASF. Byproducts or waste materials from one manufacturing process can in turn be used in another, increasing efficiency and cutting costs. The result is a highly sophisticated and flexible chain of production, where plants are effectively interlinked and resources, energy, and manpower are optimally used. In addition to reducing costs and increasing efficiency, the Verbund concept offers potential to cut energy requirements and waste. Over the years, BASF has extended the Verbund concept to other activities that can benefit from better coordination and integration. Examples include R&D, purchasing, logistics, and BASF's interaction with customers and the community at large.

The Verbund strategy is the core strength of BASF's activities worldwide. The first Asian Verbund site was built at Kuantan in Malaysia, to serve the South East Asian market. Strategically located at one of China's chemical bases in the Yangtze River delta and well supported by an extensive rail network and other logistics infrastructure, the new site in Nanjing will mainly supply local customers in the prosperous eastern coastal cities. The Nanjing project is the largest construction endeavor in BASF's history and involved an investment of USD 2.9 billion.

After gaining approval from the State Council – the first such approval for a cracker project with foreign participation – the joint venture contract between BASF and Sinopec was signed in Berlin in June 2000 at a ceremony attended by German Chancellor Gerhard Schröder and Chinese Prime Minister Zhu Rongji. The groundbreaking ceremony took place in late September 2001, at a highly uncertain time globally, with the United States still in shock from the attacks on New York and Washington. To some, it must have seemed a highly inauspicious time to be embarking on such a course, but BASF vowed to remain committed to its investment decisions and to continue to explore new projects in and for the Asia Pacific region. With hindsight, this was indeed the right decision, and the Nanjing site is now ready to start commercial production in mid 2005.

Following its policy of localizing production as far as regulations allow, BASF now has ten wholly-owned enterprises and nine joint ventures in Greater China, and operates 16 sales offices, twelve production sites, and one technical service center there (see Fig. 33.1). In 2005, the company will put the world's largest integrated production facilities for polytetrahydrofuran (PolyTHF®) and tetrahydrofuran (THF) into operation in Caojing, Shanghai. Here, butane will be converted directly to THF and subsequently to PolyTHF®. The process, which was developed by BASF, eliminates the previously necessary intermediate step of 1,4-butanediol (BDO), and Caojing will be the first site in the world to use the process on a commercial scale. The facilities will produce 60,000 metric tons of PolyTHF® and 80,000 metric tons of THF per year for the spandex fibers market, mainly for use in sportswear and garment manufacture. Also in Caojing, BASF is building an integrated isocyanates complex for 240,000 metric tons of MDI per year and 160,000 metric tons of TDI per year together with Huntsman Polyurethanes and Chinese partners. Scheduled to start operations in 2006, the complex involves an investment of USD one billion and will supply key components for the rapidly growing Chinese polyurethanes market.

Fig. 33.1 BASF in Greater China.

BASF's strategy in China certainly appears to be paying off: since 1996, the company has grown sales in Greater China by an average of more than 20 percent each year in euros (see Fig. 33.2). BASF now serves more than 3,000 customers directly or via distributors in China. BASF posted sales in Greater China of more than EUR 1.9 billion in 2004, an increase of nearly 20 percent over the previous year. The range of products produced by BASF in Greater China is extremely diverse, ranging from styrene, polystyrene, polyurethanes, engineering plastics, polymer dispersions, and coatings to pigments, finishing products for the textile and leather industries, intermediates, vitamins, and crop protection products. The greater portion of sales was in plastics, which accounted for 54 percent, followed by performance products (20 percent), chemicals (18 percent), and agricultural products and nutrition (7 percent). Greater China represents 36 percent of BASF's business in Asia, a figure that is expected to rise to almost 50 percent by 2015.

The company's workforce in Greater China has grown to around 4,000 and is expected to exceed 5,000 within a few years. BASF recruits 98 percent of its employees in Asia locally.

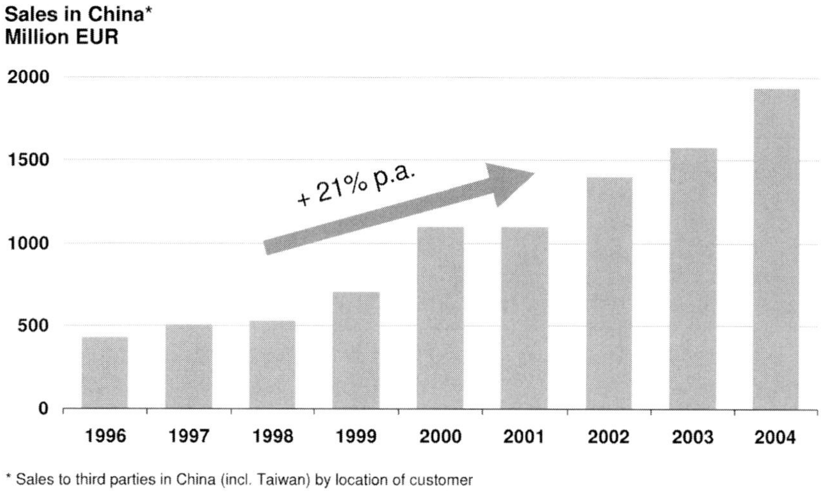

Fig. 33.2 Sales growth in China 1996 to 2004.

33.3 Sustainable Development in China

Between 2001 and 2005, BASF will have invested EUR two billion in China, but its spectrum of activities is far broader than might first be anticipated. Before retiring as chairman of BASF, Jürgen Strube visited Beijing in October 2002 and told the press there, "We are investing in more than just steel and pipes in China. Here, as in other parts of the world, our operations are guided by the principles of sustainable development."

Sustainable development is codified in BASF's Values and Principles, which are backed up by management systems. The three pillars of economic development, environmental protection, and social responsibility that underpin the sustainability principle are exemplified by BASF's activities in China. BASF sponsors activities to facilitate access to education and knowledge in the communities where its sites are located. Today's investments in this area will pay off in the long run in terms of regional competitiveness, ultimately benefiting BASF and its customers. In 1997, BASF established the Sino-German Research and Development Fund. In cooperation with this organization, BASF has supported more than 40 research projects and awarded 1,100 scholarships in China over the past seven years. In another educational project, BASF's Fine Chemicals division sponsors four-year scholarships to Shenyang Agricultural University.

In January 2004, BASF became a founding member of the China Business Council for Sustainable Development and hosted a symposium to encourage direct contact with its stakeholders. This organization brings us together with other like-minded members such as Sinopec, which is now not only our investment partner but also a partner on the journey to promoting sustainable development in China.

Another crucial aspect of future growth is attention to environmental matters, and BASF is playing a part here, too. China's rapid economic growth is associated with many challenges. Emission levels, for instance, naturally increase with the rising number of vehicles. Automobile production in China is expected to grow by an average of eight percent per year up to 2015. BASF is using its products and knowledge to assist the sustainable development of the Chinese economy. On the one hand, BASF supplies innovative fuel additives under the Keropur® brand that ensure better fuel consumption and lower emissions. On the other, it uses its knowledge to establish environmentally friendly technologies. For example, in June 2004, BASF and the Chinese Research Academy of Environmental Sciences (CRAES) signed a strategic alliance in the fuels sector. The goal of the cooperation is to improve the quality of Chinese gasoline significantly and to adapt it to the requirements of the latest engine technology. Among other things, this agreement provides for the joint creation of an engine test laboratory in Beijing that will employ internationally standardized test procedures. Other examples of BASF products that specially address environmental issues include Ecoflex®, a biodegradable polymer that eliminates the problem of plastic waste, and Cyclanon® ECO, a post-clearing agent for use in the textile industry that helps save water.

33.4 Outlook

Although BASF has a successful track record and the market potential is bright, the risks have to be kept in mind. The setback of the Asian crises is still fresh in everyone's mind.

BASF therefore aims to harness the changing business environment and achieve sustainable, profitable growth. However, operating successfully in an emerging and highly dynamic market also requires careful evaluation of the risks that could endanger that success. BASF therefore analyzes several key risk factors in China on a regular basis. The most prominent ones at present are the development of:

- the Chinese economy
- the banking sector
- the Renminbi
- the energy supply
- the WTO implementation process
- the legal framework and
- social challenges

Overall, BASF believes that the risk of missing out on opportunities is far larger than market risks. We are convinced that Greater China will continue to be an important growth engine for BASF in Asia Pacific. It represents a key success factor in ensuring that BASF remains the world's leading chemical company, that BASF remains – The Chemical Company.

33.5
Summary

- By 2010, the large and dynamically growing Chinese market is expected to be the second-largest national chemical market in the world. BASF recognized its enormous potential early, and defined a long-term strategy for Asia in the mid 1990s.
- However, the company already had a long history of involvement with China. The article traces its development from the 1880s to the present day.
- The company follows the strategy of building a strong local manufacturing base and has established a significant number of wholly-owned enterprises and joint ventures in China. This year the largest investment project in the company's history will come on stream in the form of the new petrochemical "Verbund" site (integrated site) in Nanjing.
- BASF's strategy in China appears to be paying off, with sales growth in Greater China averaging more than 20 percent each year since 1996. The company currently employs around 4,000 people in Greater China, 98 percent of them recruited locally.
- The spectrum of BASF's activities in China ranges beyond economic considerations: it is also supporting environmental and social projects, the two other pillars of sustainable development, in the country.
- BASF aims to harness the changing business environment in China and achieve sustainable, profitable growth. Nevertheless, the company is aware of the potential risks and continues to analyze key risk factors on a regular basis.

References

Abelshauser, W., von Hippel, W., Johnson, J. A., and Stokes, R.G. *German Industry and Global Enterprise, BASF: The History of a Company*, Cambridge University Press, Cambridge, **2004**, p. 96, Fig. I.5.

Sung Jae Koh, *Stages of Industrial Development in Asia; a Comparative History of the Cotton Industry in Japan, India, China, and Korea*, Philadelphia: University of Pennsylvania Press, **1996**, pp. 88–90, 190–200.

Index

Value Creation: Strategies for the Chemical Industry. 2nd Edition. F. Budde, U.-H. Felcht, H. Frankemölle (Eds.)

ISBN: 3-527-31266-8